Digital Signal Processing and Applications with the C6713 and C6416 DSK

TOPICS IN DIGITAL SIGNAL PROCESSING

C. S. BURRUS and T. W. PARKS: *DFT/FFT AND CONVOLUTION ALGORITHMS: THEORY AND IMPLEMENTATION*

JOHN R. TREICHLER, C. RICHARD JOHNSON, JR., and MICHAEL G. LARIMORE: *THEORY AND DESIGN OF ADAPTIVE FILTERS*

T. W. PARKS and C. S. BURRUS: *DIGITAL FILTER DESIGN*

RULPH CHASSAING and DARRELL W. HORNING: *DIGITAL SIGNAL PROCESSING WITH THE TMS320C25*

RULPH CHASSAING: *DIGITAL SIGNAL PROCESSING WITH C AND THE TMS320C30*

RULPH CHASSAING: *DIGITAL SIGNAL PROCESSING LABORATORY EXPERIMENTS USING C AND THE TMS320C31 DSK*

RULPH CHASSAING: *DSP APPLICATIONS USING C AND THE TMS320C6x DSK*

RULPH CHASSAING: *DIGITAL SIGNAL PROCESSING AND APPLICATIONS WITH THE C6713 AND C6416 DSK*

Digital Signal Processing and Applications with the C6713 and C6416 DSK

Rulph Chassaing

Worcester Polytechnic Institute

WILEY-
INTERSCIENCE

A JOHN WILEY & SONS, INC., PUBLICATION

Published by John Wiley & Sons, Inc., Hoboken, New Jersey.
Published simultaneously in Canada.

For general information on our other products and services please contact our Customer Care Department within the U.S. at 877-762-2974, outside the U.S. at 317-572-3993 or fax 317-572-4002.

Wiley also publishes its books in a variety of electronic formats. Some content that appears in print, however, may not be available in electronic format.

Library of Congress Cataloging-in-Publication Data:
Chassaing, Rulph.
    Digital signal processing and applications with the C6713 and C6416 DSK / by Rulph Chassaing.
        p.   cm.
    Includes bibliographical references and index.
    ISBN 0-471-69007-4
    1. Signal processing—Digital techniques.   2. Texas Instruments TMS320 series microprocessors.
    I. Title.
    TK5102.9.C47422 2004
    621.382′2—dc22
                                                                        2004050924

Printed in the United States of America.

10 9 8 7 6 5 4 3 2 1

Contents

Preface

Digital signal processors, such as the TMS320 family of processors, are used in a wide range of applications, such as in communications, controls, speech processing, and so on. They are used in cellular phones, digital cameras, high-definition television (HDTV), radio, fax transmission, modems, and other devices. These devices have also found their way into the university classroom, where they provide an economical way to introduce real-time digital signal processing (DSP) to the student.

Texas Instruments introduced the TM320C6x processor, based on the very-long-instruction-word (VLIW) architecture. This new architecture supports features that facilitate the development of efficient high-level language compilers. Throughout the book we refer to the C/C++ language simply as C. Although TMS320C6x/assembly language can produce fast code, problems with documentation and maintenance may exist. With the available C compiler, the programmer must "let the tools do the work." After that, if the programmer is not satisfied, Chapters 3 and 8 and the last few examples in Chapter 4 can be very useful.

This book is intended primarily for senior undergraduate and first-year graduate students in electrical and computer engineering and as a tutorial for the practicing engineer. It is written with the conviction that the principles of DSP can best be learned through interaction in a laboratory setting, where students can appreciate the concepts of DSP through real-time implementation of experiments and projects. The background assumed is a course in linear systems and some knowledge of C.

Most chapters begin with a theoretical discussion, followed by representative examples that provide the necessary background to perform the concluding experiments. There are a total of 105 programming examples, most using C code, with a few in assembly and linear assembly code. A list of these examples appears on page xvii. A total of 22 students' projects are also discussed. These projects cover a wide range of applications in filtering, spectrum analysis, modulation techniques, speech processing, and so on.

Programming examples are included throughout the text. This can be useful to the reader who is familiar with both DSP and C programming but who is not necessarily an expert in both. Many assignments are included at the end of Chapters 1–6.

This book can be used in the following ways:

1. For a DSP course with a laboratory component, using parts of Chapters 1–9. If needed, the book can be supplemented with some additional theoretical materials, since its emphasis is on the practical aspects of DSP. It is possible to cover Chapter 7 on adaptive filtering following Chapter 4 on finite impulse response (FIR) filtering (since there is only one example in Chapter 7 that uses materials from Chapter 5). It is my conviction that adaptive filtering should be incorporated into an undergraduate course in DSP.

2. For a laboratory course using many of the examples and experiments from Chapters 1–7 and Chapter 9. The beginning of the semester can be devoted to short programming examples and experiments and the remainder of the semester for a final project. The wide range of sample projects (for both undergraduate and graduate students) discussed in Chapter 10 can be very valuable.

3. For a senior undergraduate or first-year graduate design project course using selected materials from Chapters 1–10.

4. For the practicing engineer as a tutorial and reference, and for workshops and seminars, using selected materials throughout the book.

In Chapter 1 we introduce the tools through three programming examples. These tools include the powerful Code Composer Studio (CCS) provided with the TMS320C6713 DSP starter kit (DSK). It is essential to perform these examples before proceeding to subsequent chapters. They illustrate the capabilities of CCS for debugging, plotting in both the time and frequency domains, and other matters. Appendix H contains several programming examples using the TMS320C6416 DSK.

In Chapter 2 we illustrate input and output (I/O) with the AIC23 stereo codec on the DSK board through many programming examples. Chapter 3 covers the architecture and the instructions available for the TMS320C6x processor. Special instructions and assembler directives that are useful in DSP are discussed. Programming examples using both assembly and linear assembly are included in this chapter.

In Chapter 4 we introduce the z-transform and discuss FIR filters and the effect of window functions on these filters. Chapter 5 covers infinite impulse response (IIR) filters. Programming examples to implement real-time FIR and IIR filters are included. Appendix D illustrates MATLAB for the design of FIR and IIR filters.

Chapter 6 covers the development of the fast Fourier transform (FFT). Programming examples on FFT are included using both radix-2 and radix-4 FFT. In

Chapter 7 we demonstrate the usefulness of the adaptive filter for a number of applications with least mean squares (LMS). Programming examples are included to illustrate the gradual cancellation of noise or system identification. Students have been very receptive to applications in adaptive filtering. Chapter 8 illustrates techniques for code optimization.

In Chapter 9 we introduce DSP/BIOS and discuss a number of schemes (Visual C++, MATLAB, etc.) for real-time data transfer (RTDX) and communication between the PC and the DSK.

Chapter 10 discusses a total of 22 projects implemented by undergraduate and graduate students. They cover a wide range of DSP applications in filtering, spectrum analysis, modulation schemes, speech processing, and so on.

A CD is included with this book and contains all the programs discussed. See page xxi for a list of the folders that contain the support files for the examples and projects.

Over the last 10 years, faculty members from over 200 institutions have taken my workshops on "DSP and Applications." Many of these workshops were supported by grants from the National Science Foundation (NSF) and, subsequently, by Texas Instruments. I am thankful to NSF, Texas Instruments, and the participating faculty members for their encouragement and feedback. I am grateful to Dr. Donald Reay of Heriot-Watt University, who contributed several examples during his review of my previous book based on the TMS320C6711 DSK. I appreciate the many suggestions made by Dr. Mounir Boukadoum of the University of Quebec, Dr. Subramaniam Ganesan from Oakland University, and Dr. David Kozel from Purdue University at Calumet. I also thank Dr. Darrell Horning of the University of New Haven, with whom I coauthored my first book, *Digital Signal Processing with the TMS320C25*, for introducing me to "book writing." I thank al the students at Roger Williams University, the University of Massachusetts at Dartmouth, and Worcester Polytechnic Institute (WPI) who have taken my real-time DSP and senior design project courses, based on the TMS320 processors, over the last 20 years. The contribution of Aghogho Obi, from WPI, is very much appreciated.

The continued support of many people from Texas Instruments is also very much appreciated: Cathy Wicks and Christina Peterson, in particular, have been very supportive of this book.

Special appreciation: The laboratory assistance of Walter J. Gomes III in several workshops and during the development of many examples has been invaluable. His contribution is appreciated.

RULPH CHASSAING
Chassaing@msn.com
Chassaing@ece.wpi.edu

List of Examples

Programs/Files on Accompanying CD

A list of the folders included on the accompanying CD is shown below. The folders contain the programs/files for the examples/projects covered in the book.

Adaptc	AdaptIDFIR	AdaptIDFIRW	AdaptIDIIR	Adaptnoise
Adaptnoise_2IN	Adaptpredict	Adaptpredict_2IN	Aliasing	Am
BeatDetector	bios_4led	bios_sine_ctrl	bios_sine_intr	bpsk
Code_casm	detect_play	Dft	Dotp4	Dotp4a
dotp4clasm	Dotpintrinsic	Dotpipedfix	Dotpipedfloat	Dotpnp
Dotpnpfloat	dotpopt	Dotpp	Dotppfloat	Dsk6416
Dsk6711	Dtmf	DTMF_Bios_FFT	Dtmf_Bios_Rtdx	echo
Echo_control	Emif_lcd	Encryption	Factclasm	Factorial
FastConvo	Fastconvo_sim	FFT256c	FFT256c_poll	FFTr2
FFTr4	FFTr4_filter	FFTr4_sim	FFTsinetable	Fir
Fir3lp	FIR4types	FIR4ways	FIRcasm	FIRcasmfast
FIRcirc	FIRcirc_ext	FIRinverse	Firprn	FIRPRNbuf
Flash_sine	G722	Graphic_fft	GraphicEQ	Iir
IIR_ctrl	IIRinverse	Loop_intr	loop_poll	Loop_print
loop_stereo	Loop_store	Modulation_schemes		MuLaw
Noise_gen	Noisegen_casm	Notch2	Pll	Psk
Ramp	Ramptable	record	rtdx_lv_filter	rtdx_lv_gain
rtdx_lv_sine	rtdx_matlab_sim	rtdx_matlabFFT	rtdx_matlabFIR	rtdx_vbloop
rtdx_vbsine	rtdx_vc_FFTMatlab	rtdx_vc_FFTr4	rtdx_vc_FIR	rtdx_vc_sine
Scram16k_sw	Scram8k_DMA	Scrambler	Sin1500MATL	Sine_led_ctrl
sine_stereo	Sine2sliders	sine8_buf	sine8_LED	Sinegen_table
Sinegencasm	SinegenDE	Soundboard	speaker_recognition	
Spectrogram	speech_syn	squarewave	sum	Support
Sweep8000	SweepDE	two_tones	Twosum	Twosumfix
Twosumfloat	Twosumlasmfix	Twosumlasmfloat	Viterbi	Readme

1

DSP Development System

- Testing the software and hardware tools with Code Composer Studio
- Use of the TMS320C6713 DSK
- Programming examples to test the tools

Chapter 1 introduces several tools available for digital signal processing (DSP). These tools include the popular Code Composer Studio (CCS), which provides an integrated development environment (IDE), and the DSP starter kit (DSK) with the TMS320C6713 floating-point processor onboard and complete support for input and output. Three examples illustrate both the software and hardware tools included with the DSK. It is strongly suggested that you review these three examples before proceeding to subsequent chapters.

1.1 INTRODUCTION

Digital signal processors such as the TMS320C6x (C6x) family of processors are like fast special-purpose microprocessors with a specialized type of architecture and an instruction set appropriate for signal processing. The C6x notation is used to designate a member of Texas Instruments' (TI) TMS320C6000 family of digital signal processors. The architecture of the C6x digital signal processor is very well suited for numerically intensive calculations. Based on a very-long-instruction-word (VLIW) architecture, the C6x is considered to be TI's most powerful processor.

Digital signal processors are used for a wide range of applications, from communications and controls to speech and image processing. The general-purpose

Digital Signal Processing and Applications with the C6713 and C6416 DSK   By Rulph Chassaing
ISBN 0-471-69007-4   Copyright © 2005 by John Wiley & Sons, Inc.

1

digital signal processor is dominated by applications in communications (cellular). Applications embedded digital signal processors are dominated by consumer products. They are found in cellular phones, fax/modems, disk drives, radio, printers, hearing aids, MP3 players, high-definition television (HDTV), digital cameras, and so on. These processors have become the products of choice for a number of consumer applications, since they have become very cost-effective. They can handle different tasks, since they can be reprogrammed readily for a different application. DSP techniques have been very successful because of the development of low-cost software and hardware support. For example, modems and speech recognition can be less expensive using DSP techniques.

DSP processors are concerned primarily with real-time signal processing. Real-time processing requires the processing to keep pace with some external event, whereas non-real-time processing has no such timing constraint. The external event to keep pace with is usually the analog input. Whereas analog-based systems with discrete electronic components such as resistors can be more sensitive to temperature changes, DSP-based systems are less affected by environmental conditions. DSP processors enjoy the advantages of microprocessors. They are easy to use, flexible, and economical.

A number of books and articles address the importance of digital signal processors for a number of applications [1–22]. Various technologies have been used for real-time processing, from fiberoptics for very high frequency to DSPs very suitable for the audio-frequency range. Common applications using these processors have been for frequencies from 0 to 96 kHz. Speech can be sampled at 8 kHz (the rate at which samples are acquired), which implies that each value sampled is acquired at a rate of 1/(8 kHz) or 0.125 ms. A commonly used sample rate of a compact disk is 44.1 kHz. Analog/digital (A/D)-based boards in the megahertz sampling rate range are currently available.

The basic system consists of an analog-to-digital converter (ADC) to capture an input signal. The resulting digital representation of the captured signal is then processed by a digital signal processor such as the C6x and then output through a digital-to-analog converter (DAC). Also included within the basic system are a special input filter for anti-aliasing to eliminate erroneous signals and an output filter to smooth or reconstruct the processed output signal.

1.2 DSK SUPPORT TOOLS

Most of the work presented in this book involves the design of a program to implement a DSP application. To perform the experiments, the following tools are used:

1. *TI's DSP starter kit (DSK)*. The DSK package includes:
   (a) *Code Composer Studio* (CCS), which provides the necessary software support tools. CCS provides an integrated development environment (IDE), bringing together the C compiler, assembler, linker, debugger, and so on.

(b) A board, shown in Figure 1.1, that contains the TMS320C6713 (C6713) floating-point digital signal processor as well as a 32-bit stereo codec for input and output (I/O) support.

(c) A universal synchronous bus (USB) cable that connects the DSK board to a PC.

(d) A 5 V power supply for the DSK board.

2. *An IBM-compatible PC.* The DSK board connects to the USB port of the PC through the USB cable included with the DSK package.

3. *An oscilloscope, signal generator, and speakers.* A signal/spectrum analyzer is optional. Shareware utilities are available that utilize the PC and a sound card to create a virtual instrument such as an oscilloscope, a function generator, or a spectrum analyzer.

All the files/programs listed and discussed in this book (except some student project files in Chapter 10) are included on the accompanying CD. Most of the examples (with some minor modifications) can also run on the fixed-point C6416-based DSK. See Appendix H for the appropriate support files along with five illustrative examples. Reference 1 contains examples implemented on the C6711-based DSK (which has been discontinued). A list of all the examples is given on pages xv–xviii.

1.2.1 DSK Board

The DSK package is powerful, yet relatively inexpensive ($395), with the necessary hardware and software support tools for real-time signal processing [23–43]. It is a complete DSP system. The DSK board, with an approximate size of 5×8 in., includes the C6713 floating-point digital signal processor and a 32-bit stereo codec TLV320AIC23 (AIC23) for input and output.

The onboard codec AIC23 [37] uses a sigma–delta technology that provides ADC and DAC. It connects to a 12-MHz system clock. Variable sampling rates from 8 to 96 kHz can be set readily.

A daughter card expansion is also provided on the DSK board. Two 80-pin connectors provide for external peripheral and external memory interfaces. Two project examples in Chapter 10 illustrate the use of the external memory interface (EMIF) with light-emitting diodes (LEDs) and liquid-crystal displays (LCDs) for spectrum display.

The DSK board includes 16 MB (megabytes) of synchronous dynamic random access memory (SDRAM) and 256 kB (kilobytes) of flash memory. Four connectors on the board provide input and output: MIC IN for microphone input, LINE IN for line input, LINE OUT for line output, and HEADPHONE for a headphone output (multiplexed with line output). The status of the four user dip switches on the DSK board can be read from a program and provides the user with a feedback control interface. The DSK operates at 225 MHz. Also onboard the DSK are voltage

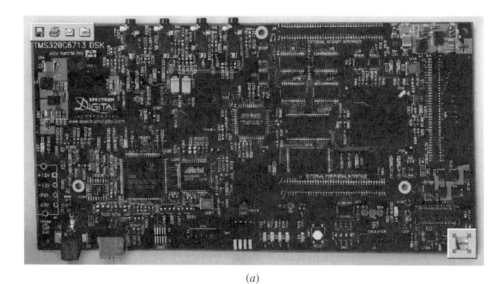

(*a*)

(*b*)

FIGURE 1.1. TMS320C6713-based DSK board: (*a*) board; (*b*) diagram. (Courtesy of Texas Instruments)

regulators that provide 1.26 V for the C6713 core and 3.3 V for its memory and peripherals.

Appendix H illustrates a DSK based on the fixed-point processor C6416.

1.2.2 TMS320C6713 Digital Signal Processor

The TMS320C6713 (C6713) is based on the VLIW architecture, which is very well suited for numerically intensive algorithms. The internal program memory is structured so that a total of eight instructions can be fetched every cycle. For example, with a clock rate of 225 MHz, the C6713 is capable of fetching eight 32-bit instructions every 1/(225 MHz) or 4.44 ns.

Features of the C6713 include 264 kB of internal memory (8 kB as L1P and L1D Cache and 256 kB as L2 memory shared between program and data space), eight functional or execution units composed of six arithmetic-logic units (ALUs) and two multiplier units, a 32-bit address bus to address 4 GB (gigabytes), and two sets of 32-bit general-purpose registers.

The C67xx (such as the C6701, C6711, and C6713) belong to the family of the C6x floating-point processors, whereas the C62xx and C64xx belong to the family of the C6x fixed-point processors. The C6713 is capable of both fixed- and floating-point processing. The architecture and instruction set of the C6713 are discussed in Chapter 3.

1.3 CODE COMPOSER STUDIO

CCS provides an IDE to incorporate the software tools. CCS includes tools for code generation, such as a C compiler, an assembler, and a linker. It has graphical capabilities and supports real-time debugging. It provides an easy-to-use software tool to build and debug programs.

The C compiler compiles a C source program with extension .c to produce an assembly source file with extension .asm. The assembler assembles an .asm source file to produce a machine language object file with extension .obj. The linker combines object files and object libraries as input to produce an executable file with extension .out. This executable file represents a linked common object file format (COFF), popular in Unix-based systems and adopted by several makers of digital signal processors [25]. This executable file can be loaded and run directly on the C6713 processor. Chapter 3 introduces the linear assembly source file with extension .sa, which is a cross between C and assembly code. A linear optimizer optimizes this source file to create an assembly file with extension .asm (similar to the task of the C compiler).

To create an application project, one can "add" the appropriate files to the project. Compiler/linker options can readily be specified. A number of debugging features are available, including setting breakpoints and watching variables; viewing memory, registers, and mixed C and assembly code; graphing results; and monitoring execution time. One can step through a program in different ways (step into, over, or out).

Real-time analysis can be performed using real-time data exchange (RTDX) (Chapter 9). RTDX allows for data exchange between the host PC and the target DSK, as well as analysis in real time without stopping the target. Key statistics and performance can be monitored in real time. Through the joint team action group (JTAG), communication with on-chip emulation support occurs to control and monitor program execution. The C6713 DSK board includes a JTAG interface through the USB port.

1.3.1 CCS Installation and Support

Use the USB cable to connect the DSK board to the USB port on the PC. Use the 5-V power supply included with the DSK package to connect to the +5-V power connector on the DSK to turn it on. Install CCS with the CD-ROM included with the DSK, preferably using the $c:\backslash C6713$ structure (in lieu of $c:\backslash ti$ as the default).

The CCS icon should be on the desktop as "C6713DSK CCS" and is used to launch CCS. The code generation tools (C compiler, assembler, linker) are used with CCS version 2.x.

CCS provides useful documentations included with the DSK package on the following (see the Help icon):

1. Code generation tools (compiler, assembler, linker, etc.)
2. Tutorials on CCS, compiler, RTDX
3. DSP instructions and registers
4. Tools on RTDX, DSP/basic input/output system (DSP/BIOS), and so on.

An extensive amount of support material (*pdf* files) is included with CCS. There are also examples included with CCS within the folder $c:\backslash C6713\backslash examples$. They illustrate the board and chip support library files, DSP/BIOS, and so on. CCS Version 2.x was used to build and test the examples included in this book. A number of files included in the following subfolders/directories within $c:\backslash C6713$ (suggested structure during CCS installation) can be very useful:

1. *myprojects*: a folder supplied only for your projects. All the folders in the accompanying book CD should be placed within this subdirectory.
2. *bin*: contains many utilities.
3. *docs*: contains documentation and manuals.
4. *c6000\cgtools*: contains code generation tools.
5. *c6000\RTDX*: contains support files for real-time data transfer.
6. *c6000\bios*: contains support files for DSP/BIOS.
7. *examples*: contains examples included with CCS.
8. *tutorial*: contains additional examples supplied with CCS.

Note that all the folders containing the programs and support files in the accompanying book CD should be transferred to the subdirectory `myprojects`. Change the properties of all the files included so that they are not read-only (all the folders can be highlighted to change the properties of their contents at once).

1.3.2 Useful Types of Files

You will be working with a number of files with different extensions. They include:

1. `file.pjt`: to create and build a project named file
2. `file.c`: C source program
3. `file.asm`: assembly source program created by the user, by the C compiler, or by the linear optimizer
4. `file.sa`: linear assembly source program. The linear optimizer uses *file.sa* as input to produce an assembly program *file.asm*
5. `file.h`: header support file
6. `file.lib`: library file, such as the run-time support library file `rts6700.lib`
7. `file.cmd`: linker command file that maps sections to memory
8. `file.obj`: object file created by the assembler
9. `file.out`: executable file created by the linker to be loaded and run on the C6713 processor
10. `file.cdb`: configuration file when using DSP/BIOS

1.4 QUICK TEST OF DSK

1. On power, a program *post.c* (Power On Self Test), stored in onboard flash memory, uses the board support library (BSL) to test the DSK. It tests the internal, external, and flash memories, the two multichannel buffered serial ports (McBSP), direct memory access (DMA), the onboard codec, and the LEDs. If all tests are successful, all four LEDs blink three times and stop (with all LEDs on). During the testing of the codec, a 1-kHz tone is generated for 1 sec.

2. Launch CCS from the icon on the desktop. A USB enumeration process takes place. Then CCS will be opened and the LEDs will turn off. Press GEL → Check DSK → Quick Test. The Quick Test can be used for confirmation of correct operation and installation. The following message is then displayed:

    Switches: 15

    Board Revision: 1

    CPLD Revision: 2

This assumes that the four dip switches $(0, 1, 2, 3)$ are all in the up position. Change the switches to $(1110)_2$ so that the first three switches $(0, 1, 2)$ are up and press the fourth switch (3) down. Repeat the procedure to select GEL → Check DSK → Quick Test and verify that the value of the switches is now 7 (with the display "Switches: 7"). You can set the value of the four user switches from 0 to 15. Within your program you can then direct the execution of your code based on these 16 values.

Alternative Quick Test of DSK

1. Open/launch CCS from the icon on the desktop if this has not been done already. Select File → Load Program. Click on the folder `sine8_LED\Debug` within myprojects to load the file `sine8_LED.out`. This loads the executable file `sine8_LED.out` into the C6713 processor. This assumes that you have already copied all the folders on the accompanying CD into your folder: `c:\c6713\myprojects`.

2. Select Debug → Run. Press the dip switch #0, which should light LED #0 on and generate a 1-kHz tone. Connect the LINE OUT (or the HEADPHONE) on the DSK board to a speaker or to an oscilloscope and verify the generation of the 1-kHz tone. The four connectors on the DSK board for I/O (MIC, LINE IN, LINE OUT, and HEADPHONE) use a 3.5-mm jack audio cable.

1.5 SUPPORT FILES

The following support files located in the folder `support` (except the library files) are used for most of the examples and projects discussed in this book:

1. `C6713dskinit.c`: contains functions to initialize the DSK, the codec, the serial ports, and for I/O. It is not included with CCS.

2. `C6713dskinit.h`: header file with function prototypes. Features such as those used to select the mic input in lieu of line input (by default), input gain, and so on are obtained from this header file (modified from a similar file included with CCS).

3. `C6713dsk.cmd`: sample linker command file. This generic file can be changed when using external memory in lieu of internal memory.

4. `Vectors_intr.asm`: a modified version of a vector file included with CCS to handle interrupts. Twelve interrupts, INT4 through INT15, are available, and INT11 is selected within this vector file. They are used for interrupt-driven programs.

5. `Vectors_poll.asm`: vector file for programs using polling.

6. `rts6700.lib, dsk6713bsl.lib, cs16713.lib`: run-time, board, and chip support library files, respectively. These files are included with CCS and are located in *C6000\cgtools\lib*, *C6000\dsk6713\lib*, and *c6000\bios\lib*, respectively.

1.6 PROGRAMMING EXAMPLES TO TEST THE DSK TOOLS

Three programming examples are introduced to illustrate some of the features of CCS and the DSK board. The primary focus is to become familiar with both the software and hardware tools. It is strongly suggested that you complete these three examples before proceeding to subsequent chapters.

Example 1.1: Sine Generation Using Eight Points with DIP Switch Control (sine8_LED)

This example generates a sinusoid using a table lookup method. More important, it illustrates some features of CCS for editing, building a project, accessing the code generation tools, and running a program on the C6713 processor. The C source program *sine8_LED.c* shown in Figure 1.2 implements the sine generation and is included in the folder *sine8_LED*.

Program Consideration

Although the purpose is to illustrate some of the tools, it is useful to understand the program *sine8_LED.c*. A table or buffer sine_table is created and filled with eight points representing $\sin(t)$, where $t = 0, 45, 90, 135, 180, 225, 270$, and 315 degrees

```
//Sine8_LED.c Sine generation with DIP switch control

#include "dsk6713_aic23.h"                    //support file for codec,DSK
Uint32 fs = DSK6713_AIC23_FREQ_8KHZ;          //set sampling rate
short loop = 0;                               //table index
short gain = 10;                              //gain factor
short sine_table[8]={0,707,1000,707,0,-707,-1000,-707};//sine values

void main()
{
 comm_poll();                                 //init DSK, codec, McBSP
 DSK6713_LED_init();                          //init LED from BSL
 DSK6713_DIP_init();                          //init DIP from BSL
 while(1)                                     //infinite loop
 {
  if(DSK6713_DIP_get(0)==0)                   //=0 if switch #0 pressed
  {
    DSK6713_LED_on(0);                        //turn LED #0 ON
    output_sample(sine_table[loop]*gain);     //output every Ts (SW0 on)
    if (++loop > 7)  loop = 0;                //check for end of table
  }
  else DSK6713_LED_off(0);                    //LED #0 off
 }                                            //end of while (1)
}                                             //end of main
```

FIGURE 1.2. Sine generation program using eight points with dip switch control (sine8_LED.c).

(scaled by 1000). Within the function `main`, another function, `comm_poll`, is called that is located in the communication and initialization support file `c6713dskinit.c`. It initializes the DSK, the AIC23 codec onboard the DSK, and the two McBSPs on the C6713 processor. Within `c6713dskinit.c`, the function `DSK6713_init` initializes the BSL file, which must be called before the two subsequent BSL functions, `DSK6713_LED_init` and `DSK6713_DIP_init`, are invoked that initialize the four LEDs and the four dip switches.

The statement `while (1)` within the function `main` creates an infinite loop. When dip switch #0 is pressed, LED #0 turns on and the sinusoid is generated. Otherwise, *DSK6713_DIP_get(0)* will be false (true if the switch is pressed) and LED #0 will be off.

The function `output_sample`, located in the communication support file `C6713dskinit.c`, is called to output the first data value in the buffer or table `sine_table[0]` = 0. The loop index is incremented until the end of the table is reached, after which it is reinitialized to zero.

Every sample period $T = 1/F_s = 1/8000 = 0.125$ ms, the value of dip switch #0 is tested, and a subsequent data value in `sine_table` (scaled by $gain = 10$) is sent for output. Within one period, eight data values (0.125 ms apart) are output to generate a sinusoidal signal. The period of the output signal is $T = 8(0.125$ ms$) = 1$ ms, corresponding to a frequency of $f = 1/T = 1$ kHz.

Create Project

In this section we illustrate how to create a project, adding the necessary files for building the project **sine8_LED**. Back up the folder *sine8_LED* (change its name) or delete its content (which can be retrieved from the book CD if needed), keeping only the C source file *sine8_LED.c* and the file *gain.gel* in order to recreate the content of that folder. Access CCS (from the desktop).

1. To create the project file *sine8_LED.pjt*. Select Project → New. Type *sine8_LED* for the project name, as shown in Figure 1.3. This project file is saved in the folder *sine8_LED* (within c:\c6713\myprojects). The .pjt file stores project information on build options, source filenames, and dependencies.

2. To add files to the project. Select Project → Add Files to Project. Look in the folder *support*, Files of type C Source Files. Double-click on the C source file *C6713dskinit.c* to add it to the project. Click on the "+" symbol to the left of the Project Files window within CCS to expand and verify that this C source file has been added to the project.

3. Repeat step 2, use the pull-down menu for Files of type, and select ASM Source Files. Double-click on the assembly source vector file *vectors_poll.asm* to add it to the project. Repeat again and select Files of type: Linker Command File, and add *C6713dsk.cmd* to the project.

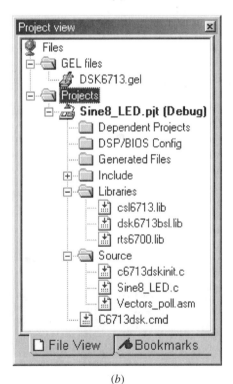

(a)

(b)

FIGURE 1.3. CCS Project windows for `sine8_LED`: (a) project creation; (b) project view files.

4. To add the library support files to the project. Repeat the previous step, but select files of type: Object and Library Files. Look in `c:\c6713\c6000\cgtools\lib` and select the run-time support library file *rts6700.lib* (which supports the C67x architecture) to add to the project. Continue this process to add the BSL file *dsk6713bsl.lib* located in c:`\c6713\c6000\dsk6713\lib`, and the chip support library (CSL) file *csl6713.lib* located in c:`\c6713\c6000\bios\lib`.

5. Verify from the Files window that the project (`.pjt`) file, the linker command (`.cmd`) file, the three library (`.lib`) files, the two C source (`.c`) files, and the assembly (`.asm`) file have been added to the project. The GEL file *dsk6713.gel* is added automatically when you create the project. It initializes the C6713 DSK invoking the BSL to use the phase-locked loop (PLL) to set the central processing unit (CPU) clock to 225 MHz (otherwise, the C6713 runs at 50 MHz by default).

6. Note that there are no "include" files yet. Select Project → Scan All File Dependencies. This adds/includes the header files *c6713dskinit.h*, along with several board and chip support header files included with CCS.

The Files window in CCS should look as in Figure 1.3*b*. Any of the files (except the library files) from CCS's Files window can be displayed by clicking on it. You should not add header or include files to the project. They are added to the project automatically when you select: Scan All File Dependencies. (They are also added when you build the project.)

It is also possible to add files to a project simply by "dragging" the file (from a different window) and dropping it into the CCS Project window.

Code Generation and Options

Various options are associated with the code generation tools: C compiler and linker to build a project.

Compiler Option

Select Project → Build Options. Figure 1.4*a* shows the CCS window Build Options for the compiler. Select the following for the compiler option with Basic (for Category): (1) c671x{-mv6710} (for Target Version), (2) Full Symbolic Debug (for Generate Debug Info), (3) Speed most critical (for Opt Speed vs. Size), and (4) None (for Opt Level and Program Level Opt). Select the Preprocessor Category and type for Define Symbols{d}: CHIP_6713, and from the Feedback Category, select for Interlisting: OPT/C and ASM{-s}. The resulting compiler option is

`-g -s`

The `-g` option is used to enable symbolic debugging information, useful during the debugging process, and is used in conjunction with the option `-s` to interlist the C

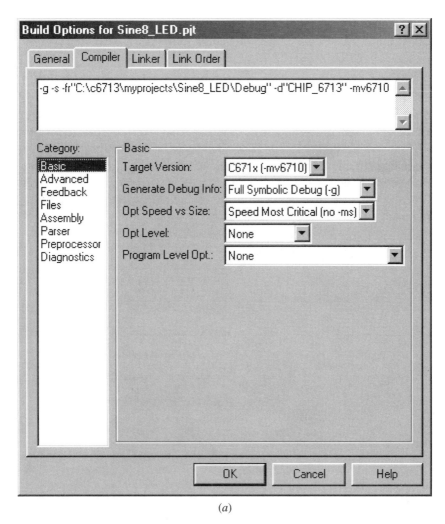

(a)

FIGURE 1.4. CCS Build options: (a) compiler; (b) linker.

source file with the assembly source file *sine8_LED.asm* generated (an additional option, -k, can be used to retain the assembly source file). The -g option disables many code optimizations to facilitate the debugging process. Press OK.

Selecting C621x or C64xx for Target Version invokes a fixed-point implementation. The C6713-based DSK can use either fixed- or floating-point processing. Most examples implemented in this book can run using fixed-point processing. Selecting C671x as Target Version invokes a floating-point implementation.

If No Debug is selected (for Generate Debug Info) and -o3:File is selected (for Opt Level), the Compiler option is automatically changed to

-s -o3

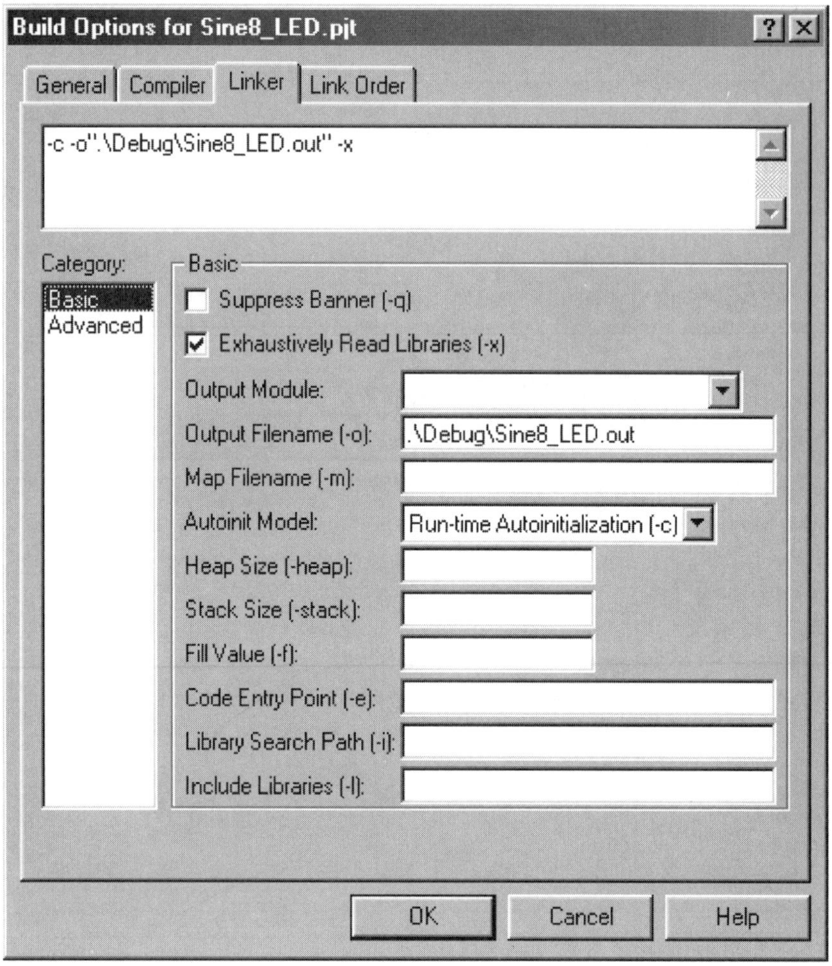

(*b*)

FIGURE 1.4. (*Continued*)

The -o3 option invokes the highest level of optimization for performance or execution speed. For now, speed is not critical (neither is debugging). Use the compiler options -gs (which you can also type directly in the compiler command window). Initially, one would not optimize for speed but to facilitate debugging. A number of compiler options are described in Ref. 28.

Linker Option

Click on Linker (from CCS Build Options). The output filename sine8_LED.out defaults to the name of the .pjt filename, and Run-time Autoinitialization defaults for Autoinit Model. The linker option should be displayed as in Figure 1.4*b*. The map file can provide useful information for debugging (memory locations of functions, etc.). The -c option is used to initialize variables at run time, and the -o option is used to name the linked executable output file sine8_LED.out. Press OK.

Note that you can/should choose to store the executable file in the subfolder "Debug," within the folder sine8_LED, especially during the debugging stage of a project.

Again, these various compiler and linker options can be typed directly within the appropriate command windows.

In lieu of adding the three library files to the project by retrieving them from their specific locations, it is more convenient to add them within the linker option window Include Libraries{-l}, typing them directly, separated by a comma. However, they will not be shown in the Files window.

Building and Running the Project

The project sine8_LED can now be built and run.

1. Build this project as **sine8_LED**. Select Project → Rebuild All or press the toolbar with the three down arrows. This compiles and assembles all the C files using cl6x and assembles the assembly file vectors_poll.asm using asm6x. The resulting object files are then linked with the library files using lnk6x. This creates an executable file sine8_LED.out that can be loaded into the C6713 processor and run. Note that the commands for compiling, assembling, and linking are performed with the Build option. A log file cc_build_Debug.log is created that shows the files that are compiled and assembled, along with the compiler options selected. It also lists the support functions that are used. Figure 1.5 shows several windows within CCS for the project sine8_LED. The building process causes all the dependent files to be included (in case one forgets to scan for all the file dependencies).

2. Select File → Load Program in order to load sine_LED.out by clicking on it (CCS includes an option to load the program automatically after a build). It should be in the folder sine8_LED\Debug. Select Debug → Run or use the toolbar with the "running man." Connect a speaker to the LINE OUT connector on the DSK. Press the dip switch #0. You should hear a tone. You can also use the headphone output at the same time.

The sampling rate F_s of the codec is set at 8 kHz. The frequency generated is $f = F_s/$(number of points) = 8 kHz/8 = 1 kHz. Connect the output of the DSK to an oscilloscope to verify a 1-kHz sinusoidal signal with an approximate amplitude of 0.8 V p-p (peak to peak).

Correcting Program Errors

1. Delete the semicolon in the statement

```
short gain = 10;
```

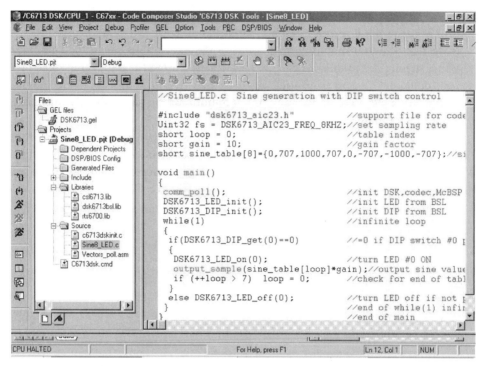

FIGURE 1.5. CCS windows for project sine8_LED.

in the C source file sine8_LED.c. If it is not displayed, double-click on it (from the Files window).

2. Select Project → Build to perform an incremental build or use the toolbar with the two (not three) arrows. The incremental build is chosen so that only the C source file sine8_LED.c is compiled. With the Rebuild option (toolbar with three arrows), files compiled and/or assembled previously would again go through this unnecessary process.

3. An error message, highlighted in red, stating that a ";" is expected, should appear in the Build window of CCS (lower left). You may need to scroll up the Build window for a better display of this error message. Double-click on the highlighted error message line. This should bring the cursor to the section of code where the error occurs. Make the appropriate correction, Build again, load, and run the program to verify your previous results.

Monitoring the Watch Window

Verify that the processor is still running (and dip switch #0 is pressed) . Note the indicator "DSP RUNNING" at the bottom left of CCS. The Watch window allows you to change the value of a parameter or to monitor a variable:

1. Select View → Quick Watch window, which should be displayed on the lower section of CCS. Type `gain`, then click on "Add to Watch." The gain value of 10 set in the program in Figure 1.2 should appear in the Watch window.

2. Change *gain* from 10 to 30 in the Watch window. Press Enter. Verify that the volume of the generated tone has increased (with the processor still running and dip switch #0 is pressed). The amplitude of the sine wave has increased from approximately 0.8 V p-p to approximately 2.5 V p-p.

3. Change *gain* to 33 (as in step 2). Verify that a higher-pitched tone exists, which implies that the frequency of the sine wave has changed just by changing its amplitude. This is not so. You have exceeded the range of the codec AIC23. Since the values in the table are scaled by 33, the range of these values is now between ±33,000. The range of output values is limited from -2^{15} to $(2^{15} - 1)$, or from −32,768 to +32,767.

   Since the AIC23 is a stereo codec, we can send data to both 16-bit channels within each sampling period. This is introduced in Chapter 2. This can be useful to experiment with the stereo effects of output signals. In Chapter 7, we use both channels for adaptive filtering where it is necessary to input one type of signal (such as noise) on one 16-bit channel and another signal (such as a desired signal) on the other 16-bit channel. In this book, we will mostly use the codec as a mono device without the need to use an adapter that is required when using both channels.

Applying the Slider Gel File

The General Extension Language (GEL) is an interpretive language similar to (a subset of) C. It allows you to change a variable such as gain, sliding through different values while the processor is running. All variables must first be defined in your source program.

1. Select File → Load GEL and open the file `gain.gel`, which you retained from the original folder, `sine8_LED` (that you backed up). Double-click on the file *gain.gel* to view it within CCS. It should be displayed in the Files window. This file is shown in Figure 1.6. By creating the slider function *gain* shown in Figure 1.6, you can start with an initial value of 10 (first value) for the variable `gain` that is set in the C program, up to a value of 35 (second value), incremented by 5 (third value).

2. Select GEL → Sine Gain → Gain. This should bring out the Slider window shown in Figure 1.7, with the minimum value of 10 set for the gain.

3. Press the up-arrow key to increase the gain value from 10 to 15, as displayed in the Slider window. Verify that the volume of the sine wave generated has increased. Press the up-arrow key again to continue increasing the slider, incrementing by 5 up to 30. The amplitude of the sine wave should be about 2.5 V p-p with a *gain* value set at 30. Now use the mouse to click directly on the Slider window and slowly increase the slider position to 31, then 32, and

```
/*gain.gel Create slider and vary amplitude (gain) of sinewave*/

menuitem "Sine Gain"

slider Gain(10,35,5,1,gain_parameter) /*incr by 5,up to 35*/
{
        gain = gain_parameter;        /*vary gain of sine*/
}
```

FIGURE 1.6. GEL file to slide through different gain values in the sine generation program (gain.gel).

FIGURE 1.7. Slider window for varying the gain of generated sine wave.

verify that the frequency generated is still 1 kHz. Increase the slider to 33 and verify that you are no longer generating a 1-kHz sine wave. The table values, scaled by the gain value, are now between ±33,000 (beyond the acceptable range by the codec).

Changing the Frequency of the Generated Sinusoid

1. Change the sampling frequency from 8 to 16 kHz by setting f_s in the C source program to *DSK6713_AIC23_FREQ_16KHZ*. Rebuild (use incremental build) the project, load and run the new executable file, and verify that the frequency of the generated sinusoid is 2 kHz. The sampling frequencies supported by the AIC23 codec are 8, 16, 24, 32, 44.1, 48, and 96 kHz.

2. Change the number of points in the lookup table to four points in lieu of eight points—for example, {0, 1000, 0, −1000}. The size of the array sine_table and the loop index also need to be changed. Verify that the generated frequency is $f = F_s/$(number of points).

   Note that the sinusoid is no longer generated if the dip switch #0 is not pressed. If a different dip switch such as switch #3 is desired (in lieu of switch #0), the BSL functions *DSK6713_DIP_get(3)*, *DSK6713_LED_on(3)*, and *DSK6713_LED_off(3)* can be substituted in the C source program.

Two sliders can readily be used, one to change the gain and the other to change the frequency. A different signal frequency can be generated by changing the loop index within the C program (e.g., stepping through every two points in the table). When you exit CCS after you build a project, all changes made to the project can be saved. You can later return to the project with the status as you left it before. For example, when returning to the project after launching CCS, select Project → Open to open an existing project such as `sine8_LED.pjt` (with all the necessary files for the project already added).

Example 1.2: Generation of the Sinusoid and Plotting with CCS (*sine8_buf*)

This example generates a sinusoid with eight points, as in Example 1.1. More important, it illustrates CCS capabilities for plotting in both time and frequency domains. The program *sine8_buf.c*, shown in Figure 1.8, implements this project. This program creates a buffer to store the output data in memory.

Create this project as *sine8_buf.pjt*, and add the necessary files to the project, as in Example 1.1 (use the C source program *sine8_buf.c* in lieu of *sine8_LED.c*). Note that the necessary header support files are added to the project by selecting Project → Scan All File Dependencies. The necessary

```
//sine8_buf Sine generation. Output buffer plotted within CCS

#include "dsk6713_aic23.h"                  //codec-DSK support file
Uint32 fs=DSK6713_AIC23_FREQ_8KHZ;          //set sampling rate
int loop = 0;                               //table index
short gain = 10;                            //gain factor
short sine_table[8]={0,707,1000,707,0,-707,-1000,-707};//sine values
short out_buffer[256];                      //output buffer
const short BUFFERLENGTH = 256;             //size of output buffer
int i = 0;                                  //for buffer count

interrupt void c_int11()                    //interrupt service routine
{
 output_sample(sine_table[loop]*gain);      //output sine values
 out_buffer[i] = sine_table[loop]*gain;     //output to buffer
 i++;                                       //increment buffer count
 if(i==BUFFERLENGTH) i=0;                   //if @ bottom reinit count
 if (++loop > 7) loop = 0;                  //check for end of table
 return;                                    //return from interrupt
}

void main()
{
  comm_intr();                              //init DSK, codec, McBSP
  while(1);                                 //infinite loop
}
```

FIGURE 1.8. Sine generation with output stored in memory as well (`sine8_buf.c`).

support files for this project, c6713dskinit.c, vectors_intr.asm and C6713dsk.cmd, are in the folder support, and the three library support files can be added using Project → Build Options and selecting the linker option (Include Libraries). Type them, separating each by a comma. Note that since this program is interrupt-driven (in lieu of polling), the vector file vectors_intr.asm (in lieu of vectors_poll.asm) is added to the project.

Within the function *main*, *comm._intr* (in lieu of *comm_poll* in Example 1.1) is called. This function resides in c6713dskinit.c to support interrupt-driven programs. The statement while(1) within the function main creates an infinite loop to wait for an interrupt to occur. On interrupt, execution proceeds to the interrupt service routine (ISR) *c_int11*. This ISR address is specified in the file vectors_intr.asm with a branch instruction to this address, using interrupt INT11. Interrupts are discussed in more detail in Chapter 3.

Within the ISR, the function output_sample, located in the communication and initialization file c6713dskinit.c, is called to output the first data value in sine_table. The loop index is incremented until the end of the table is reached; after that, it is reinitialized to zero. An output buffer is created to capture a total of 256 (specified by *BUFFERLENGTH*) sine data values. Execution returns from ISR to the while (1) infinite loop to wait for each subsequent interrupt.

Build this project as **sine8_buf**. Load and run the executable file *sine8_buf.out* and verify that a 1-kHz sinusoid is generated with the output connected to a speaker or a scope (as in Example 1.1).

Plotting with CCS

The output buffer is being updated continuously every 256 points (you can readily change the buffer size). Use CCS to plot the current output data stored in the buffer *out_buffer*.

1. Select View → Graph → Time/Frequency. Change the Graph Property Dialog so that the options in Figure 1.9*a* are selected for a time-domain plot (use the pull-down menu when appropriate). The starting address of the output buffer is out_buffer. The other options can be left as default. Figure 1.10 shows a time-domain plot of the sinusoidal signal within CCS.

2. Figure 1.9*b* shows CCS's Graph Property Display for a frequency-domain plot. Choose a fast Fourier transform (FFT) order so that the frame size is 2^{order}. Press OK and verify that the FFT magnitude plot is as shown in Figure 1.10. The spike at 1000 Hz represents the frequency of the sinusoid generated.

You can obtain many different windows within CCS. From the Build window, right-click and select Float In Main Window. To change the screen size, right-click on the Build window and deselect Allow Docking. For example, you can get the time-domain plot (separated). Right-click on the time-domain plot, select Float In Main Window, and again right-click on the same time-domain plot window and deselect Allow Docking. You can then move it.

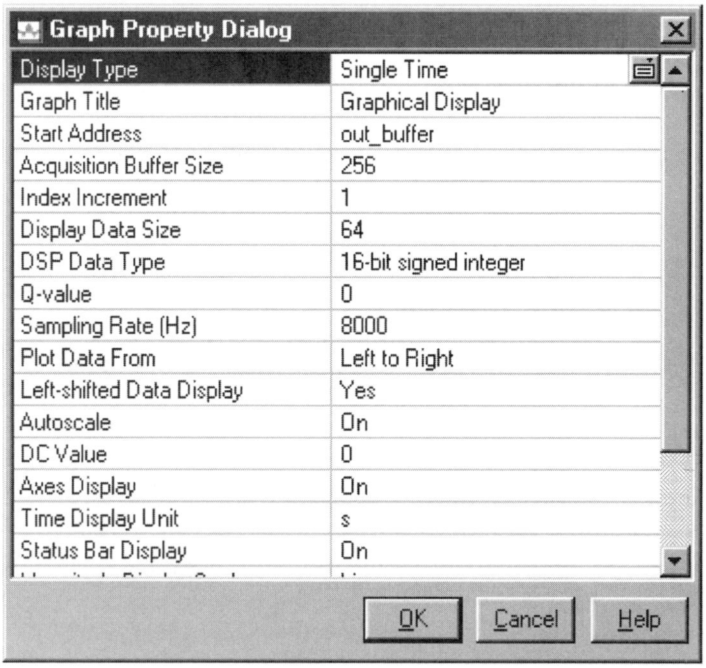

Graph Property Dialog   ×

Display Type	Single Time
Graph Title	Graphical Display
Start Address	out_buffer
Acquisition Buffer Size	256
Index Increment	1
Display Data Size	64
DSP Data Type	16-bit signed integer
Q-value	0
Sampling Rate (Hz)	8000
Plot Data From	Left to Right
Left-shifted Data Display	Yes
Autoscale	On
DC Value	0
Axes Display	On
Time Display Unit	s
Status Bar Display	On

[OK]   [Cancel]   [Help]

(*a*)

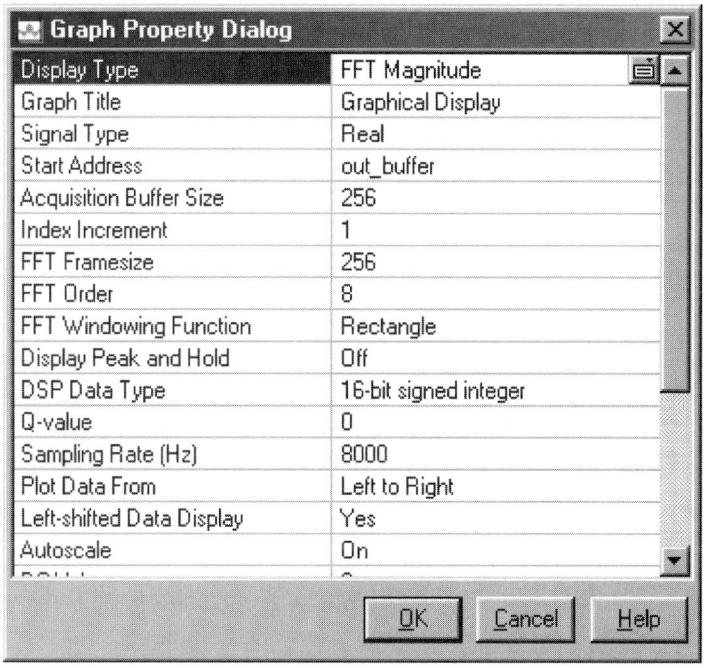

Graph Property Dialog   ×

Display Type	FFT Magnitude
Graph Title	Graphical Display
Signal Type	Real
Start Address	out_buffer
Acquisition Buffer Size	256
Index Increment	1
FFT Framesize	256
FFT Order	8
FFT Windowing Function	Rectangle
Display Peak and Hold	Off
DSP Data Type	16-bit signed integer
Q-value	0
Sampling Rate (Hz)	8000
Plot Data From	Left to Right
Left-shifted Data Display	Yes
Autoscale	On

[OK]   [Cancel]   [Help]

(*b*)

FIGURE 1.9. CCS Graph Property Dialog for `sine8_buf`: (*a*) for time-domain plot; (*b*) for frequency-domain plot.

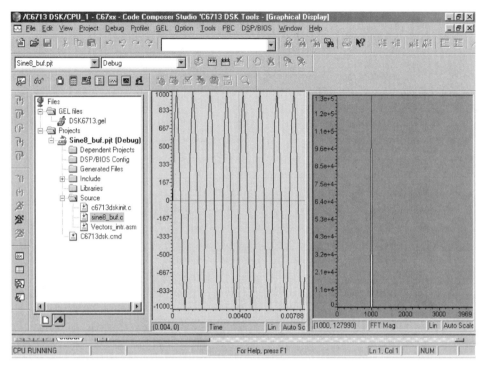

FIGURE 1.10. CCS windows for sine8_buf showing both time- and frequency-domain plots of a generated 1-kHz sine wave.

Viewing and Saving Data from Memory in a File

To view the content of that buffer, select View → Memory and specify `out_buffer` for the address, and select the 16-bit signed integer (or hex, etc.) for the format.

To save the content of the output buffer in a file, select File → Data → Save. Save the file as sine8_buf.dat (as type hex, for example) in the folder sine8_buf. From the Storing Memory window, use `out_buffer` as the buffer's address with length 256. You can then plot this data [with MATLAB for example] and verify the 1-kHz sinusoidal waveform (with 8 kHz as the sampling rate).

Example 1.3: Dot Product of Two Arrays (`dotp4`)

Operations such as addition/subtraction and multiplication are the key operations in a DSP. A very important operation is multiply/accumulate, which is useful in a number of applications requiring digital filtering, correlation, and spectrum analysis. Since the multiplication operation is executed commonly and is essential for most DSP algorithms, it is important that it executes in a single cycle. With the C6713 we can actually perform two multiply/accumulate operations within a single cycle.

This example illustrates additional features of CCS, such as single-stepping, setting breakpoints, and profiling for the benchmark. Again, the purpose here is to

```
//Dotp4.c Multiplies two arrays, each array with 4 numbers

int dotp(short *a,short *b,int ncount);//function prototype
#include <stdio.h>                      //for printf
#include "dotp4.h"                      //header file with data
#define count 4                         //# data in each array
short x[count] = {x_array};            //declaration of 1st array
short y[count] = {y_array};            //declaration of 2nd array

main()
{
  int result = 0;                       //result sum of products

  result = dotp(x, y, count);           //call dotp function
  printf("result = %d (decimal) \n", result);   //print result
}

int dotp(short *a,short *b,int ncount) //dot product function
{
  int sum = 0;                          //init sum
  int i;

  for (i = 0; i < ncount; i++)
     sum += a[i] * b[i];                //sum of products
  return(sum);                          //return sum as result
}
```

FIGURE 1.11. Sum-of-products program using C code (dotp4.c).

```
//dotp4.h  Header file with two arrays of numbers

#define x_array 1,2,3,4

#define y_array 0,2,4,6
```

FIGURE 1.12. Header file with two arrays each with four numbers (dotp4.h).

become more familiar with the tools. We invoke C compiler optimization to see how performance or execution speed can be drastically increased.

The C source file dotp4.c in Figure 1.11 takes the sum of products of two arrays, each with four numbers, contained in the header file dotp4.h in Figure 1.12. The first array contains the four numbers 1, 2, 3, and 4, and the second array contains the four numbers 0, 2, 4, and 6. The sum of products is $(1 \times 0) + (2 \times 2) + (3 \times 4) + (4 \times 6) = 40$.

The program can be readily modified to handle a larger set of data. No real-time implementation is used in this example, and no real-time I/O support files are needed. The support functions for interrupts are not needed here.

Create this project as **dotp4** and add the following files to the project (see Example 1.1):

1. `dotp4.c`: C source file
2. `vectors_poll.asm`: vector file defining the entry address *c_int00*
3. `C6713dsk.cmd`: generic linker command file
4. `rts6700.lib`: library file

Do not add any "include" files using "Add Files to Project" since they are added by selecting Project → Scan All File Dependencies. The header file `stdio.h` is needed due to the `printf` statement in the program `dotp4.c` to print the result.

Implementing a Variable Watch

1. Select Project → Options with `-gs` as the compiler option and the default linker option with no optimization.
2. Rebuild All by selecting the toolbar with the three arrows (or select Project → Rebuild All). Load the executable file *dotp4.out* within the folder *dotp4\Debug.*
3. Select View → Quick Watch. Type *sum* to watch the variable *sum* and click on "Add to Watch." The message "identifier not found" associated with *sum* is displayed (as Value) because this local variable does not exist yet.
4. Set a breakpoint at the line of code

   ```
   sum += a[i] * b[i];
   ```

   by placing the mouse cursor (clicking) on that line, then right-click and select the Toggle breakpoint. Or, preferably, with the cursor on that line of code (at the extreme left), double-click. A red circle to the left of that line of code should appear. (Note: placing the cursor on a line of code with a set breakpoint and double clicking will remove the breakpoint.)
5. Select Debug → Run (or use the "running man" toolbar). The program executes up to (excluding) the line of code with the set breakpoint. A yellow arrow will also point to that line of code.
6. Single-step using F8. Repeat or continue to single-step and observe/watch the variable *sum* in the Watch window change in value to 0, 4, 16, 40. Select Debug → Run and verify that the resulting value of `sum` is printed as

   ```
   sum = 40 (decimal)
   ```

7. Note the `printf` statement in the C program `dotp4.c` for printing the result. This statement (while excellent for debugging) should be avoided after the debugging stage, since it takes over 6000 cycles to execute.

Animating

1. Select File → Reload Program to reload the executable file `dotp4.out`. Or, preferably, select Debug → Restart. Note that after the executable file is

loaded, the entry address for execution is c_int00, as can be verified by the disassembled file.

2. The same breakpoint should be set already at the same line of code as before. Select Debug → Animate or use the equivalent toolbar in the left window (below the Halt running man). Observe the variable *sum* change in values through the Watch window. The speed of animation can be controlled by selecting Option → Customize → Animate Speed (the maximum speed is set to default at 0 second).

Benchmarking (Profiling) without Optimization

In this section we illustrate how to benchmark a section of code: in this case, the *dotp* function. Verify that the options for the compiler (-g) and linker (-c -o dotp4.out) are still set. To profile code, you must use the compiler option -g for symbolic debugging information. Remove any breakpoint by double-clicking on the line of code with the set breakpoint (or right-click and select the Toggle breakpoint).

1. Select Debug → Restart.
2. Select Profiler → Start New Session and enter dotp4 as the Profile Session Name. Then press OK.
3. Click on the icon to "Create Profile Area" (see Figure 1.13*a*). This icon is the third icon from the bottom left in Figure 1.13*b*. Figure 1.13*b* shows the added profile area for the function dotp within the C source file dotp4.c.

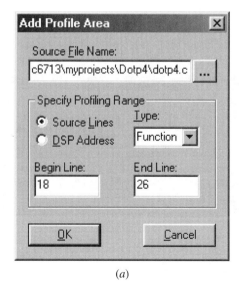

(*a*)

FIGURE 1.13. CCS display of project dotp4 for profiling: (*a*) profile area for function *dotp*; (*b*) profiling function dotp with no optimization; (*c*) profiling function dotp with level 3 optimization; (d) profiling printf.

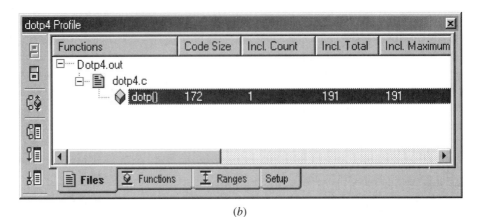

(b)

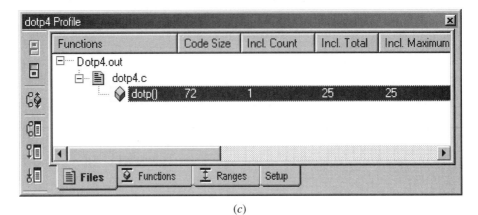

(c)

(d)

FIGURE 1.13. (*Continued*)

4. Run the program. Verify the results shown in Figure 1.13*b*. This indicates that it takes 191 cycles to execute the function dotp (with no optimization).

Benchmarking (Profiling) with Optimization

In this section we illustrate how to optimize the program using one of the optimization options, -o3. The program's execution speed can be increased using the optimizing C compiler. Change the compiler option (select Project → Build Options) to

```
-g  -o3
```

and use the same linker options as before (you can type this option directly). The option -o3 invokes the highest level of compiler optimization. Various compiler options are described in Ref. 28. Rebuild All (toolbar with three arrows) and load the executable file dotp4.out (or select File → Reload Program). Re-create the Profile Area as in Figure 1.13*a*.

Select Debug → Run. Verify that it takes now 25 cycles (from 191) to execute the dotp function, as shown in Figure 1.13*c*. This is a considerable improvement using the C compiler optimizer. The code size is reduced from 172 to 72. The dot product example can be also optimized using an intrinsic function or the code optimization techniques discussed in Chapter 8.

Profiling Printf

Again restart the program (Debug → Restart). Click on the icon *Ranges* at the bottom of the profile area. Highlight *printf* from the C source program, drag it to the profiling area window, and drop it by releasing the cursor. Verify that the code size of *printf* is 32 and that it takes 6316 cycles to execute, as shown in Figure 1.13*d*.

Note that in lieu of using Figure 1.13*a* to profile the function dotp, you can highlight it, drag it, and drop it with your mouse in the profiling area.

1.7 SUPPORT PROGRAMS/FILES CONSIDERATIONS

The following support files are used for practically all the examples in this book: (1) *c6713dskinit.c*, (2) *vectors_intr.asm* or *vectors_poll.asm*, and (3) *c6713dsk.cmd*. For now, the emphasis associated with these files should be on using them.

1.7.1 Initialization/Communication File (*c6713dskinit.c*)

Several BSL and CSL support functions are included in the initialization and communication (init/comm) file *c6713dskinit.c*. A partial listing is shown in Figure 1.14. It includes functions to initialize the DSK and provide for input and output.

```
//C6713dskinit.c Partial list of init/comm file.Includes CSL/BSL funct
...
void c6713_dsk_init()                    //dsp-peripheral init
{
DSK6713_init();                          //BSL to init DSK-EMIF,PLL
hAIC23_handle=DSK6713_AIC23_openCodec(0, &config);//handle to codec
DSK6713_AIC23_setFreq(hAIC23_handle, fs);  //set sample rate
MCBSP_config(DSK6713_AIC23_DATAHANDLE,&AIC23CfgData);//32bits interface
MCBSP_start(DSK6713_AIC23_DATAHANDLE,MCBSP_XMIT_START | MCBSP_RCV_START
 | MCBSP_SRGR_START | MCBSP_SRGR_FRAMESYNC,220); //start data channel
}

void comm_poll()                         //comm/init using polling
{
    poll = 1;                            //1 if using polling
    c6713_dsk_init();                    //init DSP and codec
}

void comm_intr()                         //for comm/init using interrupt
{
 poll = 0;                               //0 since not polling
 IRQ_globalDisable();                    //disable interrupts
 c6713_dsk_init();                       //init DSP and codec
CODECEventId=MCBSP_getXmtEventId(DSK6713_AIC23_codecdatahandle);//Xmit
 ...
 IRQ_setVecs(vectors);                   //point to the IRQ vector
 IRQ_map(CODECEventId, 11);              //map McBSP1 Xmit to INT11
 IRQ_reset(CODECEventId);                //reset codec INT 11
 IRQ_globalEnable();                     //globally enable interrupts
 IRQ_nmiEnable();                        //enable NMI interrupt
 IRQ_enable(CODECEventId);               //enable CODEC eventXmit INT11
 output_sample(0);                       //start McBSP interrup out a sample
}

void output_sample(int out_data)    //out to Left and Right channels
{
 short CHANNEL_data;
 AIC_data.uint=0;                        //clear data structure
 AIC_data.uint=out_data;                 //32-bit data -->data structure
 ...
 if(poll) while(!MCBSP_xrdy(DSK6713_AIC23_DATAHANDLE));//ready to Xmit?
   MCBSP_write(DSK6713_AIC23_DATAHANDLE,AIC_data.uint);//write data
}

void output_left_sample(short out_data)  //for output->left channel
{
 AIC_data.uint=0;                        //clear data structure
 AIC_data.channel[LEFT]=out_data;  //data->Left channel->data structure
 if(poll) while(!MCBSP_xrdy(DSK6713_AIC23_DATAHANDLE));//ready to Xmit?
 MCBSP_write(DSK6713_AIC23_DATAHANDLE,AIC_data.uint);//out->leftchannel
}

void output_right_sample(short out_data)  //for output->right channel
...
Uint32 input_sample()                    //for 32-bit input
{
 short CHANNEL_data;
```

FIGURE 1.14. Partial listing of communication/initialization support program (C6713dskinit.c).

```
 if (poll) while(!MCBSP_rrdy(DSK6713_AIC23_DATAHANDLE));//receiveready?
   AIC_data.uint=MCBSP_read(DSK6713_AIC23_DATAHANDLE); //read data
 ...
 return(AIC_data.uint);
}

short input_left_sample()                      //input to left channel
{
 if(poll) while(!MCBSP_rrdy(DSK6713_AIC23_DATAHANDLE));//receiveready?
   AIC_data.uint=MCBSP_read(DSK6713_AIC23_DATAHANDLE);//read->left chan
 return(AIC_data.channel[LEFT]);               //return left channel data
}

short input_right_sample()                     //input to right channel
 ...
```

FIGURE 1.14. (*Continued*)

The function *comm_intr()* in an interrupt-driven program or *comm_poll()* in a polling-based program calls the appropriate functions to initialize the DSK. These two functions are located in the init/comm. file. When using an interrupt-driven program, interrupt #11 (INT11) is configured and enabled (selected). The nonmaskable interrupt bit must be enabled as well as the global interrupt enable (GIE) bit. A different interrupt, such as INT12, can be selected readily by modifying slightly the init/comm. file and the vector file that contains the branching address to the corresponding ISR in the main C source program. INT11 is generated via the serial port (McBSP).

The function *input_sample()* is used to input data and the function *output_sample()* to output data. Most of the examples throughout the book utilize the AIC23 codec in a mono format, defaulting to the left channel to read or write a 16-bit data. The example loop_stereo.c in Chapter 2 illustrates the stereo capability of the codec to input 16-bit data into each (left and right) channel and output a 16-bit data from each channel. Some adaptive filtering examples in Chapter 7 use both input channels to acquire two different 16-bit input data signals.

The code input = input_sample();, casting input as a short, acquires 16-bit data through the left (default) channel. Similarly, *output_sample ((short) . . .)*; outputs 16-bit data from the left (default) channel.

A polling-based program (non-interrupt-driven) continuously polls or tests whether or not data are ready to be received or transmitted. This scheme is in general less efficient than the interrupt scheme. For input, the content of the serial port control register (SPCR) bit 1 [the second least significant bit (LSB)], as shown in Figure B.8 (Appendix B), is continuously tested to determine when data are available to be received or read. For output, the content of SPCR bit 17 is tested (Figure B.8) to determine when data are available to be transmitted. An input data value is accessed through the data receive register of the McBSP. An output data value is sent through the data transmit register of McBSP.

The MCBSP1 transmit interrupt is used and INT11 is selected in the examples throughout the book. If the program is polling-based, the McBSP is continuously tested before reading (for input) or writing (for output).

Within the function `output_sample()` used for output, in the code segment

```
If (poll) while(!MCBSP_xrdy(...)); MCBSP_write(...);
```

the first line of code continuously tests (if polling-based) the transmit ready *xrdy* register bit. If it is a 1, then the subsequent line of code is executed to write (output). If the transmit ready bit is a 0 (not ready), then the `while()` statement becomes *while (true)* and execution remains in an infinite loop until the transmit ready bit becomes a 1 (ready). If the program is not polling-based, then the transmit ready bit is not tested and writing (output) occurs every sample period.

Similarly, within the function `input_sample()` used for input, in the code segment

```
If (poll) while(!MCBSP_rrdy(...)); MCBSP_read(...);
```

the first line of code continuously tests (if polling-based) the receive ready *rrdy* register bit. If it is a 1 (ready), the subsequent line of code reads the data. If it is a 0 (not ready), the *while ( )* statement causes execution to remain in an infinite loop until the receive ready bit register becomes a 1. If the program is not polling-based, the receive ready bit is not tested and reading occurs every sample period.

The examples throughout the book use both interrupt-driven and polling-based programs. A polling-based program can be readily changed to interrupt-driven and vice versa. Interrupts are discussed further in Chapter 3.

Header File (c6713dskinit.h)

The corresponding header support file `c6713dskinit.h` contains the function prototypes as well as various register settings associated with the AIC23 codec. For example (see *c6713dskinit.h*):

1. The mic input can be set in lieu of the line input by changing the value of register 4 from the (default) value of 0x0011 to 0x0015.

2. In Chapter 2, a loop program yields an output that is the delayed input, with the same frequency but attenuated (by default). To increase the gain of the (default) left line input channel, change the value of register 0 from 0x0017 to 0x001c. This value will produce an output of the same amplitude as the input. Note that either the line input or the mic input can be made active.

1.7.2 Vector File (vectors_intr.asm/vectors_poll.asm)

To select interrupt INT11, a branch instruction to the ISR `c_int11` located in the C program (see `sine8_buf.c`) is placed at the address INT11 in `vectors_intr.asm`. A listing of the file `vectors_intr.asm` is shown in Figure 1.15. Note

```
*Vectors_intr.asm Vector file for interrupt INT11
     .global _vectors                    ;global symbols
     .global _c_int00
     .global _vector1
     .global _vector2
     .global _vector3
     .global _vector4
     .global _vector5
     .global _vector6
     .global _vector7
     .global _vector8
     .global _vector9
     .global _vector10
     .global _c_int11                    ;for INT11
     .global _vector12
     .global _vector13
     .global _vector14
     .global _vector15

     .ref _c_int00                       ;entry address

VEC_ENTRY .macro addr                    ;macro for ISR
     STW    B0,*--B15
     MVKL   addr,B0
     MVKH   addr,B0
     B      B0
     LDW    *B15++,B0
     NOP    2
     NOP
     NOP
     .endm

_vec_dummy:
  B      B3
  NOP    5

 .sect ".vecs"                           ;aligned IST section
 .align 1024
_vectors:
_vector0:    VEC_ENTRY _c_int00          ;RESET
_vector1:    VEC_ENTRY _vec_dummy        ;NMI
_vector2:    VEC_ENTRY _vec_dummy        ;RSVD
_vector3:    VEC_ENTRY _vec_dummy
_vector4:    VEC_ENTRY _vec_dummy
_vector5:    VEC_ENTRY _vec_dummy
_vector6:    VEC_ENTRY _vec_dummy
_vector7:    VEC_ENTRY _vec_dummy
_vector8:    VEC_ENTRY _vec_dummy
_vector9:    VEC_ENTRY _vec_dummy
_vector10:   VEC_ENTRY _vec_dummy
_vector11:   VEC_ENTRY _c_int11          ;ISR address
_vector12:   VEC_ENTRY _vec_dummy
_vector13:   VEC_ENTRY _vec_dummy
_vector14:   VEC_ENTRY _vec_dummy
_vector15:   VEC_ENTRY _vec_dummy
```

FIGURE 1.15. Vector file for an interrupt-driven program (vectors_intr.asm).

the underscore preceding the name of the routine or function being called. The ISR is also referenced in vectors_intr.asm using .ref _c_int11.

For a non-interrupt-driven or polling-based program, a separate file vectors_poll.asm is used, in lieu of vectors_intr.asm, by

1. Deleting the reference to the interrupt service routine (ISR) .ref _c_int11
2. Replacing the branch instruction to the ISR for interrupt INT11 by (NOP), which is a no operation instruction.

1.7.3 Linker Command File (c6713dsk.cmd)

The linker command file C6713dsk.cmd is listed in Figure 1.16. It shows that sections such as .text reside in internal RAM (IRAM), which is mapped to the internal memory of the C6713 digital signal processor. It can be used as a generic sample linker command file even though some portion of it is not necessary. Chapter 2 contains an example illustrating the use of the *pragma* directive to specify a section such as EXT_RAM in synchronous DRAM (SDRAM). SDRAM is a section in external memory that starts at the address 0x80000000. Chapter 2 contains an example illustrating the use of the onboard flash memory (burning the flash) that starts at address 0x90000000. In Chapter 4, we illustrate the implementation of a digital filter is assembly code using external memory SDRAM. Chapter 10 contains

```
/*C6713dsk.cmd Linker command file*/

MEMORY
{
   IVECS:      org=0h,          len=0x220
   IRAM:       org=0x00000220,  len=0x0002FDE0 /*internal memory*/
   SDRAM:      org=0x80000000,  len=0x00100000 /*external memory*/
   FLASH:      org=0x90000000,  len=0x00020000 /*flash memory*/
}
SECTIONS
{
   .EXT_RAM :> SDRAM
   .vectors :> IVECS      /*in vector file*/
   .text    :> IRAM
   .bss     :> IRAM
   .cinit   :> IRAM
   .stack   :> IRAM
   .sysmem  :> IRAM
   .const   :> IRAM
   .switch  :> IRAM
   .far     :> IRAM
   .cio     :> IRAM
   .csldata :> IRAM
}
```

FIGURE 1.16. Generic linker command file (C6713dsk.cmd).

two projects that utilize the EMIF 80-pin connector on the DSK, which starts at address 0xA0000000, to interface to external LEDs and LCDs.

Linker options include -heap size to specify the heap size in bytes for dynamic memory allocation (default is 1kB) and the option -stack size to specify the C system stack size in bytes. Other linker options can be found in Ref. 26.

The linker allocates the program in memory using a default location algorithm. It places the various sections into appropriate memory locations, where code and data reside. By using a linker command file with extension .cmd, one can customize the allocation process, specifying MEMORY and SECTIONS directives within the linker command file. The linker directive MEMORY (uppercase) defines a memory model and designates the origin and length of various available memory spaces. The directive SECTIONS (uppercase) allocate the output sections into defined memory and designate the various code sections to available memory spaces.

Most of the examples in the book invoke internal memory. The generic sample linker command file, shown in Figure 1.16, can be used for almost all of the examples in the book, even if neither external nor flash memory is utilized.

1.8 COMPILER/ASSEMBLER/LINKER SHELL

In previous examples the code generation tools for compiling, assembling, and linking were invoked within CCS while building a project. The tools may also be invoked directly outside CCS using a DOS shell.

1.8.1 Compiler

The compiler shell can be invoked using

```
cl6x [options] [files]
```

to compile and assemble files that can be C files with extension .c, assembly files with extension .asm, and linear assembly (introduced in Chapter 3) with extension .sa. A linear assembly program file is a cross between C and assembly that can provide a compromise between the more versatile C program and the most efficient assembly program. For example, the command

```
Cl6x -gks -o3 file1.c, file2, file3.asm, file4.sa
```

invokes the C compiler to compile *file1* and *file2* (defaults to extension .c) and generates the assembly files *file1.asm* and *file2.asm*. This also invokes the assembler optimizer to optimize *file4.sa* and create *file4.asm*. Then the assembler (invoked with the shell command cl6x) assembles the four assembly source files and creates the four object files *file1.obj, . . . , file4.obj*. The option -gs adds debugger-specific information for debugging purposes and interlists C statements into assembly files, respectively. The -k option is used to keep the assembly source files generated.

Four levels of compiler optimizations are available, with -o3 to invoke the highest level of optimization. Level 0 allocates variables to registers. Level 1 performs all level 0 optimizations, eliminates local common expressions, and removes unused assignments. Level 2 performs all the level 1 optimizations plus loop optimizations and rolling. Level 3 performs all level 2 optimizations and removes functions that are not called. There are also compiler optimizations to minimize code size (with possible degradation in execution speed).

Note that full optimization may change memory locations that can affect the functionality of a program. In such cases, these memory locations must be declared as volatile. The compiler does not optimize volatile variables. A volatile variable is allocated to an uninitialized section in lieu of a register. Volatiles can be used when memory access is to be exactly as specified in the C code.

Initially, the functionality of a program is of primary importance. One should *not* invoke any (or too-high-level) optimization option initially while debugging, since additional debugger-specific information is provided to enhance the debugging process. Such additional information suppresses the level of performance. It is also difficult to debug a program after optimization, since the lines of code are usually no longer arranged in a serial fashion. Compiler options can also be set using the environment variable with C_OPTION.

1.8.2 Assembler

An assembly-coded source file *file3.asm* can also be assembled using

```
asm6x file3.asm
```

to create file3.obj. The .asm extension is optional. The resulting object file is then linked with a run-time support library to create an executable COFF file with extension .out that can be loaded directly and run on the DSp. Examples using assembly-coded source files are introduced in Chapter 3.

1.8.3 Linker

The linker can be invoked using

```
lnk6x -c prog1.obj -o prog1.out -l rts6700.lib
```

The -c option tells the linker to use special conventions defined by the C environment for automatic variable initialization at run time (another linker option, -cr, initializes the variables at load time). The -l option invokes a library file such as the run-time support library file rts6700.lib. These options [-c (or -cr) and

-1] must be used when linking. The object file `prog1.obj` is linked with the library file(s) and creates the executable file `prog1.out`. Without the -o option, the executable file `a.out` (by default) is created.

The linker can also be invoked with the compiler shell command with the -z option

```
Cl6x -gks -o3 prog1.c prog2.asm -z -o prog.out -m prog.map
-l rts6700.lib
```

to create the executable file `prog.out`. The -m option creates a map file that provides a list of all the addresses of sections, symbols, and labels that can be useful for debugging.

The linker also links automatically a `boot` program when using C programs to initialize the run-time environment, setting the entry point to `c_int00`. The symbol `_c_int00` is defined automatically when the linker option -c (or -cr) is invoked. The function `_c_int00`, included in the run-time support library, is the entry point in the `boot` program that sets up the stack and calls `main`. The run-time library support program `boot.c` is used to auto-initialize variables. The linker option -c invokes the initialization process with `boot.c`. Note that it is defined in the vector files `vectors_intr.asm` and `vectors_poll.asm`.

The book CD contains all the main source files used in this book, located in separate folders, and some support files necessary for many examples and projects are located in the folder *support*. Other needed support files are included with CCS within *c:\C6713*.

1.9 ASSIGNMENTS

1. Write a program to generate a cosine with a frequency of 666.66 Hz. Verify your output result using LINE OUT, as well as plotting the generated cosine in both time and frequency domains.

2. Write a polling-based program so that once dip switch #3 is pressed, LED #3 turns on and a 666.66 Hz cosine is generated for approximately 5 seconds. [*Hint*: also use (incorporate) the delay associated with turning a LED on.]

3. Write a program to multiply two arrays, each containing the five numbers 1, 2, 3, 4, and 5 (i.e., $1^2 + 2^2 + 3^2 + 4^2 + 5^2$). Verify your result using a watch window and printing it within CCS in the Build window.

4. Write an interrupt-driven program to capture an input sinusoidal signal of amplitude 3 V p-p and a frequency of 1 kHz, and output that sampled signal every 0.0625 ms. Use the function `input_sample` in a similar fashion as the function `output_sample` used in Examples 1.1 and 1.2—for example,

```
input = input_sample();
```

casting input as short (16-bit). Verify that the output signal has the same frequency as the input signal but is reduced in amplitude. Increase the input signal frequency until the output is reduced drastically. What is the approximate frequency at which this occurs? This represents the bandwidth of the onboard AIC23 codec (as illustrated in Chapter 2).

REFERENCES

Note: References 23 to 43 are included with the DSK package.

1. R. Chassaing, *DSP Applications Using C and the TMS320C6x DSK*, Wiley, New York, 2002.

2. R. Chassaing, *Digital Signal Processing Laboratory Experiments Using C and the TMS320C31 DSK*, Wiley, New York, 1999.

3. R. Chassaing, *Digital Signal Processing with C and the TMS320C30*, Wiley, New York, 1992.

4. R. Chassaing and D. W. Horning, *Digital Signal Processing with the TMS320C25*, Wiley, New York, 1990.

5. N. Kehtarnavaz and M. Keramat, *DSP System Design Using the TMS320C6000*, Prentice Hall, Upper Saddle River, NJ, 2001.

6. N. Kehtarnavaz and B. Simsek, *C6x-Based Digital Signal Processing*, Prentice Hall, Upper Saddle River, NJ, 2000.

7. N. Dahnoun, *DSP Implementation Using the TMS320C6x Processors*, Prentice Hall, Upper Saddle River, NJ, 2000.

8. Steven A. Tretter, *Communication System Design Using DSP Algorithms with Laboratory Experiments for the TMS320C6701 and TMS320C6711*, Kluwer Academic, New York, 2003.

9. J. H. McClellan, R. W. Schafer, and M. A. Yoder, *DSP First: A Multimedia Approach*, Prentice Hall, Upper Saddle River, NJ, 1998.

10. C. Marven and G. Ewers, *A Simple Approach to Digital Signal Processing*, Wiley, New York, 1996.

11. J. Chen and H. V. Sorensen, *A Digital Signal Processing Laboratory Using the TMS320C30*, Prentice Hall, Upper Saddle River, NJ, 1997.

12. S. A. Tretter, *Communication System Design Using DSP Algorithms*, Plenum Press, New York, 1995.

13. A. Bateman and W. Yates, *Digital Signal Processing Design*, Computer Science Press, New York, 1991.

14. Y. Dote, *Servo Motor and Motion Control Using Digital Signal Processors*, Prentice Hall, Upper Saddle River, NJ, 1990.

15. J. Eyre, The newest breed trade off speed, energy consumption, and cost to vie for an ever bigger piece of the action, *IEEE Spectrum*, June 2001.

16. J. M. Rabaey, ed., VLSI design and implementation fuels the signal-processing revolution, *IEEE Signal Processing*, Jan. 1998.

17. P. Lapsley, J. Bier, A. Shoham, and E. Lee, *DSP Processor Fundamentals: Architectures and Features*, Berkeley Design Technology, Berkeley, CA, 1996.

18. R. M. Piedra and A. Fritsh, Digital signal processing comes of age, *IEEE Spectrum*, May 1996.

19. R. Chassaing, The need for a laboratory component in DSP education: a personal glimpse, *Digital Signal Processing*, Jan. 1993.

20. R. Chassaing, W. Anakwa, and A. Richardson, Real-time digital signal processing in education, *Proceedings of the 1993 International Conference on Acoustics, Speech and Signal Processing (ICASSP)*, Apr. 1993.

21. S. H. Leibson, DSP development software, *EDN Magazine*, Nov. 8, 1990.

22. D. W. Horning, An undergraduate digital signal processing laboratory, *Proceedings of the 1987 ASEE Annual Conference*, June 1987.

23. *TMS320C6000 Programmer's Guide*, SPRU198G, Texas Instruments, Dallas, TX, 2002.

24. *TMS320C6211 Fixed-Point Digital Signal Processor–TMS320C6711 Floating-Point Digital Signal Processor*, SPRS073C, Texas Instruments, Dallas, TX, 2000.

25. *TMS320C6000 CPU and Instruction Set Reference Guide*, SPRU189F, Texas Instruments, Dallas, TX, 2000.

26. *TMS320C6000 Assembly Language Tools User's Guide*, SPRU186K, Texas Instruments, Dallas, TX, 2002.

27. *TMS320C6000 Peripherals Reference Guide*, SPRU190D, Texas Instruments, Dallas, TX, 2001.

28. *TMS320C6000 Optimizing C Compiler User's Guide*, SPRU187K, Texas Instruments, Dallas, TX, 2002.

29. *TMS320C6000 Technical Brief*, SPRU197D, Texas Instruments, Dallas, TX, 1999.

30. *TMS320C64x Technical Overview*, SPRU395, Texas Instruments, Dallas, TX, 2000.

31. *TMS320C6x Peripheral Support Library Programmer's Reference*, SPRU273B, Texas Instruments, Dallas, TX, 1998.

32. *Code Composer Studio User's Guide*, SPRU328B, Texas Instruments, Dallas, TX, 2000.

33. *Code Composer Studio Getting Started Guide*, SPRU509, Texas Instruments, Dallas, TX, 2001.

34. *TMS320C6000 Code Composer Studio Tutorial*, SPRU301C, Texas Instruments, Dallas, TX, 2000.

35. *TLC320AD535C/I Data Manual Dual Channel Voice/Data Codec*, SLAS202A, Texas Instruments, Dallas, TX, 1999.

36. *TMS320C6713 Floating-Point Digital Signal Processor*, SPRS186, Texas Instruments, Dallas, TX.

37. *TLV320AIC23 Stereo Audio Codec, 8- to 96-kHz, with Integrated Headphone Amplifier Data Manual*, SLWS106G, Texas Instruments, Dallas, TX, 2003.

38. *TMS320C6000 DSP Phase-Locked Loop (PLL) Controller Peripheral Reference Guide*, SPRU233, Texas Instruments, Dallas, TX.

39. *Migrating from TMS320C6211/C6711 to TMS320C6713*, SPRA851, Texas Instruments, Dallas, TX, 2003.

40. *How to begin Development Today with the TMS320C6713 Floating-Point DSP*, SPRA809, Texas Instruments, Dallas, TX, 2003.

41. *TMS320C6000 DSP/BIOS User's Guide*, SPRU423, Texas Instruments, Dallas, TX, 2002.

42. *TMS320C6000 Optimizing C Compiler Tutorial*, SPRU425A, Texas Instruments, Dallas, TX, 2002.

43. *TMS320C6000 Chip Support Library API User's Guide*, SPRU401F, Texas Instruments, Dallas, TX, 2003.

44. B. W. Kernigan and D. M. Ritchie, *The C Programming Language*, Prentice Hall, Upper Saddle River, NJ, 1988.

45. G. R. Gircys, *Understanding and Using COFF*, O'Reilly & Associates, Newton, MA, 1988.

2

Input and Output with the DSK

- Input and output with the onboard AIC23 stereo codec
- Programming examples using C code

2.1 INTRODUCTION

Typical applications using DSP techniques require at least the basic system shown in Figure 2.1, consisting of analog input and output. Along the input path is an antialiasing filter for eliminating frequencies above the *Nyquist frequency*, defined as one-half of the sampling frequency F_s. Otherwise, aliasing occurs, in which case a signal with a frequency higher than one-half F_s is disguised as a signal with a lower frequency. The sampling theorem tells us that the sampling frequency must be at least twice the highest-frequency component f in a signal, so that

$$F_s > 2f$$

which is also

$$1/T_s > 2(1/T)$$

where T_s is the sampling period, or

$$T_s < T/2$$

Digital Signal Processing and Applications with the C6713 and C6416 DSK   By Rulph Chassaing
ISBN 0-471-69007-4   Copyright © 2005 by John Wiley & Sons, Inc.

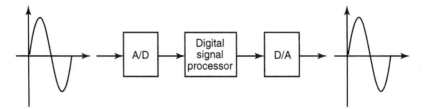

FIGURE 2.1. DSP system with input and output.

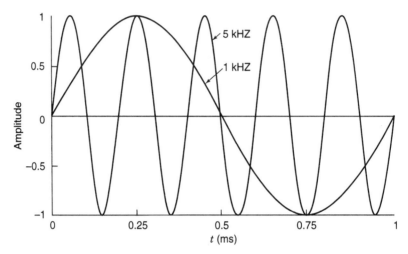

FIGURE 2.2. Aliased sinusoidal signal.

The sampling period T_s must be less than one-half the period of the signal. For example, if we assume that the ear cannot detect frequencies above 20 kHz, we can use a lowpass input filter with a bandwidth or cutoff frequency at 20 kHz to avoid aliasing. We can then sample a music signal at $F_s > 40$ kHz (typically, 44.1 or 48 kHz) and remove frequency components higher than 20 kHz. Figure 2.2 illustrates an aliased signal. Let the sampling frequency $F_s = 4$ kHz, or a sampling period of $T_s = 0.25$ ms. It is impossible to determine whether it is the 5- or 1-kHz signal that is represented by the sequence $(0, 1, 0, -1)$. A 5-kHz signal will appear as a 1-kHz signal; hence, the 1-kHz signal is an aliased signal. Similarly, a 9-kHz signal would also appear as a 1-kHz aliased signal.

2.2 TLV320AIC23 (AIC23) ONBOARD STEREO CODEC FOR INPUT AND OUTPUT

The DSK board includes the TLV320AIC23 (AIC23) codec for input and output. The ADC circuitry on the codec converts the input analog signal to a digital representation to be processed by the DSP. The maximum level of the input signal to be converted is determined by the specific ADC circuitry on the codec, which is 6 V p-p with the onboard codec. After the captured signal is processed, the result needs to be sent to the outside world. Along the output path in Figure 2.1 is a DAC, which performs the reverse operation of the ADC. An output filter smooths out or reconstructs the output signal. ADC, DAC, and all required filtering functions are performed by the single-chip codec AIC23 on board the DSK.

The AIC23 is a stereo audio codec based on sigma–delta technology [1–5]. The functional block diagram of the AIC23 codec is shown in Figure 2.3. It performs all the functions required for ADC and DAC, lowpass filtering, oversampling, and so on. The AIC23 codec contains specifications for data transfer of words with length 16, 20, 24, and 32 bits. A diagram of the AIC23 codec interfaced to the C6713 DSK is shown in *6713_dsk_schem.pdf*, included with the CCS package.

Sigma–delta converters can achieve high resolution with high oversampling ratios but with lower sampling rates. They belong to a category in which the sampling rate can be much higher than the Nyquist rate. Sample rates of 8, 16, 24, 32, 44.1, 48, and 96 kHz are supported and can be readily set in the program.

A digital interpolation filter produces the oversampling. The quantization noise power in such devices is independent of the sampling rate. A modulator is included to shape the noise so that it is spread beyond the range of interest. The noise spectrum is distributed between 0 and $F_s/2$, so that only a small amount of noise is within the signal frequency band. Therefore, within the actual band of interest, the noise power is considerably lower. A digital filter is also included to remove the out-of-band noise.

A 12-MHz crystal supplies the clocking to the AIC23 codec (as well as to the DSP and the USB interface). Using this 12-MHz master clock, with oversampling rates of $250 F_s$ and $272 F_s$, an exact audio sample rate of 48 kHz (12 MHz/250) and a CD rate of 44.1 kHz (12 MHz/272) can be obtained. The sampling rate is set by the codec's register SAMPLERATE.

The ADC converts an input signal into discrete output digital words in a 2's-complement format that corresponds to the analog signal value. The DAC includes an interpolation filter and a digital modulator. A decimation filter reduces the digital data rate to the sampling rate. The DAC's output is first passed through an internal lowpass reconstruction filter to produce an output analog signal. Low noise performance for both ADC and DAC is achieved using oversampling techniques with noise shaping provided by sigma–delta modulators.

Communication with the AIC23 codec for input and output uses two multi-channel buffered serial ports McBSPs on the C6713. McBSP0 is used as a unidirectional channel to send a 16-bit control word to the AIC23. McBSP1 is used as a bidirectional channel to send and receive audio data.

Alternative I/O daughter cards can be used for input and output. Such cards can plug into the DSK through the external peripheral interface 80-pin connector J3 on the DSK board.

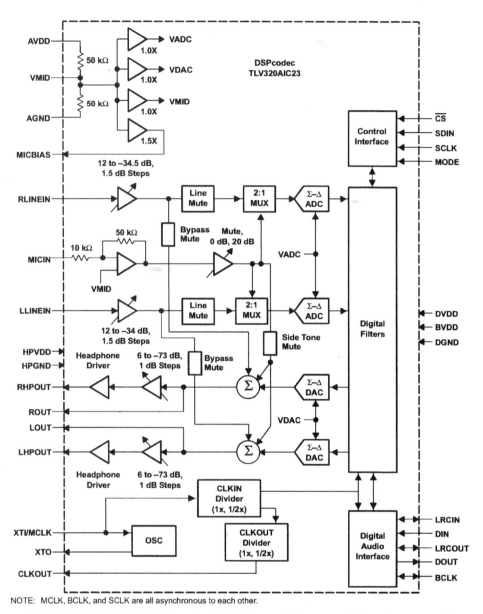

FIGURE 2.3. TLV320AIC23 codec block diagram (Courtesy of Texas Instruments).

2.3 PROGRAMMING EXAMPLES USING C CODE

Several examples follow to illustrate input and output with the DSK. They are included to familiarize you with both the hardware and software tools and provide some background to implement a specific application. The example *sine2sliders* illustrates the use of two sliders, an echo example demonstrates the effects of a variable-length buffer on an echo, a noise generator example is used in Chapter 4 as the input to a digital filter, an example illustrates the use of onboard flash memory, and so on. A list of all the examples included in this book appears on pages xv–xviii.

Example 2.1: Loop Program Using Interrupt (loop_intr)

This example illustrates input and output with the AIC23 codec. Figure 2.4 shows the C source program *loop_intr.c*, which implements the loop program. It is interrupt-driven using INT11, as in Example 1.2.

This program example is very important since it can be used as a base program to build on. For example, to implement a digital filter, one would need to insert the appropriate algorithm between the input and output functions. The two functions input_sample and output_sample, as well as the function comm_intr, are included in the communication support file *C6713dskinit.c*. This is done so that the C source program is kept as small as possible. The file *C6713dskinit.c* can be used as a "black box program" since it is used in many examples throughout this book.

After the initialization and selection/enabling of an interrupt, execution waits within the infinite while loop until an interrupt occurs. Upon interrupt, execution proceeds to the ISR c_int11, as specified in the vector file vectors_intr.asm. An interrupt occurs every sample period $T_s = 1/F_s = 1/(8\,\text{kHz}) = 0.125\,\text{ms}$, at which time an input sample value is read from the codec's ADC and then sent as output to the codec's DAC.

Execution returns from interrupt to the while(1) statement waiting for a subsequent interrupt. [Note that in lieu of waiting within the while(1) infinite loop,

```
//Loop_intr.c Loop program using interrupt.Output=delayed input

#include "dsk6713_aic23.h"              //codec-DSK support file
Uint32 fs=DSK6713_AIC23_FREQ_8KHZ;     //set sampling rate

interrupt void c_int11()               //interrupt service routine
{
    short sample_data;

    sample_data = input_sample();      //input data
    output_sample(sample_data);        //output data
    return;
}

void main()
{
    comm_intr();                       //init DSK, codec, McBSP
    while(1);                          //infinite loop
}
```

FIGURE 2.4. Loop program using interrupt (loop_intr.c).

one could be processing code.] Upon interrupt, execution proceeds to ISR, "services" the necessary task dictated by ISR, then returns to the calling function waiting for the occurrence of a subsequent interrupt.

1. Within the function *output_sample*, support functions from the BSL are included to write data using the two serial ports: McBSP0 for control and McBSP1 for data transfer (*MCBSP_write*). Most of the programs in the book will output using 16 bits. In this fashion, output_sample is made to default to the left 16-bit channel and no adapter need be used (see the comments in C6713dskinit.c). Otherwise, one would need to use output_right_sample.

2. Within the function *comm_intr*, the following tasks are performed.

   (a) Initialize the DSK.

   (b) Configure/select INT11.

   (c) Enable the specific interrupt.

   (d) Enable the global enable interrupt (GIE) bit and the nonmaskable interrupt.

   (e) Initiate communication.

The interrupt functions called for the tasks above are within the board and chip support files included with CCS.

Create and build this project as **loop_intr**. The main C source file is in the folder *loop_intr*. Use the same support files as in Example 1.2: the vector file for the interrupt-driven and linker command file located in the folder *support*, and the run-time support, board support, and chip support library files that can be added with the building option for the linker.

Input a sinusoidal waveform to the LINE IN connector on the DSK, with an amplitude of approximately 2 V p-p and a frequency between approximately 1 and 3 kHz. Connect the output of the DSK, LINE OUT to a speaker or to an oscilloscope and verify a tone of the same input frequency, but attenuated to approximately 0.8 V p-p. Using an oscilloscope, the output is a delayed version of the input signal.

Increase the amplitude of the input sinusoidal waveform beyond 6 V p-p and observe that the output signal becomes distorted.

Input with Gain

To adjust the gain of the left line-input channel, the corresponding header support file *c6713dskinit.h* of the communication/init "black box" file needs to be modified slightly. First, copy this header file AND *c6713dskinit.c from* the *support* folder into the folder *loop_intr* so that you do not modify the original header file. Remove the init file from the project and replace it with the one in the folder *loop_intr*. This will keep the original init support file unchanged in the folder *support*. Modify the set-up register 0, which controls the left input volume, from 0x0017 to 0x001c in order to increase the left line-input volume.

Rebuild the project, making sure that you are adding *c6713dskinit.c* from the folder *loop_intr* (and not from the folder *support*). In this fashion, the corresponding header file *c6713dskinit.h* that will be included will come from that same folder.

Load/run the executable file *loop_intr.out*, and verify that the output amplitude is not attenuated and is the same as the input amplitude of 2 V p-p. Values for the set-up register 0 from 0x0018 to 0x001c will cause the output amplitude to increase from 0.8 to 2 V p-p.

The left input channel was selected since *input_sample* and *output_sample* default to the left channel. Otherwise, if the right line-input volume is to be increased by modifying the set-up register 1, an adapter/connector with two inputs and one single-ended output connections would be needed. See Example 2.3 (*loop_stereo/ sine_stereo*).

Input from a Microphone

To select an input from a microphone in lieu of line input, modify the header file set-up register 4 from 0x0011 to 0x0015 (third LSB as a 1) so that the ADC gets its input from MIC IN. The microphone input and line input are multiplexed, and only one is active at a time. Rebuild the project to verify your output, with the input to the MIC IN connector.

Example 2.2: Loop Program Using Polling (loop_poll)

This example implements a polling-based loop program to illustrate the input and output of a sample value every sample period T_s. Note that the program loop_intr.c in Example 2.1 is an interrupt-driven program. The C source program loop_poll.c shown in Figure 2.5 implements this loop program. The

```
//loop_poll.c Loop program using polling.Output=delayed input


#include "DSK6713_AIC23.h"          //codec-DSK file support
Uint32 fs=DSK6713_AIC23_FREQ_8KHZ;  //set sampling rate

void main()
{
  short sample_data;

  comm_poll();                      //init DSK, codec, McBSP
  while(1)                          //infinite loop
  {
    sample_data = input_sample();   //input sample
    output_sample(sample_data);     //output sample
  }
}
```

FIGURE 2.5. Loop program using polling (loop_poll.c).

polling technique uses a continuous procedure of testing when the data are ready. Although it is simpler than the interrupt technique, it is less efficient since the input and output data need to be continuously tested to determine when they are ready to be received or transmitted.

1. The input to the ADC is from the data receive register (DRR) of the McBSP1. Since this is a polling-driven program, the SPCR bit 1, which is the receive ready register (RRDY), is first tested to determine if it is a 1 or enabled (see Figure B.8). Within `input_sample`, execution of the statement

   ```
   While (!MCBSP_rrdy())
   ```

   remains in an infinite loop until RRDY becomes 1 or enabled. Execution then proceeds to read/receive the data.

2. Within the function `output_sample`, the MCBSP1 *writes* the output from the DAC to the data transmit register (DXR) of McBSP1. Since this is a polling-driven program, the transmit ready register (XRDY) bit 17 of SPCR (see Figure B.8) is first tested to see if it is a 1 or enabled. Within `output_sample`, execution of the statement

   ```
   While (!MCBSP_xrdy())
   ```

   remains in an infinite loop until the transmit ready register becomes 1 or enabled. Execution then proceeds to transmit/write the data.

The same support files as in Example 1.1 are used: the "black box" communication/init file *c6713dskinit.c*, the vector file `vectors_poll.asm`, the linker command file `c6713dsk.cmd` (all three from the folder `support`), and the three library-support files.

Create and build this project as **loop_poll**. Use the same input as in Example 2.1 and verify the same results.

Example 2.3: Stereo Input and Stereo Output (loop_stereo/sine_stereo)

Loop Program with Stereo Input and Stereo Output (loop_stereo)

This example demonstrates input and output using the stereo capability of the onboard AIC23 codec. It requires the use of an adapter with two inputs and one output that connects to the DSK. Such an adapter has one input connector white (or silver) that represents the left channel and another input connector red (or gold) that represents the right channel. This adapter becomes essential for some of the examples on adaptive filtering that require two separate input signals, processing each input separately. Figure 2.6 shows the loop program *loop_stereo* to illustrate.

```
//Loop_stereo.c   Stereo input and output with both channels

#include "dsk6713_aic23.h"          //codec-DSK support file
Uint32 fs=DSK6713_AIC23_FREQ_8KHZ;  //set sampling rate

#define    LEFT  0                   //reversed in init file
#define    RIGHT 1
union {Uint32 combo; short channel[2];} AIC23_data;

interrupt void c_int11()            //interrupt service routine
{
AIC23_data.combo = input_sample();  //input 32-bit sample

output_left_sample(AIC23_data.channel[LEFT]);//left channels for I/O

return;
}

void main()                         //main function
{
     comm_intr();                   //init DSK, codec, McBSP
     while(1);                      //infinite loop
}
```

FIGURE 2.6. Loop program with stereo input and output (`loop_stereo.c`).

Within the function `input_sample`, support functions from the BSL are included to read a 32-bit data. The function *input_sample* captures 32-bit data, 16 bits from the left input channel and 16 bits from the right input channel. The *union* statement is used to process each channel independently. The union of *AIC23_data* and *combo* contains these 32-bit input data. The line of code for output is from the left channel (by default) to output 16-bit data from the left input channel.

Build and run this project as **loop_stereo** using the support files as in Example 1.2 for an interrupt-driven program. The main C source file *loop_stereo.c* is contained in the folder `loop_stereo`. Connect a 1 kHz (with approximate amplitude of 2 V p-p) sine wave into the left input channel and a 2-kHz sine wave into the right input channel. Verify that the left (default) output channel has the same input signal frequency of 1 kHz, but reduced in amplitude (as expected). You do not need a second adapter for the output side since the output defaults to the left channel. Change the output line of code to

output_left_sample(AIC23_data.channel[RIGHT]);

and verify that the output is the 2-KHz sine wave from the right input channel. With the line of code

output_right_sample(AIC23_data.channel[RIGHT]);

```
//Sine_stereo.c Sine generation with output to both channels

#include "dsk6713_aic23.h"           //codec-DSK support file
Uint32 fs=DSK6713_AIC23_FREQ_8KHZ;   //set sampling rate

#define LEFT  0                       //reversed in init file
#define RIGHT 1
union {Uint32 combo; short channel[2];} AIC23_data;

short loop = 0, gain = 10;
short sine_table[8]={0,707,1000,707,0,-707,-1000,-707};//sine values

interrupt void c_int11()             //interrupt service routine
{
AIC23_data.channel[RIGHT]=sine_table[loop]*gain; //for right channel
AIC23_data.channel[LEFT]=sine_table[loop]*gain;  //for left channel

output_sample(AIC23_data.combo);     //output to both channels

if (++loop > 7) loop = 0;            //reint index if @ end of table
}

void main()
{
    comm_intr();                     //init DSK, codec, McBSP
    while(1) ;                       //infinite loop
}
```

FIGURE 2.7. Sine generation with stereo outputs (sine_stereo.c).

two adapters are required to verify that the output from the right channel is the 2-kHz sine wave from the right input channel. You can also use one adapter at the input side to capture the two different signals and one stereo cable at the output side.

Experiment with this project, inputting different signals into each channel and outputting from each channel using adapters and stereo cable. Verify that you can select each input and output channel independently.

Sine Generation with Stereo Output (sine_stereo)

Figure 2.7 shows the C source file *sine_stereo.c*, included in the folder *sine_stereo*, to illustrate further the codec as a stereo device. Build and run this project as **sine_stereo**. Verify that the generated 1 kHz sinusoid is through both output channels, using an adapter or stereo cable at the output side of the DSK.

Example 2.4: Sine Generation with Two Sliders for Amplitude and Frequency Control (sine2sliders)

The polling-based program *sine2sliders.c* in Figure 2.8 generates a sine wave. Two sliders are used to vary both the amplitude (gain) and the frequency of the sinusoid

```
//Sine2sliders.c Sine generation with different # of points

#include "DSK6713_AIC23.h"              //codec-DSK interface support
Uint32 fs=DSK6713_AIC23_FREQ_8KHZ;     //set sampling rate
short loop = 0;
short sine_table[32]={0,195,383,556,707,831,924,981,1000,
      981,924,831,707,556,383,195,0,-195,-383,-556,-707,-831,-924,
      -981,-1000,-981,-924,-831,-707,-556,-383,-195};   //sine data
short gain = 1;                        //for gain slider
short frequency = 2;                   //for frequency slider

void main()
{
 comm_poll();                          //init DSK,codec,McBSP
 while(1)                              //infinite loop
 {
 output_sample(sine_table[loop]*gain);//output scaled value
 loop += frequency;                    //incr frequency index
 loop = loop % 32;                     //modulo 32 to reinit index
 }
}
```

FIGURE 2.8. Sine generation making use of two sliders to control the amplitude and frequency of the sine wave generated (sine2sliders.c).

```
/*Sine2sliders.gel Two sliders to vary gain and frequency*/

menuitem "Sine Parameters"

slider Gain(1,8,1,1,gain_parameter)              /*incr by 1,up to 8*/
      {
        gain = gain_parameter;                   /*vary gain*/
      }

slider Frequency(2,8,2,2,frequency_parameter) /*incr by 2,up to 8*/
      {
        frequency = frequency_parameter;         /*vary frequency*/
      }
```

FIGURE 2.9. GEL file with two slider functions to control the amplitude and frequency of the sine wave generated (sine2sliders.gel).

generated. Using a lookup table with 32 points, the variable frequency is obtained by selecting a different number of points per cycle. The gain slider scales the volume/amplitude of the waveform signal. The appropriate GEL file *sine2sliders.gel* is shown in Figure 2.9.

The 32 sine data values in the table or buffer correspond to $\sin(t)$, where $t = 0$, $11.25, 22.5, 33.75, 45, \ldots, 348.75$ degrees (scaled by 1000). The frequency slider takes on a value from 2 to 8, incremented by 2. The modulo operator is used to test when the end of the buffer that contains the sine data values is reached. When the loop index reaches 32, it is reinitialized to zero. For example, with the frequency slider at position 2, the loop or frequency index steps through every other value in the table. This corresponds to 16 data values within one cycle.

Build this project as **sine2sliders**. Use the appropriate support files for a polling-driven program. The main C source file *sine2sliders.c* is contained in the folder **sine2sliders**. Verify that the frequency generated is $f = F_s/16 = 500\,\text{Hz}$. Increase the slider position (the use of a slider was introduced in Example 1.1) to 4, 6, 8 and verify that the signal frequencies generated are 1000, 1500, and 2000 Hz, respectively. Note that when the slider is at position 4, the loop or frequency index steps through the table selecting the eight values (per cycle): sin[0], sin[4], sin[8], ..., sin[28] that correspond to the data values 0, 707, 1000, 707, 0, −707, −1000, and −707. The resulting frequency generated is then $f = F_s/8 = 1\,\text{kHz}$ (as in Example 1.1).

Example 2.5: Loop Program with Input Data Stored in Memory (loop_store)

The program *loop_store.c* in Figure 2.10 is an interrupt-based program and is included in the folder **loop_store**. Each time an interrupt INT11 occurs, a sample is read from the codec's ADC and written to the codec's DAC. Furthermore, each sample is written to a 512-element circular buffer implemented using an array *buffer* and an index i that is incremented after each sample is stored. The index is reset to zero when it reaches the end of the buffer. Consequently, the array always contains the 512 most recent sample values.

```
//Loop_store.c Data acquisition.Input data stored also into buffer

#include "DSK6713_AIC23.h"            //codec-DSK interface support
Uint32 fs=DSK6713_AIC23_FREQ_8KHZ;   //set sampling rate
#define BUFFER_SIZE 512               //buffer size
short buffer[BUFFER_SIZE];            //buffer where data is stored
int i = 0;

interrupt void c_int11()              //interrupt service routine
{
 output_sample((short)input_sample());//output acquired data
 buffer[i] =((short)input_sample());//store input data into buffer
 i++;                                 //increment buffer index
 if (i==BUFFER_SIZE) i = 0;           //reinit index if buffer full
 return;                              //return from ISR
}

void main()
{
 comm_intr();                         //init DSK, codec, McBSP
 while(1);                            //infinite loop
}
```

FIGURE 2.10. Loop program with input data stored in memory (loop_store.c).

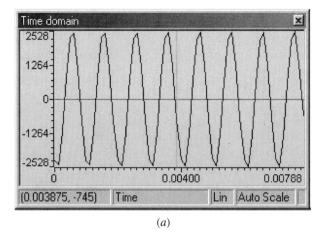

(a)

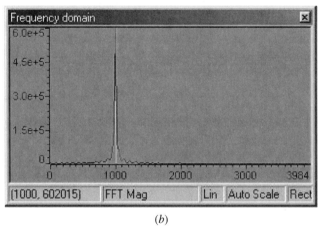

(b)

FIGURE 2.11. CCS graphs with the `loop_store` program: (a) time-domain plot of stored input data representing a 1-kHz sine wave; (b) FFT magnitude of stored data representing a 1-kHz sine wave.

Build this project as **loop_store**. Input a sinusoidal signal with amplitude of approximately 1/2 V p-p and a frequency of 1 kHz. Run and verify your output results.

Use CCS to plot the stored input data in both the time and frequency domains (see also Example 1.2). Select View → Graph → Time/Frequency. For the time-domain plot, specify a starting address "buffer," 512 points for the acquisition buffer size, 64 points for the data size display (for a clearer plot), a 16-bit signed integer for the data type, and 8000 for the sampling rate. Verify the 1-kHz time-domain sine-wave plot within CCS, as shown in Figure 2.11a.

Select View → Graph → Time/Frequency again and FFT magnitude for display to obtain a frequency-domain plot of the stored input data. Specify a display data size of 512 with an FFT order of $M = 9$, where $2^M = 512$. The spike at 1 kHz in Figure 2.11b represents the 1-kHz sine wave plot within CCS.

```
//Loop_print.c Data acquisition.Loop with data printed to a file

#include "DSK6713_AIC23.h"          //codec-DSK support file
Uint32 fs=DSK6713_AIC23_FREQ_8KHZ;  //set sampling rate
#include <stdio.h>
#define BUFFER_SIZE 64              //buffer size
int i=0, j=0;
short buffer[BUFFER_SIZE];          //buffer for data
FILE *fptr;                         //file pointer

interrupt void c_int11()            //ISR
{
 buffer[i]=((short)input_sample()); //store data in buffer
 i++;                               //increment buffer count
 if (i==BUFFER_SIZE - 1)            //if buffer full
   {
    fptr = fopen("sine.dat","w");   //create output data file
    for (j=0; j<BUFFER_SIZE; j++)
      fprintf(fptr,"%d\n",buffer[j]);//write buffer data to file
    fclose(fptr);                   //close file
    i = 0;                          //initialize buffer count
    puts("done");                   //finished storing to file
   }
 output_sample((short)input_sample());  //output data
 return;                            //return from ISR
}

void main()
{
 comm_intr();                       //init DSK, codec, McBSP
 puts("start\n");                   //print "start" indicator
 while(1);                          //infinite loop
}
```

FIGURE 2.12. Loop program to store I/O data in memory and in a file (`loop_print.c`).

Example 2.6: Loop with Data in a Buffer Printed to a File (`loop_print`)

This example extends the preceding loop program so that the acquired input data are stored in a memory buffer and then printed to a file. Figure 2.12 shows the C source program `loop_print.c` (included in the folder **loop_print**) that implements this example. It takes a long time (more than 3000 cycles) to execute the printf statement in the program (see Example 1.3). This can be reduced to about 30 cycles using DSP/BIOS, introduced in Chapter 9.

After initialization of the DSK, the *puts* statement prints the word start as an indicator within the CCS command window; then execution proceeds to the infinite while loop. Upon each interrupt, execution proceeds to ISR, and a newly acquired data value is stored in a buffer of size 64.

The buffer index i is incremented to store each new sampled data value. When the end of the buffer is reached, indicating that the buffer is full, a file sine.dat is "opened" and the contents of the buffer are written into that file. Then the indicator done is printed within the CCS window. This process is repeated continuously so that a new set of 64 data points is acquired, and the done indicator is again displayed (after each set of data fills the buffer and is written to sine.dat).

Build and run this project as ***loop_print***. Input a sine-wave signal of approximately $\frac{1}{2}$ V p-p with a 1-kHz frequency. Halt execution after the indicator done is displayed. The buffer of 64 input data representing the sine wave can be retrieved from the file *sine.dat* in the same folder loop_print\Debug. Note that the third set of 64 points will be stored in the buffer and printed in the file *sine.dat* if execution of the program is halted after the third done indicator. A plot program or MATLAB can be used to plot sine.dat and verify a 1-kHz sine wave. You can also verify your results by plotting the content of the buffer within CCS, as in the previous example. Note that the output is not displayed appropriately in real time due to the slow execution of the print statement. You can comment the section of code that is associated with printing the input data into a file to verify that a loop program is also implemented.

Example 2.7: Square-Wave Generation Using a Lookup Table (squarewave)

This example generates a square wave using a lookup table. Figure 2.13 shows a listing of the program *squarewave.c* (located in the folder **squarewave**) that implements this project example. A buffer of size 64 is created. Within *main*, the buffer table is loaded with data: the first half with $(2^{15} - 1) = 32{,}767$ and the second half with $-2^{15} = -32{,}768$. Upon each interrupt that occurs every sample period T_s, one data value from the buffer is sent for output. After each data value from the table is output, execution returns to the infinite while loop, waiting for the next interrupt to occur and output the subsequent value in the table. When the end of the buffer (table) is reached, the buffer index is reinitialized to the beginning of the buffer.

Build and run this project as **squarewave**. Verify a square-wave output signal of approximately 3 V p-p. Note that the valid input data to the codec are between -2^{15} and $(2^{15} - 1)$ or between $-32{,}768$ and $32{,}767$. Change the values in the first half of the table using 0x8000 = 32,768 in lieu of 0x7FFF = 32,767. Rebuild/run and verify that the square-wave signal is no longer generated.

Note that increasing the number of points in the table produces a more pronounced charging/discharging effect (since it is AC coupled) due to the output capacitor (see the block diagram of the AIC23 codec). For example, with 64 points, the fundamental frequency is at 8 kHz/64 = 125 Hz. Doubling the number of points

```
//Squarewave.c Generates a squarewave using a look-up table

#include "dsk6713_aic23.h"          //codec-DSK interface support
Uint32 fs=DSK6713_AIC23_FREQ_8KHZ;  //set sampling rate
#define table_size (int)0x40        //size of table=64
short data_table[table_size];       //data table array
int i;

interrupt void c_int11()            //interrupt service routine
{
 output_sample(data_table[i]);      //output value each Ts
 if (i < table_size) ++i;           //if table size is reached
 else i = 0;                        //reinitialize counter
 return;                            //return from interrupt
}

main()
{
for(i=0; i<table_size/2; i++)       //set 1st half of buffer
   data_table[i] = 0x7FFF;          //with max value (2^15)-1
for(i=table_size/2;i<table_size;i++) //set 2nd half of buffer
   data_table[i] = -0x8000;         //with -(2^15)
i = 0;                              //reinit counter
comm_intr();                        //init DSK, codec, McBSP
while (1);                          //infinite loop
}
```

FIGURE 2.13. Square-wave generation program (squarewave.c).

will double the period of the square wave, and the discharging effect will be more pronounced (time constant reduced relative to one-half of the period of the square wave). Change the sampling frequency to 16 or 24 kHz and verify that the charching/discharging effect of the capacitor is less pronounced.

Example 2.8: Ramp Generation Using a Lookup Table (ramptable)

Figure 2.14 shows a listing of the program ramptable.c, which generates a ramp using a lookup table. A buffer of size 1024 is created. Within main, the buffer table is loaded with 1024 values: 0, 0x20, 0x40, ..., or 0, 32, 64, ..., 32,736 in decimal.

Build and run this project as **ramptable**. Verify that a ramp with an approximate peak value of 1.5 V is generated. The ramp has a negative slope due to the 2's-complement format of the AIC23 codec.

Replace the value 0x20 with -0x20 and verify that a ramp is generated with a positive slope with a peak value of approximately 1.5 V.

Note that reducing the number of points will result in a "cleaner" ramp due to AC coupling, as in the previous square wave generation example.

```
//Ramptable.c Generates a ramp using a look-up table

#include "dsk6713_aic23.h"              //codec-dsk support file
Uint32 fs=DSK6713_AIC23_FREQ_8KHZ;     //set sampling rate
#define table_size (int)0x400          //size of table=1024
short data_table[table_size];          //data table array
int i;

interrupt void c_int11()               //interrupt service routine
{
 output_sample(data_table[i]);         //ramp value for each Ts
 if (i < table_size-1) i++;            //if table size is reached
 else i = 0;                           //reinitialize counter
 return;                               //return from interrupt
}

main()
{
  for(i=0; i < table_size; i++)
    {
     data_table[i] = 0x0;              //clear each buffer location
     data_table[i] = i * 0x20;         //set to 0,32,64,96,..,32736
    }
  i = 0;                               //reinit counter
  comm_intr();                         //init DSK, codec, McBSP
  while (1);                           //infinite loop
}
```

FIGURE 2.14. Ramp generation program using a lookup table (ramptable.c).

Example 2.9: Ramp Generation without a Lookup Table (ramp)

Example 2.8 is based on loading a table with a set of values, then outputting each value in the table every sample period, wrapping around when the end of the table is reached. Figure 2.15 shows a listing of the program ramp.c, which generates a ramp using an alternative approach to Example 2.8. Starting with an initial output value of 0, the output value is incremented by 0x20 every sample period T_s. The values sent for output are then 0, 32, 64, 96, . . . , 32,736.

Build and run this project as **ramp**. Verify that a ramp with a negative slope and an approximate peak value of 3 V is generated. To obtain a ramp with a positive slope, change output to

```
output -= 0x20;
```

so that the output becomes 0, –32, –64, . . . , –32,736. Also change the if statement to reinitialize output, or

```
if  (output == -0x7FFF)
```

```
//Ramp.c Generates a ramp

#include "dsk6713_aic23.h"          //codec-DSK support file
Uint32 fs=DSK6713_AIC23_FREQ_8KHZ; //set sampling frequency
short output;

interrupt void c_int11()           //interrupt service routine
{
 output_sample(output);            //output each sample period
 output += 0x20;                   //incr output value
 if (output >= 0x7FFF)             //if peak is reached
 output = 0;                       //reinitialize
 return;                           //return from interrupt
}

 void main()
{
 output = 0;                       //init output to zero
 comm_intr();                      //init DSK, codec, McBSP
 while(1);                         //infinite loop
}
```

FIGURE 2.15. Ramp generation program without a lookup table (`ramp.c`).

Rebuild, and verify that the output is now a ramp with a positive slope with an approximate peak voltage of 3 V.

Example 2.10: Echo (echo)

Figure 2.16 shows a listing of the program *echo.c*, which echoes an input signal. The length or size of the buffer determines the echo effect. A buffer size of 2000 barely generates a clear echo, while a size of 16,000 produces too much delay and the effect is more of a repeat. The output consists of a newly acquired sample added to the oldest sample already stored in the buffer. If the buffer size is too small, the time delay between the newest and oldest samples is too small to create an audible echo effect. The oldest sample is attenuated to enhance the echo effect.

After a new sample is acquired and stored at memory location x, the output becomes the sum of the new sample and the oldest sample stored at memory location x+1, where x = 0, 1, 2, . . . , 2998. When the buffer index reaches the end of the buffer (`buffer[2999]`), where a newly acquired sample is stored, the oldest sample is at the beginning of the buffer.

Build and run this project as **echo**. A wave file, *Theforce.wav* (included on the CD), can be used as input. Play this file continuously with loop-around. The shareware utility Goldwave (described in Appendix E) can be used to play this file.

```
//Echo.c Echo effect changed with size of buffer (delay)

#include "dsk6713_aic23.h"              //codec-DSK support file
Uint32 fs=DSK6713_AIC23_FREQ_8KHZ;      //set sampling rate
short input, output;
short bufferlength = 3000;              //buffer size for delay
short buffer[3000];                     //create buffer
short i = 0;
short amplitude = 5;                    //to vary amplitude of echo

interrupt void c_int11()                //ISR
{
 input = input_sample();                //newest input sample data
 output=input + 0.1*amplitude*buffer[i];//newest + oldest samples
 output_sample(output);                 //output sample

 buffer[i] = input;                     //store newest input sample
 i++;                                   //increment buffer count
 if (i >= bufferlength) i = 0;          //if end of buffer reinit
}

main()
{
 comm_intr();                           //init DSK, codec, McBSP
 while(1);                              //infinite loop
}
```

FIGURE 2.16. Echo generation (echo.c).

Vary the size of the buffer from 1000 to 8000 and observe that a larger buffer size produces a greater delay between the newest and oldest samples.

A fading effect is obtained if the output (in lieu of the input) is stored in the buffer, using

```
buffer[i] = output;
```

Rebuild/run and verify this fading echo effect.

Example 2.11: Echo with Control for Different Effects (echo_control)

This example extends Example 2.10 to incorporate additional echo effects. Three sliders are used: one to vary the amplitude of the oldest sample, one to change the buffer size for different amounts of delay, and one to create a fading effect. The program echo_control.c, listed in Figure 2.17, implements this project using a polling-driven program (the previous example is interrupt-driven). Use the same

```
//Echo_control.c  Echo effects with fading
//3 sliders to control effects: buffer size, amplitude, fading

#include "DSK6713_AIC23.h"            //codec-DSK file support
Uint32 fs=DSK6713_AIC23_FREQ_8KHZ;   //set sampling rate
short input, output;
short buffer[8000];                  //max size of buffer
short bufferlength = 1000;           //initial buffer size
short i = 0;                         //buffer index
short delay = 3;                     //determines size of buffer
short delay_flag = 1;                //flag if buffer size changes
short amplitude = 5;                 //amplitude control by slider
short echo_type = 1;                 //1 for fading(0 with no fading)

main()
{
 short new_count;                    //count for new buffer

 comm_poll();                        //init DSK, codec, McBSP
 while(1)                            //infinite loop
 {
 output=input+0.1*amplitude*buffer[i];//newest + oldest samples
 if (echo_type == 1)                 //if fading is desired
  {
   new_count = (i-1) % bufferlength; //previous buffer location
   buffer[new_count] = output;       //to store most recent output
  }
 output_sample(output);              //output delayed sample

 input = input_sample();             //newest input sample data
 if (delay_flag != delay)            //if delay has changed
  {                                  //new buffer size
   delay_flag = delay;               //reint for future change
   bufferlength = 1000*delay;        //new buffer length
   i = 0;                            //reinit buffer count
  }
 buffer[i] = input;                  //store input sample
 i++;                                //increment buffer index
 if (i == bufferlength) i=0;         //if @ end of buffer reinit
 }
}
```

FIGURE 2.17. Echo generation with controls for different effects (echo_control.c).

.wav file, *Theforce.wav* (on the CD), for input, as in Example 2.10. The output is the sum of the newest input sample plus the oldest sample. Note that for fading, the output is stored in the buffer.

1. Build and run this project as **echo_control**.
2. Access the three sliders: amplitude, delay, and type. The GEL file *echo_control.gel* is shown in Figure 2.18. Set the amplitude slider to posi-

```
//Echo_control.gel Sliders vary time delay,amplitude,and type of echo

menuitem "Echo with Fading"

slider Amplitude(1,8,1,1,amplitude_parameter) /*incr by 1, up to 8*/
  {
   amplitude = amplitude_parameter;           /*vary amplit of echo*/
  }
slider Delay(1,8,1,1,delay_parameter)         /*incr by 1, up to 8*/
  {
   delay = delay_parameter;                    /*vary buffer size*/
  }
slider Type(0,1,1,1,echo_typeparameter)       /*incr by 1, up to 1*/
  {
   echo_type = echo_typeparameter;             /*echo type for fading*/
  }
```

FIGURE 2.18. GEL file for echo control of amplitude, delay, and fading (echo_control.gel).

tion 5, and set the delay slider to position 3. Since the delay is not equal to delay_flag, the size of the buffer has changed. The new buffer size is bufferlength = 1000 × 3 = 3000. These two slider settings correspond to the same conditions as in Example 2.10. The delay slider can take on the values 1, 2, . . . , 8, allowing for buffer lengths of 1000, 2000, 3000, . . . , 8000. Increase the delay slider to position 4, and then to position 5, to produce a longer time delay between the newest and oldest samples and listen to the echo effects.

3. The slider "type" in position 1 creates/adds a fading effect, since the output becomes the most recent output. A clearer fading effect is produced just after you stop "playing" the input .wav file.

Experiment with the three sliders for different echo effects.

Example 2.12: Sine Generation with Table Values Generated within the Program (*sinegen_table*)

This example creates one period of sine data values for a table. Then these values are output for generating a sine wave. Figure 2.19 shows a listing of the program *sinegen_table.c*, which implements this project. The frequency generated is $f = F_s/$(number of points) = 8000/10 = 800 Hz.

Build and run this project as **sinegen_table.** Verify a sine wave generated with a frequency of 800 Hz. Change the number of points to generate a 400-Hz sine wave (only table_size needs to be changed).

```
//Sinegen_table.c Generates sinusoid with generated values

#include "DSK6713_AIC23.h"            //codec-DSK support file
Uint32 fs=DSK6713_AIC23_FREQ_8KHZ;    //set sampling rate
#include <math.h>
#define table_size (short)10          //set table size
short sine_table[table_size];         //sine table array
int i;

interrupt void c_int11()              //interrupt service routine
{
 output_sample(sine_table[i]);        //output each sine value
 if (i < table_size - 1) ++i;         //incr index until end of table
 else i = 0;                          //reinit index if end of table
 return;                              //return from interrupt
}

void main()
{
float pi=3.14159;

for(i = 0; i < table_size; i++)
    sine_table[i] = 10000*sin(2.0*pi*i/table_size); //scaled values

i = 0;
comm_intr();                          //init DSK, codec, McBSP
while(1);                             //infinite loop
}
```

FIGURE 2.19. Sine-wave generation program using the table generated within a program (sinegen_table.c).

Example 2.13: Sine Generation with a Table Created by MATLAB (sin1500MATL)

This example illustrates the generation of a sinusoid using a lookup table created with MATLAB. Figure 2.20 shows a listing of the MATLAB program sin1500.m, which generates a file with 128 data points over 24 cycles. The sine-wave frequency generated is

$$f = F_s \,(\text{number of cycles})/(\text{number of points}) = 1500\,\text{Hz}$$

Run sin1500.m within MATLAB and verify the header file sin1500.h with 128 points, as shown in Figure 2.21. Different numbers of points representing sinusoidal signals of different frequencies can readily be obtained with minor changes in the MATLAB program sin1500.m.

Figure 2.22 shows a listing of the C source file sin1500MATL.c, which implements this project in real time. This program includes the header file generated by MATLAB. See also Example 2.12, which generates the table within the main C source program in lieu of using MATLAB.

```
%sin1500.m Generates 128 points representing sin(1500) Hz
%Creates file sin1500.h
for i=1:128
  sine(i) = round(1000*sin(2*pi*(i-1)*1500/8000)); %sin(1500)
end

fid = fopen('sin1500.h','w');          %open/create file
fprintf(fid,'short sin1500[128]={');   %print array name,"={"
fprintf(fid,'%d, ' ,sine(1:127));      %print 127 points
fprintf(fid,'%d' ,sine(128));          %print 128th point
fprintf(fid,'};\n');                   %print closing bracket
fclose(fid);                           %close file
```

FIGURE 2.20. MATLAB program to generate a lookup table for sine-wave data (sin1500.m).

```
short sin1500[128]={0, 924, 707, -383, -1000, -383, 707, 924, 0, -924,
-707, 383, 1000, 383, -707, -924, 0, 924, 707, -383, -1000, -383, 707,
924, 0, -924, -707, 383, 1000, 383, -707, -924, 0, 924, 707, -383,
-1000, -383, 707, 924, 0, -924, -707, 383, 1000, 383, -707, -924, 0,
924, 707, -383, -1000, -383, 707, 924, 0, -924, -707, 383, 1000, 383,
-707, -924, 0, 924, 707, -383, -1000, -383, 707, 924, 0, -924, -707,
383, 1000, 383, -707, -924, 0, 924, 707, -383, -1000, -383, 707, 924,
0, -924, -707, 383, 1000, 383, -707, -924, 0, 924, 707, -383, -1000,
-383, 707, 924, 0, -924, -707, 383, 1000, 383, -707, -924, 0, 924, 707,
-383, -1000, -383, 707, 924, 0, -924, -707, 383, 1000, 383, -707,
-924};
```

FIGURE 2.21. Sine lookup-table header file generated by MATLAB (sin1500.h).

```
//Sin1500MATL.c Generates sine from table created with MATLAB

#include "DSK6713_AIC23.h"              //codec-DSK support file
Uint32 fs=DSK6713_AIC23_FREQ_8KHZ;     //set sampling rate
#include "sin1500.h"                    //created with MATLAB
int i=0;

interrupt void c_int11()               //ISR
{
 output_sample(sin1500[i]);            //output each sine value
 if (i < 127) ++i;                     //incr until end of table
 else i = 0;
 return;                               //return from interrupt
}

void main()
{
 comm_intr();                          //init DSK, codec, McBSP
 while(1);                             //infinite loop
}
```

FIGURE 2.22. Sine generation program using a header file with sine data values generated by MATLAB (sin1500MATL.c).

Build and run this project as **sin1500MATL**. Verify that the output is a 1500-Hz sine-wave signal. Within CCS, be careful when you view the header file sin1500.h so as not to truncate it.

Example 2.14: Amplitude Modulation (AM)

This example illustrates an amplitude modulation (AM) scheme. Figure 2.23 shows a listing of the program AM.c, which generates an AM signal. The buffer *baseband* contains 20 points and represents a baseband cosine signal with a frequency of $f = F_s/20 = 400\,\text{Hz}$. The buffer *carrier* also contains 20 points and represents a carrier signal with a frequency of $f = F_s$ (number of cycles)/(number of points) = F_s/(number points per cycle) = 2 kHz. The output equation shows the baseband signal being modulated by the carrier signal. The variable amp is used to vary the modulation. The polling-driven C source program AM.c implements this project.

Build and implement this project as **AM**. Verify that the output consists of the 2-kHz carrier signal and two sideband signals. The sideband signals are at the frequency of the carrier signal + or − the frequency of the sideband signal, or at 1600 and 2400 Hz.

Load the GEL file AM.gel, increase the variable amp, and verify the baseband signal being modulated (modulation index controlled by amp,). Note that the

```
//AM.c AM using table for carrier and baseband signals

#include "DSK6713_AIC23.h"                   //codec-dsk support file
Uint32 fs=DSK6713_AIC23_FREQ_8KHZ;           //set sampling rate
short amp = 1;                               //index for modulation

void main()
{
 short baseband[20]={1000,951,809,587,309,0,-309,-587,-809,-951,
       -1000,-951,-809,-587,-309,0,309,587,809,951};//400Hz baseband
 short carrier[20] ={1000,0,-1000,0,1000,0,-1000,0,1000,0,
       -1000,0,1000,0,-1000,0,1000,0,-1000,0};       //2kHz carrier
 short output[20];
 short k;

 comm_poll();                               //init DSK, codec, McBSP
 while(1)                                   //infinite loop
  {
   for (k=0; k<20; k++)
    {
     output[k]= carrier[k] + ((amp*baseband[k]*carrier[k]/10)>>12);
     output_sample(20*output[k]);           //scale output
    }
  }
}
```

FIGURE 2.23. Amplitude modulation program (AM.c).

product of the carrier and baseband signals (within the output equation) is scaled by 2^{12} (shifted right by 12). Projects on modulation are included in Chapter 10.

Example 2.15: Sweep Sinusoid Using a Table with 8000 Points (*sweep8000*)

Figure 2.24 shows a listing of the program *sweep8000.c*, which generates a sweeping sinusoidal signal using a table lookup with 8000 points. The header file *sine8000_table.h* contains the 8000 data points that represent a one-cycle sine wave. Since the output rate is $F_s = 8\,\text{kHz}$, 8000 points are chosen to represent a 1-second interval. The file *sine8000_table.h* (in the folder sweep8000) is generated with MATLAB using

```
1000*sin(2*pi*i*start_freq/8000)
```

```
//Sweep8000.c Sweep sinusoid using table with 8000 points

#include "DSK6713_AIC23.h"          //codec-DSK support file
Uint32 fs=DSK6713_AIC23_FREQ_8KHZ;  //set sampling rate
#include "sine8000_table.h"         //one cycle with 8000 points
short start_freq = 100;             //initial frequency
short stop_freq = 3500;             //maximum frequency
short step_freq = 200;              //increment/step frequency
short amp = 30;                     //amplitude
short delay_msecs = 1000;           //# of msec at each frequency
short freq;                         //variable frequency
short t;
short i = 0;

void main()
{
 comm_poll();                               //init DSK, codec, McBSP
 while(1)                                    //infinite loop
  {
   for(freq=start_freq;freq<=stop_freq;freq+=step_freq)
    {                                        //step thru freqs
     for(t=0; t<8*delay_msecs; t++)    //output 8*delay_msecs samples
      {                                      // at each freq
       output_sample(amp*sine8000[i]);//output
       i = (i + freq) % 8000;          //next freq sample
      }
    }
  }
}
```

FIGURE 2.24. Program to generate a sweeping sinusoid using a lookup table with 8000 points (sweep8000.c).

```
//sine8000_table.h Sine table with 8000 points generated with MATLAB

short sine8000[8000]=
{0, 1, 2, 2, 3, 4, 5, 5,
6, 7, 8, 9, 9, 10, 11, 12,
13, 13, 14, 15, 16, 16, 17, 18,
19, 20, 20, 21, 22, 23, 24, 24,
25, 26, 27, 27, 28, 29, 30, 31,
31, 32, 33, 34, 35, 35, 36, 37,
38, 38, 39, 40, 41, 42, 42, 43,
44, 45, 46, 46, 47, 48, 49, 49,
50, 51, 52, 53, 53, 54, 55, 56,
57, 57, 58, 59, 60, 60, 61, 62,
63, 64, 64, 65, 66, 67, 67, 68,
69, 70, 71, 71, 72, 73, 74, 75,
75, 76, 77, 78, 78, 79, 80, 81,
82, 82, 83, 84, 85, 86, 86, 87,
88, 89, 89, 90, 91, 92, 93, 93,
94, 95, 96, 96, 97, 98, 99, 100,
100, 101, 102, 103, 103, 104, 105, 106,
107, 107, 108, 109, 110, 111, 111, 112,
.
.
.
-13, -12, -11, -10, -9, -9, -8, -7,
-6, -5, -5, -4, -3, -2, -2, -1};
```

FIGURE 2.25. Partial listing of a sine with 8000 data points (sine8000_table.h).

Figure 2.25 shows a partial listing of the file *sine8000_table.h*.

The initial frequency is set at 100 Hz and increments every 200 Hz until a stop frequency of 3500 Hz is reached. The frequencies generated are 100, 300, 500, . . . , 3500 Hz, and each frequency is generated for 1 second.

Increase delay_msecs from 1000 to 2000 for a slower sweep, since each frequency would be generated for 2 seconds rather than 1 second. If *step_freq* is increased to 700, the frequencies generated would be 100, 800, 1500, 2200, and 2900 Hz.

The index i is incremented by $i+freq$, which determines the values chosen from the table (see also Example 2.4, *sine2sliders*). For example, to generate 100 Hz, every 100th value in the table is selected to output 80 data points, corresponding to 1 cycle, that is, 8000 points over 100 cycles. With this scheme, 8000 points are always used to generate each frequency over x cycles per second.

Build and run this project as **sweep8000**. Verify the output as a sweeping sinusoid. Note that the source program *sweep8000.c* is polling-driven (use the appropriate vector file). A slider can be used to control the amplitude of the frequency generated with the variable *amp*, the duration at each frequency with *delay_msecs* (sweep speed), and the incremental frequency with *step_freq*.

```
//Noise_gen.c Pseudo-random sequence generation

#include "DSK6713_AIC23.h"           //codec-DSK support file
Uint32 fs=DSK6713_AIC23_FREQ_8KHZ;//set sampling rate
#include "noise_gen.h"               //header file for noise sequence
short fb;
shift_reg sreg;                      //shift reg structure

interrupt void c_int11()             //interrupt service routine
{
 short prnseq;                       //for pseudo-random sequence

 if(sreg.bt.b0)                      //sequence{1,-1}based on bit b0
     prnseq = -8000;                 //scaled negative noise level
 else
     prnseq = 8000;                  //scaled positive noise level
 fb  =(sreg.bt.b0)^(sreg.bt.b1);   //XOR bits 0,1
 fb ^=(sreg.bt.b11)^(sreg.bt.b13);//with bits 11,13 ->fb
 sreg.regval<<=1;                    //shift register 1 bit to left
 sreg.bt.b0 = fb;                    //close feedback path

 output_sample(prnseq);              //output scaled sequence
 return;                             //return from interrupt
}

void main()
{
 sreg.regval = 0xFFFF;               //set shift register
 fb = 1;                             //initial feedback value
 comm_intr();                        //init DSK, codec, McBSP
 while (1);                          //infinite loop
}
```

FIGURE 2.26. Pseudorandom noise sequence generation program (noise_gen.c).

Example 2.16: Pseudorandom Noise Sequence Generation (noise_gen)

The program noise_gen.c, shown in Figure 2.26, generates a pseudorandom noise sequence. It uses a software-based implementation of a maximal-length sequence technique for generating a pseudorandom sequence. An initial 16-bit seed is assigned to a register. Bits b0, b1, b11, and b13 are XORed, and the result is placed in a feedback variable. The register with the initial seed value is then shifted 1 bit to the left. The feedback variable is then assigned to bit b0 of the register. A scaled minimum or maximum is assigned to *prnseq*, depending on whether the register's bit b0 is 0 or 1. This scaled value corresponds to the noise-level amplitude. The header file *noise_gen.h* (on the CD) defines the shift register bits.

Build and run this project as **noise_gen**. You can view the noise in the time domain or hear it. Increase the noise-level amplitude for a scaled value of ±16,000 (in lieu of ±8000) and verify that the noise generated is louder. Connect the output to a spectrum analyzer. Verify that the output spectrum is relatively flat until the cutoff frequency of approximately 3800 Hz, which represents the bandwidth of the antialiasing filter on the codec AIC23.

```
//Sine_led_ctrl.c Sine generation with DIP Switch control

#include "dsk6713_aic23.h"              //codec-dsk support file
Uint32 fs=DSK6713_AIC23_FREQ_8KHZ;    //set sampling rate
short sine_table[8]={0,707,1000,707,0,-707,-1000,-707};//sine values
short loop=0, gain=10;
short j=0, k = 0;                      //delay counter
short flag = 0;                        //for LED on
short const delay = 800;               //for delay loop
short on_time = 2;                     //led is on for on-time secs

void main()
{
 comm_poll();                          //init BSL
 DSK6713_LED_init();                   //init LEDs
 DSK6713_DIP_init();                   //init DIP SWs
 while(1)                              //infinite loop
   {
    if(DSK6713_DIP_get(0)==0 &&(k<=(on_time*5))) //if SW0 pressed
     {
      if(flag==0) DSK6713_LED_toggle(0);   //LED0 toggles
      else        DSK6713_LED_off(0);      //turn LED0 off

      output_sample(sine_table[loop]*gain);//output with gain
      if (loop < 7) loop++;            //increment loop index
      else   loop = 0;                 //reinit if end of table
      if (j < delay) ++j;              //delay counter
      else
        {
         j = 0;                        //reset delay counter
         if (flag == 1)
           {
            flag = 0;                  //if flag=1 toggle LED
            k++;
           }
         else flag = 1;                //toggle flag
        }
     }
    else
      {
       DSK6713_LED_off(0);             //turn off LED0
       if(DSK6713_DIP_get(0)==1) k=0;//If LED0 off reset counter
      }
   }
}
```

FIGURE 2.27. Sine generation with a dip switch control program (sine_led_ctrl.c).

Example 2.17: Sine Generation with DIP Switch Control (sine_led_ctrl)

The program sine_led_ctrl.c, shown in Figure 2.27, implements a sine generation using a DIP switch to control how long the sine wave is generated. When switch #0 is pressed, LED #0 toggles and a 1-kHz sine wave is generated, with the duration determined by the variable in the program *on-time*. A slider is used to vary *on-time* between 1 and 10 seconds. Unlike Example 1.1, after switch #0 is pressed, the sine is generated but only for *on-time* seconds.

Build and run this project as **sine_led_ctrl**. Press switch #0 and verify that both LED #0 toggles and a 1-kHz sine wave is generated for 2 seconds, with *on-time* set at 2 sec in the program. Load the gel file *sine_led_ctrl.gel* (on the CD) to obtain the slider. Increase the slider value to 8. Is the sine wave generated and LED #0 toggles for approximately 8 seconds after switch #0 is pressed?

Note that the delay associated with turning on the LED is incorporated within *delay* in determining the value of the delay loop set in the program.

Example 2.18: Use of External Memory to Record Voice (record)

This example illustrates the use of the *pragma* directive in a C source program to store data in external (in lieu of internal) memory. The DSK board includes 16 MB of SDRAM external memory that provides a larger section of memory than the on-chip internal memory. Figure 2.28 shows the C source program *record.c* that implements this project example. It defines a buffer size of 2400000 allowing approximately $(2.4 \times 10^6)/8000 = 300$ seconds = 5 minutes of speech to be recorded and stored in external memory, sampling at 8 kHz.

The *pragma* directive specifies a section called buffer to reside in a memory section specified by .EXTRAM. The following lines need to be added in the linker command file. Within SECTIONS, add

$$.EXTRAM :> SRAM\_EXT$$

and within MEMORY, add

$$SRAM\_EXT:\quad org = 0x80000000,\quad len = 0x01000000$$

Note that SDRAM could have been specified in lieu of *SRAM_EXT*. External memory starts at the address 0x80000000.

To use voice as input with a microphone into the DSK, the header file c6713dskinit.h needs to be changed so that register 4 is 0x0015 in lieu of 0x0011, as discussed in Chapter 1 and illustrated in Example 2.1. If you have a microphone with the appropriate preamplification, you can connect it directly into the line input in lieu of the mic input (not with most microphones).

Build this project as **record**. Load/run the program.

1. Press switch #3 and input voice for approximately a few seconds. Verify that LED #3 is turned on to indicate that the input voice is being recorded and stored in a buffer. Release switch #3 to stop recording.

2. Connect the output to a speaker. Press switch #0 and verify that the input voice (stored in external memory) is replayed. LED #0 should turn on to indicate that. Release LED #0 to stop replaying.

```
//Record.c Illustrates use of external memory with voice as input

#include "dsk6713_aic23.h"                 //codec-DSK support file
Uint32 fs=DSK6713_AIC23_FREQ_8KHZ;         //set sampling rate
#define N   2400000                        //large buffer size
long i; short var;                         //flag when recording/playing
short buffer[N];
#pragma DATA_SECTION(buffer,".EXTRAM")     //buffer->external memory

void main()
{
 comm_poll();                              //init DSK, codec, McBSP
 DSK6713_DIP_init();
 DSK6713_LED_init();
 while(1)                                  //infinite loop
   {
    if(DSK6713_DIP_get(3) == 0)            //if SW#3 is pressed
      {
       DSK6713_LED_on(3);                  //turn on LED#3
       for (i = 0; i<N; i++)
          buffer[i] = input_sample();      //input data
       DSK6713_LED_off(3);                 //LED#3 off when buffer full
       break;
      }
   };
 var = 0;
 while(1)
   {
    if((DSK6713_DIP_get(0)==0)&&(var==0))  //if SW#0 pressed/var=0
      {
       DSK6713_LED_on(0);                  //turn on LED#0
       for (i = 0; i<N; i++)
          output_sample(buffer[i]*10);     //play back
       var=1;
       DSK6713_LED_off(0);                 //LED#0 off when finished
      }
    if(DSK6713_DIP_get(0)==1) var=0;       //toggle flag
   };
}
```

FIGURE 2.28. C source program to illustrate the use of input voice stored in external memory (record.c).

```
//Flash_sine.c Sine generation-illustrates use of flash memory

#include "dsk6713_aic23.h"              //codec-DSK support file
Uint32 fs=DSK6713_AIC23_FREQ_8KHZ;     //set sampling rate
short loop = 0,  gain = 10;
short sine_table[8] = {0,707,1000,707,0,-707,-1000,-707};//sine values

interrupt void c_int11()               //interrupt service routine
{
 output_sample(sine_table[loop]*gain);//output with gain
 if (++loop > 7) loop = 0;             //if end of buffer,reinit index
 return;
}

void main()
{
 comm_intr();                          //init DSK, codec, McBSP
 while(1);                             //infinite loop
}
```

FIGURE 2.29. C source program to program the onboard flash (*flash_sine.c*).

Example 2.19: Use of Flash Memory—Programming the Onboard Flash (*Flash_sine*)

This example illustrates the use of the onboard flash memory to run an application. It illustrates the steps used to invoke the flash utilities, erase the onboard flash, and reprogram it for a specific application. A 1-kHz sine generation program is used as an application example. Figure 2.29 shows the C source sine generation program *flash_sine.c* that implements this project example.

1. Build this project as **flash_sine**. Add the necessary support files to this project: the initialization "black box" file, the appropriate vector file, the linker command file, and the three library support files. Load and run the executable (.out) file, and verify that a 1-kHz sine wave is generated.

2. Remove (delete) the vector and the linker command files from the project to replace them with the following files:

   (a) vecs_int_flash.asm—a modified version of the vector file included with CCS to copy the code from flash to internal memory upon boot up. It includes a starting address for flash and the code size.

   (b) c6713dsk_flash.cmd—a new linker command file. It sets up a section called *bootload* starting at 0x200 with a length of 0x200.

   These two files are included on the CD. Rebuild the project and verify again that the 1-kHz tone is generated using the new executable (.out) file.

Creating .hex File
The executable (.out) file needs to be converted from a *COFF* to a hex file format that can then be loaded into flash. The *COFF*-to-hex converter file hex6x.exe

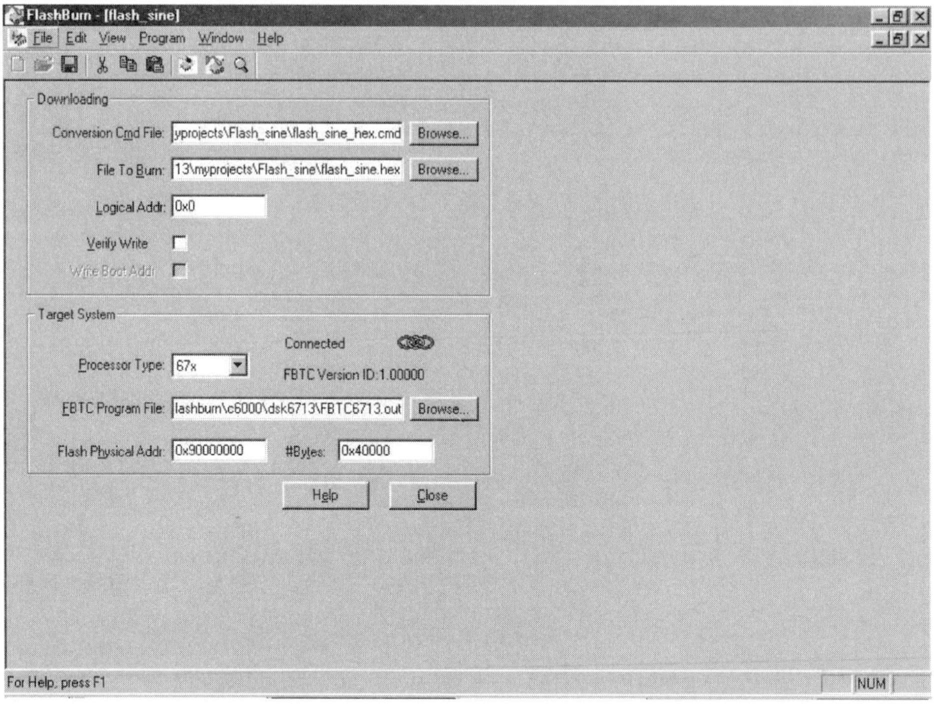

FIGURE 2.30. Flashburn utility (`.cdd`) configuration.

is included with CCS in the directory: c:\c6713\c6000\cgtools\bin. To invoke
hex6x.exe within DOS, an appropriate path needs to be created. This path can be
set in the autoexec.exe file. Alternatively, copy hex6x.exe into the folder
flash_sine. Access DOS, and from the folder *flash_sine*, type

```
hex6x    flash_sine_hex.cmd
```

to create flash_sine.hex. Within the linker file flash_sine_hex.cmd (included
on the CD), the executable *flash_sine.out* is specified as input (within the debug subfolder) and the resulting file as *flash_sine.hex*. A flash length of 0x40000 is specified,
which should be at least the length of the actual code (it can be found in the .map
file). If this length is not large enough, you will be prompted to increase it.

Configuring the Flash Burn (.cdd) Utility
Within CCS, select Tools → FlashBurn to invoke the flash burn utility. Then select
File → New, to configure the flash burn utility and create *flash_sine.cdd*, as shown
in Figure 2.30, with the following fields:

   1. *Conversion Cmd File*: c:\c6713\myprojects\flash_sine\flash_sine_hex.cmd
   2. *File to Burn*: c:\c6713\myprojects\flash_sine\flash_sine.hex

3. *FBTC   Program   File*:   c:\c6713\bin\utilities\flashburn\c6000\dsk6713\ FBTC6713.out

Save this file as *flash_sine.cdd* (already on the CD within the folder *flash_sine*).

Erasing and Programming the Flash Memory

The flash program *post* runs when the DSK is powered. It performs a number of tasks to check the memory and the LEDs, generates a 1-kHz tone for 1 second, and so on.

Note: Continuing this step erases (kills) the flash program *post* and replaces it with the *flash_sine* program. Be assured that it will be brought back to life readily.

Within the Flashburn utility (.cdd file) shown in Figure 2.30, select Program → Erase Flash. This erases any program (*post.hex*) stored in the flash memory. Still, within the (.cdd) Flashburn utility, select Program → Program Flash. This loads *flash_sine.hex* created initially into the flash memory.

To verify that the sine generation program is stored in flash memory, close the (.cdd) Flashburn utility, exit CCS, and unplug the power to the DSK. Turn back on the power to the DSK. The *post* program no longer runs. The LEDs are not turned on, and the 1-kHz sinusoid is not generated for 1 second. Instead, verify from the DSK output that the 1-kHz sine generation program now runs.

Recovering the Post Program

Launch CCS: Select Tools → FlashBurn → File → Open. Look in *c:\c6713\ examples\dsk6713\bsl\post* to open the Flashburn utility configuration file `post.cdd`. Note that there is no entry in the Conversion Cmd File. Select Program → Erase Flash to erase any program (*flash_sine.hex*) stored in the flash memory. Then select Program → Program Flash to download *post.hex* into the flash memory (bringing it back to life when the DSK is powered again).

2.4 ASSIGNMENTS

1. Write a loop program, interrupt-driven using INT12, incorporating directly into the program the BSL/CSL support functions to read and write without using the init/comm "black box" file *C6713dskinit.c*.

2. Write a loop program, polling-based, incorporating directly into the program the BSL/CSL support functions to read and write without using the init/comm "black box" file *C6713dskinit.c*.

3. Write a program to generate a square wave using a sine wave as input. The program should test if the input is greater than or equal to a variable *accum*, in which case the output is a positive scaled value, where *accum* accumulates with a positive step size value. If not, the output is a negative scaled value and *accum* accumulates with a negative step size value. Choose a step size value of $2*pi*f/F_s$, using $f = 1\,\text{kHz}$ (1 V p-p) as the input signal frequency and $F_s = 96\,\text{kHz}$ as the sampling frequency. Cast the step size and *accum* as floating-point values. Verify the output square-wave signal.

4. In the echo program example, add a repetitive echo effect so that if a repetitive echo effect is desired, buffer [i] = output (or input for no repetitive effect).

5. Implement an AM scheme to obtain an AM signal using an external input as the sideband signal and a 2-kHz carrier signal from a lookup table. Test your results using a sampling frequency of 8 kHz, and a sinusoidal input signal with amplitude below 0.35 V and frequency less than 2 kHz (a higher input signal frequency will cause aliasing). Such a small input signal as the sideband yields a more stable output.

6. Write a program to generate a cosine signal of frequency 666 Hz when dip switch #0 (SW0) is pressed (turning on LED #0), 1.33 kHz when SW1 is pressed, 2 kHz when SW2 is pressed, and 2.66 kHz when SW3 is pressed. The program could take the following shape: Start with a 12-point table representing the cosine values, sampling at 8 kHz. Build and test this program as **sine_4freq**.

   (a) Test if SW0 is pressed. If so, turn on LED #0. Then, while SW0 is still pressed, the cosine is generated and a loop index is preincrementing to step through every point in the table until the end of the table is reached. This should generate the 666 Hz.

   (b) Test if SW1 is pressed. If so, preincrement the loop index until the end of the table is reached. Now, every other point is selected so that the generated frequency is $f = F_s/6 = 1.33$ kHz.

   (c) Repeat the previous step for SW2 so that four points are selected and $f = F_s/4 = 2$ kHz, and repeat the previous step for SW3 so that three points are selected and $f = F_s/3 = 2.66$ kHz. Note that to generate the 2.66 kHz, all four switches need to be pressed.

REFERENCES

1. *TLV320AIC23 Stereo Audio Codec, 8- to 96-kHz, with Integrated Headphone Amplifier Data Manual*, SLWS106G, Texas Instruments, Dallas, TX, 2003.

2. S. Norsworthy, R. Schreier, and G. Temes, *Delta–Sigma Data Converters: Theory, Design and Simulation*, IEEE Press, Piscataway, NJ, 1997.

3. P. M. Aziz, H. V. Sorensen, and J. Van Der Spiegel, An overview of sigma delta converters, *IEEE Signal Processing*, Jan. 1996.

4. J. C. Candy and G. C. Temes, eds., *Oversampling Delta–Sigma Data Converters: Theory, Design and Simulation*, IEEE Press, Piscataway, NJ, 1992.

5. C. W. Solomon, Switched-capacitor filters, *IEEE Spectrum*, June 1988.

6. *PCM3002/PCM3003 16-/20-Bit Single-Ended Analog Input/Output Stereo Audio Codecs*, SBAS079, Texas Instruments, Dallas, TX, 2000.

7. *TMS320C6000 McBSP: AC'97 Codec Interface*, SPRA528, Texas Instruments, Dallas, TX, 1999.

3

Architecture and Instruction Set of the C6x Processor

- Architecture and instruction set of the TMS320C6x processor
- Addressing modes
- Assembler directives
- Linear assembler
- Programming examples using C, assembly, and linear assembly code

3.1 INTRODUCTION

Texas Instruments introduced the first-generation TMS32010 DSP in 1982, the TMS320C25 in 1986 [1], and the TMS320C50 in 1991. Several versions of each of these processors—C1x, C2x, and C5x—are available with different features, such as faster execution speed. These 16-bit processors are all fixed-point processors and are code-compatible.

In a von Neumann architecture, program instructions and data are stored in a single memory space. A processor with a von Neumann architecture can make a read or a write to memory during each instruction cycle. Typical DSP applications require several accesses to memory within one instruction cycle. The fixed-point processors C1x, C2x, and C5x are based on a modified Harvard architecture with separate memory spaces for data and instructions that allow concurrent accesses.

Digital Signal Processing and Applications with the C6713 and C6416 DSK   By Rulph Chassaing
ISBN 0-471-69007-4   Copyright © 2005 by John Wiley & Sons, Inc.

Quantization error or round-off noise from an ADC is a concern with a fixed-point processor. An ADC uses only a best-estimate digital value to represent an input. For example, consider an ADC with a word length of 8 bits and an input range of ±1.5 V. The steps represented by the ADC are: input range/2^8 = 3/256 = 11.72 mV. This produces errors that can be up to ±(11.72 mV)/2 = ±5.86 mV. Only a best estimate can be used by the ADC to represent input values that are not multiples of 11.72 mV. With an 8-bit ADC, 2^8 or 256 different levels can represent the input signal. An ADC with a larger word length, such as a 16-bit ADC (or larger, currently very common), can reduce the quantization error, yielding a higher resolution. The more bits an ADC has, the better it can represent an input signal.

The TMS320C30 floating-point processor was introduced in the late 1980s. The C31, the C32, and the more recent C33 are all members of the C3x family of floating-point processors [2,3]. The C4x floating-point processors, introduced subsequently, are code-compatible with the C3x processors and are based on the modified Harvard architecture [4].

The TMS320C6201 (C62x), announced in 1997, is the first member of the C6x family of fixed-point digital signal processors. Unlike the previous fixed-point processors, C1x, C2x, and C5x, the C62x is based on a VLIW architecture, still using separate memory spaces for instructions and data, as with the Harvard architecture. The VLIW architecture has simpler instructions, but more are needed for a task than with a conventional DSP architecture.

The C62x is not code-compatible with the previous generation of fixed-point processors. Subsequently, the TMS320C6701 (C67x) floating-point processor was introduced as another member of the C6x family of processors. The instruction set of the C62x fixed-point processor is a subset of the instruction set of the C67x processor. Appendix A contains a list of instructions available on the C6x processors. A more recent addition to the family of the C6x fixed-point processors is the C64x. The C64x is introduced in Appendix H.

An application-specific integrated circuit (ASIC) has a DSP core with customized circuitry for a specific application. A C6x processor can be used as a standard general-purpose DSp programmed for a specific application. Specific-purpose digital signal processors are the modem, echo canceler, and others.

A fixed-point processor is better for devices that use batteries, such as cellular phones, since it uses less power than does an equivalent floating-point processor. The fixed-point processors, C1x, C2x, and C5x, are 16-bit processors with limited dynamic range and precision. The C6x fixed-point processor is a 32-bit processor with improved dynamic range and precision. In a fixed-point processor, it is necessary to scale the data. Overflow, which occurs when an operation such as the addition of two numbers produces a result with more bits than can fit within a processor's register, becomes a concern.

A floating-point processor is generally more expensive since it has more "real estate" or is a larger chip because of additional circuitry necessary to handle integer as well as floating-point arithmetic. Several factors, such as cost, power consumption, and speed, come into play when choosing a specific DSp. The C6x processors are particularly useful for applications requiring intensive computations. Family members of the C6x include both fixed-point (e.g., C62x, C64x) and floating-point (e.g., C67x) processors. Other DSp's are also available from companies such as Motorola and Analog Devices [5].

Other architectures include the Super Scalar, which requires special hardware to determine which instructions are executed in parallel. The burden is then on the processor more than on the programmer, as in the VLIW architecture. It does not necessarily execute the same group of instructions, and as a result, it is difficult to time. Thus, it is rarely used in DSP.

3.2 TMS320C6x ARCHITECTURE

The TMS320C6713 onboard the DSK is a floating-point processor based on the VLIW architecture [6–10]. Internal memory includes a two-level cache architecture with 4 kB of level 1 program cache (L1P), 4 kB of level 1 data cache (L1D), and 256 kB of level 2 memory shared between program and data space. It has a glueless (direct) interface to both synchronous memories (SDRAM and SBSRAM) and asynchronous memories (SRAM and EPROM). Synchronous memory requires clocking but provides a compromise between static SRAM and dynamic DRAM, with SRAM being faster but more expensive than DRAM.

On-chip peripherals include two McBSPs, two timers, a host port interface (HPI), and a 32-bit EMIF. It requires 3.3 V for I/O and 1.26 V for the core (internal). Internal buses include a 32-bit program address bus, a 256-bit program data bus to accommodate eight 32-bit instructions, two 32-bit data address buses, two 64-bit data buses, and two 64-bit store data buses. With a 32-bit address bus, the total memory space is $2^{32} = 4$ GB, including four external memory spaces: CE0, CE1, CE2, and CE3. Figure 3.1 shows a functional block diagram of the C6713 processor included with CCS.

Independent memory banks on the C6x allow for two memory accesses within one instruction cycle. Two independent memory banks can be accessed using two independent buses. Since internal memory is organized into memory banks, two loads or two stores of instructions can be performed in parallel. No conflict results if the data accessed are in different memory banks. Separate buses for program, data, and direct memory access (DMA) allow the C6x to perform concurrent program fetches, data read and write, and DMA operations. With data and instructions residing in separate memory spaces, concurrent memory accesses are possible. The C6x has a byte-addressable memory space. Internal memory is organized as separate program and data memory spaces, with two 32-bit internal ports (two 64-bit ports with the C64x) to access internal memory.

The C6713 on the DSK includes 264 kB of internal memory, which starts at 0x00000000, and 16 MB of external SDRAM, mapped through CE0 starting at 0x80000000. The DSK also includes 512 kB of Flash memory (256 kB readily available to the user), mapped through CE1 starting at 0x90000000. Figure 3.2 shows

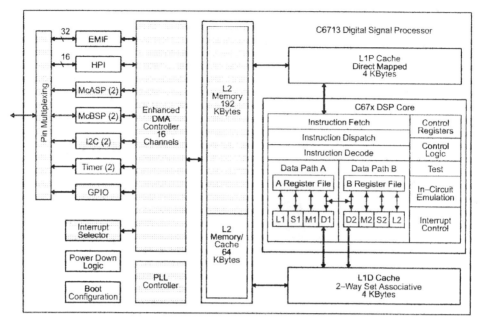

FIGURE 3.1. Functional block diagram of TMS320C6713 (Courtesy of Texas Instruments).

the L2 internal memory configuration, included with CCS [7]. Table 3.1 shows the memory map, also included with CCS [7]. A schematic diagram of the DSK is included with CCS (*6713dsk_schem.pdf*).

With the DSK operating at 225 MHz, one can ideally achieve two multiplies and accumulates per cycle, for a total of 450 million multiplies and accumulates (MACs) per second. With six of the eight functional units in Figure 3.1 (not the .D units described below) capable of handling floating-point operations, it is possible to perform 1350 million floating-point operations per second (MFLOPS). Operating at 225 MHz, this translates into 1800 million instructions per second (MIPS) with a 4.44-ns instruction cycle time.

3.3 FUNCTIONAL UNITS

The CPU consists of eight independent functional units divided into two data paths, A and B, as shown in Figure 3.1. Each path has a unit for multiply operations (.M), for logical and arithmetic operations (.L), for branch, bit manipulation, and arithmetic operations (.S), and for loading/storing and arithmetic operations (.D). The .S and .L units are for arithmetic, logical, and branch instructions. All data transfers make use of the .D units.

The arithmetic operations, such as subtract or add (SUB or ADD), can be performed by all the units, except the .M units (one from each data path). The eight

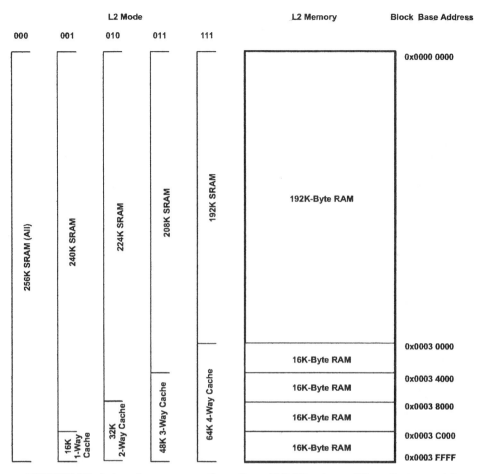

FIGURE 3.2. Internal memory configuration of L2 (Courtesy of Texas Instruments).

functional units consist of four floating/fixed-point ALUs (two .L and two .S), two fixed-point ALUs (.D units), and two floating/fixed-point multipliers (.M units). Each functional unit can read directly from or write directly to the register file within its own path. Each path includes a set of sixteen 32-bit registers, A0 through A15 and B0 through B15. Units ending in 1 write to register file A, and units ending in 2 write to register file B.

Two cross-paths (1x and 2x) allow functional units from one data path to access a 32-bit operand from the register file on the opposite side. There can be a maximum of two cross-path source reads per cycle. Each functional unit side can access data from the registers on the opposite side using a cross-path (i.e., the functional units on one side can access the register set from the other side). There are 32 general-purpose registers, but some of them are reserved for specific addressing or are used for conditional instructions.

TABLE 3.1   Memory Map

Memory Block Description	Block Size (Bytes)	Hex Address Range
Internal RAM (L2)	192 K	0000 0000–0002 FFFF
Internal RAM/cache	64 K	0003 0000–0003 FFFF
Reserved	24 M–256 K	0004 0000–017F FFFF
External memory interface (EMIF) registers	256 K	0180 0000–0183 FFFF
L2 registers	128 K	0184 0000–0185 FFFF
Reserved	128 K	0186 0000–0187 FFFF
HPI registers	256 K	0188 0000–018B FFFF
McBSP 0 registers	256 K	018C 0000–018F FFFF
McBSP 1 registers	256 K	0190 0000–0193 FFFF
Timer 0 registers	256 K	0194 0000–0197 FFFF
Timer 1 registers	256 K	0198 0000–019B FFFF
Interrupt selector registers	512	019C 0000–019C 01FF
Device configuration registers	4	019C 0200–019C 0203
Reserved	256 K–516	091C 0204–019F FFFF
EDMA RAM and EDMA registers	256 K	01A0 0000–01A3 FFFF
Reserved	768 K	01A4 0000–01AF FFFF
GPIO registers	16 K	01B0 0000–01B0 3FFF
Reserved	240 K	01B0 4000–01B3 FFFF
I2C0 registers	16 K	01B4 0000–01B4 3FFF
I2C1 registers	16 K	01B4 4000–01B4 7FFF
Reserved	16 K	01B4 8000–01B4 BFFF
McASP0 registers	16 K	01B4 C000–01B4 FFFF
McASP1 registers	16 K	01B5 0000–01B5 3FFF
Reserved	160 K	01B5 4000–01B7 BFFF
PLL registers	8 K	01B7 C000–01B7 DFFF
Reserved	264 K	01B7 E000–01BB FFFF
Emulation registers	256 K	01BC 0000–01BF FFFF
Reserved	4 M	01C0 0000–01FF FFFF
QDMA registers	52	0200 0000–0200 0033
Reserved	16 M–52	0200 0034–02FF FFFF
Reserved	720 M	0300 0000–2FFF FFFF
McBSP0 data port	64 M	3000 0000–33FF FFFF
McBSP1 data port	64 M	3400 0000–37FF FFFF
Reserved	64 M	3800 0000–3BFF FFFF
McASP0 data port	1 M	3C00 0000–3C0F FFFF
McASP1 data port	1 M	3C10 0000–3C1F FFFF
Reserved	1 G + 62 M	3C20 0000–7FFF FFFF
EMIF CE0*	256 M	8000 0000–8FFF FFFF
EMIF CE1*	256 M	9000 0000–9FFF FFFF
EMIF CE2*	256 M	A000 0000–AFFF FFFF
EMIF CE3*	256 M	B000 0000–BFFF FFFF
Reserved	1 G	C000 0000–FFFF FFFF

* The number of EMIF address pins (EA[21:2]) limits the maximum addressable memory (SDRAM) to 128 MB per CE space.

Source: Courtesy of Texas Instruments.

FIGURE 3.3. One FP with three EPs showing the "p" bit of each instruction.

3.4 FETCH AND EXECUTE PACKETS

The architecture VELOCITI, introduced by TI, is derived from the VLIW architecture. An execute packet (EP) consists of a group of instructions that can be executed in parallel within the same cycle time. The number of EPs within a fetch packet (FP) can vary from one (with eight parallel instructions) to eight (with no parallel instructions). The VLIW architecture was modified to allow more than one EP to be included within an FP.

The least significant bit of every 32-bit instruction is used to determine if the next or subsequent instruction belongs in the same EP (if 1) or is part of the next EP (if 0). Consider an FP with three EPs: EP1, with two parallel instructions, and EP2 and EP3, each with three parallel instructions, as follows:

```
        Instruction A
||      Instruction B

        Instruction C
||      Instruction D
||      Instruction E

        Instruction F
||      Instruction G
||      Instruction H
```

EP1 contains the two parallel instructions A and B; EP2 contains the three parallel instructions C, D, and E; and EP3 contains the three parallel instructions F, G, and H. The FP would be as shown in Figure 3.3. Bit 0 (LSB) of each 32-bit instruction contains a "p" bit that signals whether it is in parallel with a subsequent instruction. For example, the "p" bit of instruction B is zero, denoting that it is not within the same EP as the subsequent instruction C. Similarly, instruction E is not within the same EP as instruction F.

3.5 PIPELINING

Pipelining is a key feature in a DSp to get parallel instructions working properly, requiring careful timing. There are three stages of pipelining: program fetch, decode, and execute.

1. The *program fetch* stage is composed of four phases:
    (a) *PG*: program address generate (in the CPU) to fetch an address
    (b) *PS*: program address send (to memory) to send the address
    (c) *PW*: program address ready wait (memory read) to wait for data
    (d) *PR*: program fetch packet receive (at the CPU) to read opcode from memory
2. The *decode stage* is composed of two phases:
    (a) *DP*: to dispatch all the instructions within an FP to the appropriate functional units
    (b) *DC*: instruction decode
3. The *execute stage* is composed of 6 phases (with fixed point) to 10 phases (with floating point) due to delays (latencies) associated with the following instructions:
    (a) Multiply instruction, which consists of two phases due to one delay
    (b) Load instruction, which consists of five phases due to four delays
    (c) Branch instruction, which consists of six phases due to five delays

Table 3.2 shows the pipeline phases, and Table 3.3 shows the pipelining effects. The first row in Table 3.3 represents cycle $1, 2, \ldots, 12$. Each subsequent row represents an FP. The rows represented PG, PS, ... illustrate the phases associated with each FP. The program generate (PG) of the first FP starts in cycle 1, and the PG of the second FP starts in cycle 2, and so on. Each FP takes four phases for program fetch and two phases for decoding. However, the execution phase can take from 1

TABLE 3.2 Pipeline Phases

Program Fetch				Decode		Execute
PG	PS	PW	PR	DP	DC	E1–E6 (E1–E10 for double precision)

TABLE 3.3 Pipelining Effects

					Clock Cycle						
1	2	3	4	5	6	7	8	9	10	11	12
PG	PS	PW	PR	DP	DC	E1	E2	E3	E4	E5	E6
	PG	PS	PW	PR	DP	DC	E1	E2	E3	E4	E5
		PG	PS	PW	PR	DP	DC	E1	E2	E3	E4
			PG	PS	PW	PR	DP	DC	E1	E2	E3
				PG	PS	PW	PR	DP	DC	E1	E2
					PG	PS	PW	PR	DP	DC	E1
						PG	PS	PW	PR	DP	DC

to 10 phases (not all execution phases are shown in Table 3.3). We are assuming that each FP contains one EP.

For example, at cycle 7, while the instructions in the first FP are in the first execution phase E1 (which may be the only one), the instructions in the second FP are in the decoding phase, the instructions in the third FP are in the dispatching phase, and so on. All seven instructions are proceeding through the various phases. Therefore, at cycle 7, "the pipeline is full."

Most instructions have one execute phase. Instructions such as multiply (MPY), load (LDH/LDW), and branch (B) take two, five, and six phases, respectively. Additional execute phases are associated with floating-point and double-precision types of instructions, which can take up to 10 phases. For example, the double-precision multiply operation (MPYDP), available on the C67x, has nine delay slots, so that the execution phase takes a total of 10 phases.

The *functional unit latency*, which represents the number of cycles that an instruction ties up a functional unit, is 1 for all instructions except double-precision instructions, available with the floating-point C67x. Functional unit latency is different from a delay slot. For example, the instruction MPYDP has four functional unit latencies but nine delay slots. This implies that no other instruction can use the associated multiply functional unit for four cycles. A store has no delay slot but finishes its execution in the third execution phase of the pipeline.

If the outcome of a multiply instruction such as MPY is used by a subsequent instruction, a NOP (no operation) must be inserted after the MPY instruction for the pipelining to operate properly. Four or five NOPs are to be inserted in case an instruction uses the outcome of a load or a branch instruction, respectively.

3.6 REGISTERS

Two sets of register files, each set with 16 registers, are available: register file A (A0 through A15) and register file B (B0 through B15). Registers A0, A1, B0, B1, and B2 are used as conditional registers. Registers A4 through A7 and B4 through B7 are used for circular addressing. Registers A0 through A9 and B0 through B9 (except B3) are temporary registers. Any of the registers A10 through A15 and B10 through B15 used are saved and later restored before returning from a subroutine.

A 40-bit data value can be contained across a register pair. The 32 least significant bits (LSBs) are stored in the even register (e.g., A2), and the remaining 8 bits are stored in the 8 LSBs of the next-upper (odd) register (A3). A similar scheme is used to hold a 64-bit double-precision value within a pair of registers (even and odd).

These 32 registers are considered general-purpose registers. Several special-purpose registers are also available for control and interrupts: for example, the address mode register (AMR) used for circular addressing and interrupt control registers, as shown in Appendix B.

3.7 LINEAR AND CIRCULAR ADDRESSING MODES

Addressing modes determine how one accesses memory. They specify how data are accessed, such as retrieving an operand indirectly from a memory location. Both linear and circular modes of addressing are supported. The most commonly used mode is the indirect addressing of memory.

3.7.1 Indirect Addressing

Indirect addressing can be used with or without displacement. Register R represents one of the 32 registers A0 through A15 and B0 through B15 that can specify or point to memory addresses. As such, these registers are pointers. Indirect addressing mode uses a "*" in conjunction with one of the 32 registers. To illustrate, consider R as an address register.

1. *R. Register R contains the address of a memory location where a data value is stored.
2. *R++(d). Register R contains the memory address (location). After the memory address is used, R is postincremented (modified) such that the new address is the current address offset by the displacement value d. If d = 1 (by default), the new address is R + 1, or R is incremented to the next higher address in memory. A double minus (−−) instead of a double plus would update or postdecrement the address to R − d.
3. *++R(d). The address is preincremented or offset by d, such that the current address is R + d. A double minus would predecrement the memory address so that the current address is R − d.
4. *+R(d). The address is preincremented by d, such that the current address is R + d (as with the preceding case). However, in this case, R preincrements without modification. Unlike the previous case, R is not updated or modified.

3.7.2 Circular Addressing

Circular addressing is used to create a circular buffer. This buffer is created in hardware and is very useful in several DSP algorithms, such as in digital filtering or correlation algorithms where data need to be updated. An example in Chapter 4 illustrates the implementation of a digital filter in assembly code using a circular buffer to update the "delay" samples. Implementing a circular buffer using C code is less efficient.

The C6x has dedicated hardware to allow a circular type of addressing. This addressing mode can be used in conjunction with a circular buffer to update samples by shifting data without the overhead created by shifting data directly. As a pointer reaches the end or "bottom" location of a circular buffer that contains the last element in the buffer, and is then incremented, the pointer is automatically wrapped around or points to the beginning or "top" location of the buffer that contains the first element.

Two independent circular buffers are available using BK0 and BK1 within the AMR. The eight registers A4 through A7 and B4 through B7, in conjunction with the two .D units, can be used as pointers (all registers can be used for linear addressing). The following code segment illustrates the use of a circular buffer using register B2 (only side B can be used) to set the appropriate values within AMR:

```
MVKL    .S2   0x0004,B2    ;lower 16 bits to B2. Select A5 as pointer
MVKH    .S2   0x0005,B2    ;upper 16 bits to B2. Select BK0, set N = 5
MVC     .S2   B2,AMR       ;move 32 bits of B2 to AMR
```

The two move instructions MVKL and MVKH (using the .S unit) move 0x0004 into the 16 LSBs of register B2 and 0x0005 into the 16 most significant bits (MSBs) of B2. The MVC (move constant) instruction is the only instruction that can access the AMR and the other control registers (shown in Appendix B) and executes only on the B side in conjunction with the functional units and registers on side B. A 32-bit value is created in B2, which is then transferred to AMR with the instruction MVC to access AMR [6].

The value $0x0004 = (0100)_b$ into the 16 LSBs of AMR sets bit 2 (the third bit) to 1 and all other bits to 0. This sets the mode to 01 and selects register A5 as the pointer to a circular buffer using block BK0 (see Figure B.1).

Table 3.4 shows the modes associated with registers A4 through A7 and B4 through B7. The value $0x0005 = (0101)_b$ into the 16 MSBs of AMR sets bits 16 and 18 to 1 (other bits to 0). This corresponds to the value of N used to select the size of the buffer as $2^{N+1} = 64$ bytes using BK0. For example, if a buffer size of 128 is desired using BK0, the upper 16 bits of AMR are set to $(0110)_b = 0x0006$. If assembly code is used for the circular buffer, as execution returns to a calling C function, AMR needs to be reinitialized to the default linear mode. Hence the pointer's address must be saved.

TABLE 3.4   AMR Mode and Description

Mode	Description
0 0	For linear addressing (default on reset)
0 1	For circular addressing using BK0
1 0	For circular addressing using BK1
1 1	Reserved

3.8 TMS320C6x INSTRUCTION SET

3.8.1 Assembly Code Format

An assembly code format is represented by the field

```
Label   ||   [ ]   Instruction   Unit   Operands   ;comments
```

A label, if present, represents a specific address or memory location that contains an instruction or data. The label must be in the first column. The parallel bars (||) are there if the instruction is being executed in parallel with the previous instruction. The subsequent field is optional to make the associated instruction conditional. Five of the registers—A1, A2, B0, B1, and B2—are available to use as conditional registers. For example, [A2] specifies that the associated instruction executes if A2 is not zero. On the other hand, with [!A2], the associated instruction executes if A2 is zero. All C6x instructions can be made conditional with the registers A1, A2, B0, B1, and B2 by determining when the conditional register is zero. The instruction field can be either an assembler directive or a mnemonic. An assembler directive is a command for the assembler. For example,

```
.word value
```

reserves 32 bits in memory and fill with the specified *value*. A mnemonic is an actual instruction that executes at run time. The instruction (mnemonic or assembler directive) cannot start in column 1. The Unit field, which can be one of the eight CPU units, is optional. Comments starting in column 1 can begin with either an asterisk or a semicolon, whereas comments starting in any other columns must begin with a semicolon.

Code for the floating-point processors C3x/C4x is not compatible with code for the fixed-point processors C1x, C2x, and C5x/C54x. However, the code for the fixed-point processors C62x is compatible with the code for the floating-point C67x. C62x code is actually a subset of C67x code. Additional instructions to handle double-precision and floating-point operations are available only on the C67x processor. Also, some additional instructions are available only on the fixed-point C64x processor.

Several code segments are presented to illustrate the C6x instruction set. Assembly code for the C6x processors is very similar to C3x/C4x code. Single-task types of instructions available for the C6x make it easier to program than either the previous generation of fixed- or floating-point processors. This contributes to an efficient compiler. Additional instructions available on the C64x (but not on the C62x) resemble the multitask types of instructions for C3x/C4x processors. It is very instructive to read the comments in the programs discussed in this book. Appendix A contains a list of the instructions for the C62x/C67x processors.

3.8.2 Types of Instructions

The following illustrates some of the syntax of assembly code. It is optional to specify the eight functional units, although this can be useful during debugging and for code efficiency and optimization, discussed in Chapter 8.

1. *Add/Subtract/Multiply*
   (a) The instruction

   ```
   ADD    .L1    A3,A7,A7    ;add A3 + A7 → A7 (accum in A7)
   ```

   adds the values in registers A3 and A7 and places the result in register A7. The unit .L1 is optional. If the destination or result is in B7, the unit would be .L2.

   (b) The instruction

   ```
   SUB    .S1    A1,1,A1    ;subtract 1 from A1
   ```

   subtracts 1 from A1 to decrement it using the .S unit.

   (c) The parallel instructions

   ```
      MPY    .M2   A7,B7,B6    ;multiply 16 LSBs of A7, B7 → B6
   || MPYH   .M1   A7,B7,A6    ;multiply 16 MSBs of A7, B7 → A6
   ```

   multiplies the lower or least significant 16 bits (LSBs) of both A7 and B7 and places the product in B6, in parallel (concurrently within the same execution packet) with a second instruction that multiplies the higher or most significant 16 bits (MSBs) of A7 and B7 and places the result in A6. In this fashion, two MAC operations can be executed within a single instruction cycle. This can be used to decompose a sum of products into two sets of sum of products: one set using the lower 16 bits to operate on the first, third, fifth, ... number and another set using the higher 16 bits to operate on the second, fourth, sixth, ... number. Note that the parallel symbol is not in column 1.

2. *Load/Store*
   (a) The instruction

   ```
      LDH    .D2   *B2++,B7    ;load (B2) → B7, increment B2
   || LDH    .D1   *A2++,A7    ;load (A2) → A7, increment A2
   ```

   loads into B7 the half-word (16 bits) whose address in memory is specified/pointed to by B2. Then register B2 is incremented (postincremented) to point at the next higher memory address. In parallel is another indirect addressing mode instruction to load into A7 the content in memory whose address is specified by A2. Then A2 is incremented to point at the next higher memory address.

The instruction LDW loads a 32-bit word. Two paths using .D1 and .D2 allow for the loading of data from memory to registers A and B using the instruction LDW. The double-word load floating-point instruction LDDW on the C6713 can simultaneously load two 32-bit registers into side A and two 32-bit registers into side B.

(b) The instruction

```
STW    .D2    A1,*+A4[20]        ;store A1→(A4) offset by 20
```

stores the 32-bit word A1 in memory whose address is specified by A4 offset by 20 words (32 bits) or 80 bytes. The address register A4 is pre-incremented with offset, but it is not modified (two plus signs are used if A4 is to be modified).

3. *Branch/Move.* The following code segment illustrates branching and data transfer:

```
Loop  MVKL  .S1   x,A4          ;move 16 LSBs of x address →A4
      MVKH  .S1   x,A4          ;move 16 MSBs of x address →A4
      .
      .
      .
      SUB   .S1   A1,1,A1       ;decrement A1
[A1]  B     .S2   Loop          ;branch to Loop if A1 # 0
      NOP         5             ;five no-operation instructions
      STW   .D1   A3,*A7        ;store A3 into (A7)
```

The first instruction moves the lower 16 bits (LSBs) of address x into register A4. The second instruction moves the higher 16 bits (MSBs) of address x into A4, which now contains the full 32-bit address of x. One must use the instructions MVKL/MVKH in order to get a 32-bit constant into a register.

Register A1 is used as a loop counter. After it is decremented with the SUB instruction, it is tested for a conditional branch. Execution branches to the label or address Loop if A1 is not zero. If A1 = 0, execution continues and data in register A3 are stored in memory whose address is specified (pointed) by A7.

3.9 ASSEMBLER DIRECTIVES

An assembler directive is a message for the assembler (not the compiler) and is not an instruction. It is resolved during the assembling process and does not occupy memory space, as an instruction does. It does not produce executable code. Addresses of different sections can be specified with assembler directives. For example, the assembler directive `.sect "my_buffer"` defines a section of code or data named `my_buffer`. The directives `.text` and `.data` indicate a section for text and data, respectively. Other assembler directives, such as `.ref` and `.def`, are used for undefined and defined symbols, respectively. The assembler creates several sections indicated by directives such as `.text` for code and `.bss` for global and static variables.

Other commonly used assembler directives are:

1. `.short`: to initialize a 16-bit integer.
2. `.int`: to initialize a 32-bit integer (also `.word` or `.long`). The compiler treats a long data value as 40 bits, whereas the C6x assembler treats it as 32 bits.
3. `.float`: to initialize a 32-bit IEEE single-precision constant.
4. `.double`: to initialize a 64-bit IEEE double-precision constant.

Initialized values are specified by using the assembler directives `.byte`, `.short`, or `.int`. Uninitialized variables are specified using the directive `.usect`, which creates an uninitialized section (like the `.bss` section), whereas the directive `.sect` creates an initialized section. For example, `.usect "variable", 128` designates an uninitialized section named `variable` with a section size of 128 in bytes.

3.10 LINEAR ASSEMBLY

An alternative to C, or assembly code, is linear assembly. An assembler optimizer (in lieu of a C compiler) is used in conjunction with a linear assembly-coded source program (with extension `.sa`) to create an assembly source program (with extension `.asm`) in much the same way that a C compiler optimizer is used in conjunction with a C-coded source program. The resulting assembly-coded program produced by the assembler optimizer is typically more efficient than one resulting from the C compiler optimizer. The assembly-coded program resulting from either a C-coded source program or a linear-assembly source program must be assembled to produce an object code.

Linear assembly code programming provides a compromise between coding effort and coding efficiency. The assembler optimizer assigns the functional unit and register to use (optional to be specified by the user), finds instructions that can execute in parallel, and performs software pipelining for optimization (discussed in Chapter 8). Two programming examples at the end of this chapter illustrate a C program calling a linear assembly function. Parallel instructions are not valid in a linear assembly program. Specifying the functional unit is optional in a linear assembly program as well as in an assembly program.

In recent years, the C compiler optimizer has become more and more efficient. Although C code is less efficient (speed performance) than assembly code, it typically involves less coding effort than assembly code, which can be hand-optimized to achieve 100 percent efficiency but with much greater coding effort.

It is interesting to note that the C6x assembly code syntax is not as complex as that of the C2x/C5x or the C3x family of processors. It is actually simpler to "program" the C6x in assembly. For example, the C3x instruction

```
DBNZD   AR4,LOOP
```

decrements (due to the first D) a loop counter AR4 and branches (B) conditionally (if AR4 is nonzero) to the address specified by LOOP, with delay (due to the second D). The branch instruction with delay effectively allows the branch instruction to execute in a single cycle (due to pipelining). Such multitask instructions are not available on the C62x and C67x processors, although they were recently introduced on the C64x processor. In fact, C6x types of instructions are simpler. For example, separate instructions are available for decrementing a counter (with a SUB instruction) and branching. The simpler types of instructions are more amenable for a more efficient C compiler.

However, although it is simpler to program in assembly code to perform a desired task, this does not imply or translate into an efficient assembly-coded program. It can be relatively difficult to hand-optimize a program to yield a totally efficient (and meaningful) assembly-coded program.

Linear assembly code is a cross between assembly and C. It uses the syntax of assembly code instructions such as ADD, SUB, and MPY, but with operands/registers as used in C. In some cases this provides a good compromise between C and assembly.

Linear assembler directives include

```
.cproc
.endproc
```

to specify a C-callable procedure or section of code to be optimized by the assembler optimizer. Another directive, .reg, is used to declare variables and use descriptive names for values that will be stored in registers. Programming examples with C calling an assembly function or a linear assembly function are illustrated later in this chapter.

3.11 ASM STATEMENT WITHIN C

Assembly instructions and directives can be incorporated within a C program using the *asm* statement. The *asm* statement can provide access to hardware features that cannot be obtained using C code only. The syntax is

```
asm ("assembly code");
```

The assembly line of code within the set of quotation marks has the same format as a valid assembly statement. Note that if the instruction has a label, the first character of the label must start after the first quotation mark so that it is in column 1. The assembly statement should be valid since the compiler does not check it for syntax error but copies it directly into the compiled output file. If the assembly statement has a syntax error, the assembler would detect it.

Avoid using *asm* statements within a C program, especially within a linear assembly program. This is because the assembler optimizer could rearrange lines of code near the asm statements that may cause undesirable results.

3.12 C-CALLABLE ASSEMBLY FUNCTION

Programming examples are included later in this chapter to illustrate a C program calling an assembly function. Register B3 is preserved and is used to contain the return address of the calling function.

An external declaration of an assembly function called within a C program using extern is optional. For example,

```
extern int func();
```

is optional with the assembly function func returning an integer value.

3.13 TIMERS

Two 32-bit timers can be used to time and count events or to interrupt the CPU. A timer can direct an external ADC to start conversion or the DMA controller to start a data transfer. A timer includes a time period register, which specifies the timer's frequency; a timer counter register, which contains the value of the incrementing counter; and a timer control register, which monitors the timer's status.

3.14 INTERRUPTS

An interrupt can be issued internally or externally. An interrupt stops the current CPU process so that it can perform a required task initiated by the interrupt. The program flow is redirected to an ISR. The source of the interrupt can be an ADC, a timer, and so on. On an interrupt, the conditions of the current process must be saved so that they can be restored after the interrupt task is performed. On interrupt, registers are saved and processing continues to an ISR. Then the registers are restored.

There are 16 interrupt sources. They include two timer interrupts, four external interrupts, four McBSP interrupts, and four DMA interrupts. Twelve CPU interrupts (INT4–INT11) are available. An interrupt selector is used to choose among the 12 interrupts.

3.14.1 Interrupt Control Registers

The interrupt control registers (Appendix B) are as follows:

1. CSR (control status register): contains the global interrupt enable (GIE) bit and other control/status bits
2. IER (interrupt enable register): enables/disables individual interrupts
3. IFR (interrupt flag register): displays the status of interrupts
4. ISR (interrupt set register): sets pending interrupts
5. ICR (interrupt clear register): clears pending interrupts
6. ISTP (interrupt service table pointer): locates an ISR
7. IRP (interrupt return pointer)
8. NRP (nonmaskable interrupt return pointer)

Interrupts are prioritized, with Reset having the highest priority. The reset interrupt and nonmaskable interrupt (NMI) are external pins that have the first and second highest priority, respectively. The interrupt enable register (IER) is used to set a specific interrupt and can check if and which interrupt has occurred from the interrupt flag register (IFR).

NMI is nonmaskable, along with Reset. NMI can be masked (disabled) by clearing the nonmaskable interrupt enable (NMIE) bit within CSR. It is set to zero only upon reset or upon a nonmaskable interrupt. If NMIE is set to zero, all interrupts INT4 through INT15 are disabled. The interrupt registers are shown in Appendix B.

The reset signal is an active-low signal used to halt the CPU, and the NMI signal alerts the CPU to a potential hardware problem. Twelve CPU interrupts with lower priorities are available, corresponding to the maskable signals INT4 through INT15. The priorities of these interrupts are: INT4, INT5, . . . , INT15, with INT4 having the highest priority and INT15 the lowest priority. For an NMI to occur, the NMIE bit must be 1 (active high). On reset (or after a previously set NMI), the NMIE bit is cleared to zero so that a reset interrupt may occur.

To process a maskable interrupt, the GIE bit within the control status register (CSR) and the NMIE bit within the IER are set to 1. GIE is set to 1 with bit 0 of CSR set to 1, and NMIE is set to 1 with bit 1 of IER set to 1. Note that CSR can be ANDed with −2 (using 2's complement, the LSB is 0, while all other bits are 1's) to set the GIE bit to 0 and disable maskable interrupts globally.

The interrupt enable (IE) bit corresponding to the desirable maskable interrupt is also set to 1. When the interrupt occurs, the corresponding IFR bit is set to 1 to show the interrupt status. To process a maskable interrupt, the following apply:

1. The GIE bit is set to 1.
2. The NMIE bit is set to 1.
3. The appropriate IE bit is set to 1.
4. The corresponding IFR bit is set to 1.

TABLE 3.5 **Interrupt Service Table**

Interrupt	Offset
RESET	000h
NMI	020h
Reserved	040h
Reserved	060h
INT4	080h
INT5	0A0h
INT6	0C0h
INT7	0E0h
INT8	100h
INT9	120h
INT10	140h
INT11	160h
INT12	180h
INT13	1A0h
INT14	1C0h
INT15	1E0h

For an interrupt to occur, the CPU must not be executing a delay slot associated with a branch instruction.

The interrupt service table (IST) shown in Table 3.5 is used when an interrupt begins. Within each location is an FP associated with each interrupt. The table contains 16 FPs, each with eight instructions. The addresses on the right side correspond to an offset associated with each specific interrupt. For example, the FP for interrupt INT11 is at a base address plus an offset of 160h. Since each FP contains eight 32-bit instructions (256 bits) or 32 bytes, each offset address in the table is incremented by 20h = 32.

The reset FP must be at address 0. However, the FPs associated with the other interrupts can be relocated. The relocatable address can be specified by writing this address to the interrupt service table base (ISTB) register of the interrupt service table pointer (ISTP) register, shown in Figure B.7. On reset, ISTB is zero. For relocating the vector table, the ISTP is used; the relocatable address is ISTB plus the offset.

3.14.2 Interrupt Acknowledgment

The signals IACK and INUMx (INUM0 through INUM3) are pins on the C6x that acknowledge that an interrupt has occurred and is being processed. The four INUMx signals indicate the number of the interrupt being processed. For example,

INUM3 = 1 (MSB), INUM2 = 0, INUM1 = 1, INUM0 = 1 (LSB)

correspond to $(1011)_b = 11$, indicating that INT11 is being processed.

The IE11 bit is set to 1 to enable INT11. The IFR can be read to verify that bit IF11 is set to 1 (INT11 enabled). Writing a 1 to a bit in the interrupt set register (ISR) causes the corresponding interrupt flag to be set in IFR, whereas a 0 to a bit in the interrupt clear register (ICR) causes the corresponding interrupt to be cleared.

All interrupts remain pending while the CPU has a pending branch instruction. Since a branch instruction has five delay slots, a loop smaller than six cycles is non-interruptible. Any pending interrupt will be processed as long as there are no pending branches to be completed. Additional information can be found in Ref. 6.

3.15 MULTICHANNEL BUFFERED SERIAL PORTS

Two McBSPs are available. They provide an interface to inexpensive (industry standard) external peripherals. McBSPs have features such as full-duplex communication, independent clocking and framing for receiving and transmitting, and direct interface to AC97 and IIS compliant devices. They allow several data sizes between 8 and 32 bits. Clocking and framing associated with the McBSPs for input and output are discussed in Ref. 7.

External data communication can occur while data are being moved internally. Figure 3.4 shows an internal block diagram of a McBSP. The data transmit (DX) and data receive (DR) pins are used for data communication. Control information (clocking and frame synchronization) is through CLKX, CLKR, FSX, and FSR. The CPU or DMA controller reads data from the data receive register (DRR) and writes data to be transmitted to the data transmit register (DXR). The transmit shift register (XSR) shifts these data to DX. The receive shift register (RSR) copies the data received on DR to the receive buffer register (RBR). The data in RBR are then copied to DRR to be read by the CPU or the DMA controller.

Other registers—the serial port control register (SPCR), receive/transmit control register (RCR/XCR), receive/transmit channel enable register (RCER/XCER), pin control register (PCR), and sample rate generator register (SRGR)—support further data communication [7].

The two McBSPs are used for input and output through the onboard codec. McBSP0 is used for control and McBSP1 for transmitting and receiving data.

3.16 DIRECT MEMORY ACCESS

Direct memory access (DMA) allows for the transfer of data to and from internal memory or external devices without intervention from the CPU [7]. Sixteen enhanced DMA channels (EDMA) can be configured independently for data transfer. DMA can access on-chip memory and the EMIF, as well as the HPI. Data of different sizes can be transferred: 8-bit bytes, 16-bit half-words, and 32-bit words.

A number of DMA registers are used to configure the DMA: address (source and destination), index, count reload, DMA global data, and control registers. The source and destination addresses can be from internal program memory, internal

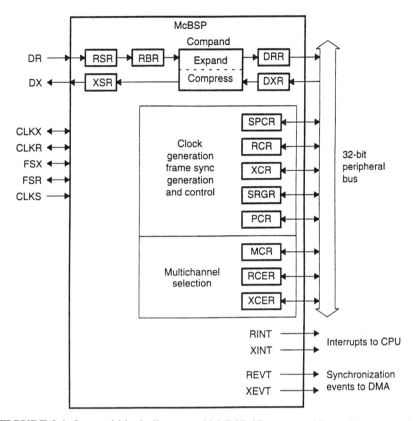

FIGURE 3.4. Internal block diagram of McBSP (Courtesy of Texas Instruments).

data memory, an external memory interface, and an internal peripheral bus. DMA transfers can be triggered by interrupts from internal peripherals as well as from external pins.

For each resource, each DMA channel can be programmed for priorities with the CPU, with channel 0 having the highest priority. Each DMA channel can be made to start initiating block transfer of data independently. A block can contain a number of frames. Within each frame can be many elements. Each element is a single data value. The DMA count reload register contains the value to specify the frame count (16 MSBs) and the element count (16 LSBs).

3.17 MEMORY CONSIDERATIONS

3.17.1 Data Allocation

Blocks of code and data can be allocated in memory within sections specified in the linker command file. These sections can be either initialized or uninitialized. The initialized sections are:

1. .cinit: for global and static variables
2. .const: for global and static constant variables
3. .switch: contains jump tables for large switch statements
4. .text: for executable code and constants

The uninitialized sections are:

1. .bss: for global and static variables
2. .far: for global and static variables declared far
3. .stack: allocates memory for the system stack
4. .sysmem: reserves space for dynamic memory allocation used by the malloc, calloc, and realloc functions

The linker can be used to place sections such as text in fast internal memory for most efficient operation.

3.17.2 Data Alignment

The C6x always accesses aligned data that allow it to address bytes, half-words, and words (32 bits). The data format consists of four byte boundaries, two half-word boundaries, and one word boundary. For example, to assign a 32-bit load with LDW, the address must be aligned with a word boundary so that the lower 2 bits of the address are zero. Otherwise, incorrect data can be loaded. A double-word (64 bits) also can be accessed. Both .S1 and .S2 can be used to execute the double-word instruction LDDW to load two 64-bit double words, for a total of 128 bits per cycle.

3.17.3 Pragma Directives

The pragma directives tell the compiler to consider certain functions. Pragmas include DATA_ALIGN, DATA_SECTION, and so on. The DATA_ALIGN pragma has the syntax

```
#pragma DATA_ALIGN (symbol,constant);
```

that aligns *symbol* to a boundary. The constant is a power of 2. This pragma directive is used later in several examples (such as in FFT program examples) to align data in memory.

The DATA_SECTION pragma has the following syntax:

```
#pragma DATA_SECTION (symbol, "my_section");
```

which allocates space for *symbol* in the section named *my_section*. This pragma directive is useful to allocate a section in external memory. For example,

```
#pragma DATA_SECTION (buffer, ".extRAM")
```

is used to place `buffer` in the section extRAM. In the linker command file, the following is specified within SECTIONS:

```
.extRAM   :  >   SDRAM
```

and within MEMORY, the following is specified:

```
SDRAM: org = 0x80000000, len = 0x01000000
```

where 0x80000000 is the address in external memory (CE0 space). Another useful `pragma` directive,

```
#pragma MUST_ITERATE (20,20)
```

tells the compiler that the loop following will execute 20 times (a minimum and maximum of 20 times).

3.17.4 Memory Models

The compiler generates a small memory model code by default. Every data object is handled as if declared *near* unless it is specifically declared *far*. If the DATA_SECTION pragma is used, the object is specified as a *far* variable.

How run-time support functions are called can be controlled by the option −mr0 with the run-time support data and calls *near*, or by the option −mr1 with the run-time support data and calls *far*. Using the *far* method to call functions does not imply that those functions must reside in off-chip memory.

Large-memory models can be generated with the linker options −mlx (x = 0 to 4). If no level is specified, data and functions default to *near*. These models can be used for calling a function that is more than 1 M word away.

3.18 FIXED- AND FLOATING-POINT FORMAT

Some fixed-point considerations are reviewed in Appendix C.

3.18.1 Data Types

Some data types are:

1. *short*: of size 16 bits represented as 2's complement with a range from -2^{15} to $(2^{15} - 1)$

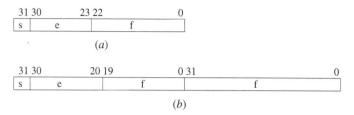

FIGURE 3.5. Data format: (*a*) single precision; (*b*) double precision.

2. *int* or *signed int*: of size 32 bits represented as 2's complement with a range from -2^{31} to $(2^{31} - 1)$

3. *float*: of size 32 bits represented as IEEE 32-bit with a range from $2^{-126} = 1.175494 \times 10^{-38}$ to $2^{+128} = 3.40282346 \times 10^{38}$

4. *double*: of size 64 bits represented as IEEE 64-bit with a range from $2^{-1022} = 2.22507385 \times 10^{-308}$ to $2^{+1024} = 1.79769313 \times 10^{+308}$

Data types such as short for fixed-point multiplication can be more efficient (fewer cycles) than using int. Use of const can also increase code performance. Notations such as Uint16 and Uint32 are supported for casting 16- and 32-bit unsigned integers, respectively.

3.18.2 Floating-Point Format

With a much wider dynamic range in a floating-point processor, scaling is not an issue. A floating-point number can be represented using single precision with 32 bits or double precision with 64 bits, as shown in Figure 3.5. In single-precision format, bit 31 represents the sign bit, bits 23 through 30 represent the exponent bits, and bits 0 through 22 represent the fractional bits, as shown in Figure 3.5*a*. Numbers as small as 10^{-38} and as large as 10^{+38} can be represented. In double-precision format, more exponent and fractional bits are available, as shown in Figure 3.5*b*. Since 64 bits are represented, a pair of registers is used. Bits 0 through 31 of the first register pair represent the fractional bits. Bits 0 through 19 of the second register pair also represent the fractional bits, with bits 20 through 30 representing the exponent bits and bit 31 the sign bit. As a result, numbers as small as 10^{-308} and as large as 10^{+308} can be represented.

Instructions ending in either SP or DP represent single and double precision, respectively. Some of the floating-point instructions (available on the C67x floating-point processor) have more latencies than do fixed-point instructions. For example, the fixed-point multiplication *MPY* requires one delay or *NOP*, whereas the single-precision *MPYSP* requires three delays and the double-precision instruction *MPYDP* requires nine delays.

The single-precision floating-point instructions ADDSP and MPYSP have three delay slots and take four cycles to complete execution. The double-precision instructions ADDDP and MPYDP have six and nine delay slots, respectively. However, the floating-point double-word load instruction LDDW (with four delay slots, as with the fixed-point LDW) can load 64 bits. Two LDDW instructions can execute in parallel through both units .S1 and .S2 to load a total of 128 bits per cycle.

A single-precision floating-point value can be loaded into a single register, whereas a double-precision floating-point value is a 64-bit value that can be loaded into a register pair such as A1:A0, A3:A2, ... , B1:B0, B3:B2, ... The least significant 32 bits are loaded into the even register pair, and the most significant 32 bits are loaded into the odd register pair.

One may need to weigh the pros and cons of dynamic range and accuracy with possible degradation in speed when using floating-point types of instructions.

3.18.3 Division

The floating-point C6713 processor has a single-precision reciprocal instruction RCPSP. A division operation can be performed by taking the reciprocal of the denominator and multiplying the result by the numerator [6]. There are no fixed-point instructions for division. Code is available to perform a division operation by using the fixed-point processor to implement a Newton–Raphson equation.

3.19 CODE IMPROVEMENT

Several code optimization schemes are discussed in Chapter 8 using both fixed- and floating-point implementations and ASM code.

3.19.1 Intrinsics

C code can be optimized further by using many of the available *intrinsics* in the run-time library support file. Intrinsic functions are similar to run-time support library functions. *Intrinsics* are available to multiply, to add, to find the reciprocal of a square root, and so on. For example, in lieu of using the asterisk operator to multiply, the intrinsic *_mpy* can be used. *Intrinsics* are special functions that map directly to inline C6x instructions. For example,

```
int _mpy()
```

is equivalent to the assembly instruction MPY to multiply the 16 LSBs of two numbers. The intrinsic function

```
int _mpyh()
```

is equivalent to the assembly instruction MPYH to multiply the 16 MSBs of two numbers.

3.19.2 Trip Directive for Loop Count

The linear assembly directive . trip is used to specify the number of times a loop iterates. If the exact number is known and used, the linear assembler optimizer can produce pipelined code (discussed in Chapter 8) and redundant loops are not generated. This can improve both code size and execution time. A . trip count specification, even if it is not the exact value, may improve performance: for example, when the actual number of iterations is a multiple of the specified value. The intrinsic function _nassert() can be used in a C program in lieu of . trip. Example 3.1 illustrates the use of _nassert() in the dot product example.

3.19.3 Cross-Paths

Data and address cross-path instructions are used to increase code efficiency. The instruction

```
MPY  .M1x  A2,B2,A4
```

illustrates a data cross-path that multiplies the two sources A2 and B2 from two different sides, A and B, with the result in A4. If the result is in the B register file, a 2x cross-path is used with the instruction

```
MPY  .M2x  A2,B2,B4
```

with the result in B4. The instruction

```
LDW  .D1T2  *A2,B2
```

illustrates an address cross-path. It loads the content in register A2 (from a register file A) into register B2 (register file B). Only two cross-paths are available on the C6x, so no more than two instructions using cross-paths are allowed within a cycle.

3.19.4 Software Pipelining

Software pipelining uses available resources to obtain efficient pipelining code. The aim is to use all eight functional units within one cycle. However, substantial coding effort can be required when the software pipelining technique is used for more complex programs. There are three stages to a pipelined code:

1. Prolog
2. Loop kernel (or loop cycle)
3. Epilog

The first stage, prolog, contains instructions to build the second-stage loop cycle, and the epilog stage (last stage) contains instructions to finish all loop iterations. Software pipelining is used by the compiler when the optimization option level −o2 or −o3 is invoked. The most efficient software pipelined code has loop trip counters that count down: for example,

```
for (i = N; i != 0; i--)
```

A dot product example with word-wide hand-coded pipelined code results in ($N/2$) + 8 cycles to obtain the sum of two arrays, with N numbers in each array. This translates to 108 cycles to find the sum of products of 200 numbers, as illustrated in Chapter 8. This efficiency is obtained using instructions such as LDW to load a 32-bit word and multiplying the lower and higher 16-bit numbers separately with the two instructions *mpy* and *mpyh*, respectively.

Removing the epilog section can also reduce the code size. The available options −msn ($n = 0, 1, 2$) directs the compiler to favor code size reduction over performance. Hand-coded software pipelined code can be produced by first drawing a dependency graph and setting up a scheduling table [8]. In Chapter 8 we discuss software pipelining in conjunction with code efficiency.

3.20 CONSTRAINTS

3.20.1 Memory Constraints

Internal memory is arranged through various banks of memory so that loads and stores can occur simultaneously. Since each bank of memory is single-ported, only one access to each bank is performed per cycle. Two memory accesses per cycle can be performed if they do not access the same bank of memory. If multiple accesses are performed to the same bank of memory (within the same space), the pipeline will stall. This causes additional cycles for execution to complete.

3.20.2 Cross-Path Constraints

Since there is one cross-path in each side of the two data paths, there can be at most two instructions per cycle using cross-paths. The following code segment is valid since both available cross-paths are used:

```
    ADD  .L1x  A1,B1,A0
||  MPY  .M2x  A2,B2,B3
```

whereas the following is not valid since one cross-path is used for both instructions:

```
    ADD .L1x A1,B1,A0
||  MPY .M1x A2,B2,A3
```

The x associated with the functional unit designates a cross-path.

3.20.3 Load/Store Constraints

The address register to be used must be on the same side as the .D unit. The following code segment is valid:

```
    LDW  .D1  *A1,A2
||  LDW  .D2  *B1,B2
```

whereas the following is not valid:

```
    LDW  .D1  *A1,A2
||  LDW  .D2  *A3,B2
```

Furthermore, loading and storing cannot be from the same register file. A load (or store) using one register file in parallel with another load (or store) must use a different register file. For example, the following code segment is valid:

```
    LDW  .D1  *A0,B1
||  STW  .D2  A1,*B2
```

The following is also valid:

```
    LDW  .D1  *A0,B1
||  LDW  .D2  *B2,A1
```

However, the following is not valid:

```
    LDW  .D1  *A0,A1
||  STW  .D2  A2,*B2
```

3.20.4 Pipelining Effects with More Than One EP within an FP

Table 3.3 shows a previous pipeline operation representing eight instructions in parallel within one FP. Table 3.6 shows the pipeline operation when there is more than one EP within an FP.

Consider the operation of six FPs (FP1 through FP6) through the pipeline. FP1 contains three execute packets, and FP2, FP3, . . . , FP6 each contains one EP. In cycles 2 through 5, FP2 through FP5, each FP starts its program fetch phase. When the CPU detects that FP1 contains more than one EP, it forces the pipeline to stall so that EP2 and EP3, within FP1, can each start its dispatching (DP) phase in cycles 6 and 7, respectively. Each instruction within an FP has a "p" bit to specify whether

TABLE 3.6   Pipelining with Stalling Effects

					Clock Cycle						
1	2	3	4	5	6	7	8	9	10	11	12
PG	PS	PW	PR	DP	DC	E1	E2	E3	E4	E5	E6
					DP	DC	E1	E2	E3	E4	E5
						DP	DC	E1	E2	E3	E4
	PG	PS	PW	PR	X	X	DP	DC	E1	E2	E3
		PG	PS	PW	X	X	PR	DP	DC	E1	E2
			PG	PS	X	X	PW	PR	DP	. DC	E1
				PG	X	X	PS	PW	PR	DP	DC
					X	X	PG	PS	PW	PR	DP

that instruction is in parallel with a subsequent instruction (if a 1, as shown in Figure 3.3). With a 0 in the LSB of an instruction, the chain is broken, and the subsequent instructions are placed in the next execute packet.

During clock cycles 1 through 4, a program fetch phase occurs. The three EPs within the same FP cause a stall in the pipeline. This allows the DP phase to start at cycle 6 (not at cycle 5) for EP2 and at cycle 7 for EP3. The subsequent FP (FP2) with only one EP (with all eight instructions in parallel) is stalled so that each of the three EPs in the previous FP (FP1) can go through the DP phase. As a result, while the fetch phase for FP2 starts at cycle 2, its DP phase does not start until cycle 8. The third FP (FP3), also with only one EP, starts its fetch stage at cycle 3, but its DP phase does not start until cycle 9, due to the pipeline stall.

The pipeline then stalls in cycles 6 and 7, as indicated with an "X." Once EP3 (within FP1) continues onto its decoding phase in cycle 8, the pipeline is released. FP2 can now continue to its DP phase in cycle 8. Since FP3 through FP6 also were stalled, each can now resume its program fetch phase in cycle 8.

Hence, with the three EPs within one FP, the pipeline stalls for two cycles. Table 3.6 illustrates the stalling pipeline effects. A pipeline stall would also take place if the first FP had four EPs, each with two parallel instructions.

3.21 PROGRAMMING EXAMPLES USING C, ASSEMBLY, AND LINEAR ASSEMBLY

Several programming examples are discussed in this section. The first example illustrates use of the intrinsic function _nassert_ to increase the efficiency of the dot product in Example 1.3. The remaining examples illustrate both assembly code and linear assembly code implementation: a C program calling an assembly function, a C program calling a linear assembly function, and an assembly-coded program calling an assembly-coded function. The focus here is on illustrating the syntax of both assembly and linear assembly code, not necessarily to produce optimized code.

We discuss further optimization techniques in Chapter 8 in conjunction with code efficiency and software pipelining.

Example 3.1: Efficient Dot Product (dotpopt)

This example uses the intrinsic function _nassert in the dot product example introduced in Example 1.3. Figure 3.6 shows a listing of the program dotpopt.c, which calls the C function dotpfunc.c listed in Figure 3.7. This function produces more efficient code, with _nassert used for alignment of the incoming pointers as constant pointers, along with a compiler option –pm. This provides additional information to the compiler about the loop.

Build and run this project example as **dotpopt**. Verify the following: (1) with a compiler option –g , the number of cycles associated with profiling the function

```
//dotpopt.c Optimized dot product of two arrays

#include    <stdio.h>
#include    <dotp4.h>                    //header file with data
#define     count 4
short x[count] = {x_array};              //declare 1st array
short y[count] = {y_array};              //declare 2nd array
volatile int result = 0;                 //result

main()
{
  result = dotpfunc(x,y,count);          //call optimized function
  printf("result = %d decimal \n", result);//print result
}
```

FIGURE 3.6. Dot product program calling a function with _nassert intrinsic (dotpopt.c).

```
//dotpfunc.c  Optimized dot product function

int dotpfunc(const short *a, const short *b, int ncount)
{
    int sum = 0;
    int i;
    _nassert((int)(a)%4 == 0);
    _nassert((int)(b)%4 == 0);
    _nassert((int)(ncount)%4 == 0);
    for ( i = 0; i < ncount; i++)
      {
        sum += (a[i] * b[i]);            //sum of products
      }
    return (sum);                        //return sum as result
}
```

FIGURE 3.7. C-called function for a dot product using _nassert (dotpfunc.c).

dotpfunc.c is 206 with a code size of 176; (2) with compiler options *-g -o3*, the number of cycles is reduced from 206 to 66, but the code size is increased to 484; (3) with options *-g -pm -o3*, the number of cycles is further reduced to 25 with a code size of 72. The *-pm* option uses program level optimization, with the source files compiled into one intermediate file. The results obtained with this option can be compared to the results obtained with the function dotp in Example 1.3. Note that the function dotpfunc.c can be readily profiled by creating a profiling area, dragging (with your mouse) the function into the profiling area.

In Chapter 8 we use optimization techniques associated with the dot product example, using two arrays each with N numbers. We show that the number of cycles can be reduced to $7 + (N/2) + 1$ with a fixed-point implementation, or 108 cycles using 200 numbers in each array. For a floating-point implementation, we obtain 124 cycles (see Table 8.4).

Example 3.2: Sum of n + (n – 1) + (n – 2) + ... + 1, Using C Calling an Assembly Function (sum)

This example illustrates a C program calling an assembly function. The C source program *sum.c* shown in Figure 3.8 calls the assembly-coded function *sumfunc.asm* shown in Figure 3.9. It implements the sum of $n + (n - 1) + (n - 2) + \ldots + 1$. The value of n is set in the main C program. It is passed through register A4 (by convention). For example, the address of more than one value can be passed to the assembly function through A4, B4, A6, B6, and so on. The resulting sum from the assembly (asm) function is returned to *result* in the C program, which then prints this resulting sum.

The assembly function's name is preceded by an underscore (by convention). The value n in register A4 in the asm function is moved to register A1 to set A1 as a loop counter since only A1, A2, B0, B1, and B2 can be used as conditional registers. A1 is then decremented. A loop section of code starts with the label or address

```
//Sum.c Finds n+(n-1)+...+1. Calls ASM function sumfunc

#include <stdio.h>

main()
{
  short n=6;                 //set value
  short result;             //result from asm function
  result = sumfunc(n);      //call ASM function sumfunc
  printf("sum = %d", result); //print result from asm function
}
```

FIGURE 3.8. C program that calls an ASM function to find $n + (n - 1) + (n - 2) + \ldots + 1$ (sum.c).

```
;Sumfunc.asm Assembly function to find n + (n-1) + ... + 1

            .def    _sumfunc            ;function called from C
_sumfunc:   MV    .L1   A4,A1          ;setup n as loop counter
            SUB   .S1   A1,1,A1        ;decrement n
LOOP:       ADD   .L1   A4,A1,A4       ;accumulate in A4
            SUB   .S1   A1,1,A1        ;decrement loop counter
     [A1]   B     .S2   LOOP           ;branch to LOOP if A1#0
            NOP         5              ;five NOPs for delay slots
            B     .S2   B3             ;return to calling routine
            NOP         5              ;five NOPs for delay slots
            .end
```

FIGURE 3.9. ASM function called from C in the project sum (sumfunc.asm).

LOOP and ends with the first branch statement B. The first addition adds $n +(n-1)$ with the result in A4. A1 is again decremented to $(n-2)$. The branch statement is conditional based on register A1, and since A1 is not zero, branching takes place and execution returns to the instruction at the address LOOP, where $A4 = n + (n-1)$ is added to $A1 = (n-2)$. This process continues until register $A1 = 0$.

The second branch instruction is to the returning address B3 (by convention) of the C calling program. The resulting sum is contained or accumulated in A4, which is passed to *result* in the C program. The five NOPs (no operation) are used to account for the five delay slots associated with a branch instruction.

The functional units .S and .L selected are shown but are not required in the program. They can be useful for debugging and analyzing which of the functional units are used in order to improve the efficiency of the program. Similarly, the two colons after the label LOOP and the function name are not required.

Build and run this project as **sum**. With a value of n set to 6 in the C program, verify that *sum* and its value of 21 are printed.

Example 3.3: Factorial of a Number Using C Calling an Assembly Function (factorial)

This example finds the factorial of a number $n \le 7$ with $n! = n(n-1)(n-2)\dots$ (1). It further illustrates the syntax of assembly code. It is very similar to Example 3.2. The value of n is set in the C source program *factorial.c*, shown in Figure 3.10, which calls the assembly function *factfunc.asm*, shown in Figure 3.11. It is instructive to read the comments.

Register A1 is again set as a loop counter. Within the loop section of code starting with the address LOOP, the first multiply is $n(n-1)$ and accumulates in register A4. The initial value of n is passed to the asm function through A4. The MPY instruction has one delay slot, which accounts for the NOP following it. Processing continues within the loop section of code until $A1 = 0$. Note that the functional units

```
//Factorial.c Finds factorial of n. Calls function factfunc

#include <stdio.h>                //for print statement

void main()
{
 short n = 7;                     //set value
 short result;                    //result from asm function
 result = factfunc(n);           //call ASM function factfunc
 printf("factorial = %d", result);//print result from asm function
}
```

FIGURE 3.10. C program that calls an ASM function to find the factorial of a number (factorial.c).

```
;Factfunc.asm Assembly function called from C to find factorial

          .def  _factfunc      ;ASM function called from C
_factfunc: MV    A4,A1          ;setup loop count in A1
          SUB   A1,1,A1        ;decrement loop count
LOOP:     MPY   A4,A1,A4       ;accumulate in A4
          NOP                  ;for 1 delay slot with MPY
          SUB   A1,1,A1        ;decrement for next multiply
    [A1]  B     LOOP           ;branch to LOOP if A1 # 0
          NOP   5              ;five NOPs for delay slots
          B     B3             ;return to calling routine
          NOP   5              ;five NOPs for delay slots
          .end
```

FIGURE 3.11. ASM function called from C that finds the factorial of a number (factfunc.asm).

are not specified in this program. The resulting factorial is returned to the calling C program through A4.

Build and run this project as **factorial**. Verify that *factorial* and its value of 5040 (7!) are printed. Note that the maximum value of *n* is 7, since result is casted as a short and 8! is greater than 2^{15}.

Example 3.4: 32-bit Pseudorandom Noise Generation Using C Calling an Assembly Function (Noisegen_casm)

The C source program *Noisegen_casm.c* in Figure 3.12 calls the function *noisefunc* located in the file *Noisegen_casmfunc.asm* (Figure 3.13) to generate a 32-bit pseudo-random noise sequence using the following scheme:

1. A 32-bit seed value such as 0×7E521603 is chosen.
2. A modulo 2 summation is performed between bits 17, 28, 30, and 31.

```
//Noisegen_casm.c Pseudorandom noise generation calling ASM function

#include "dsk6713_aic23.h"                  //codec-DSK support file
Uint32 fs=DSK6713_AIC23_FREQ_48KHZ;         //set sampling rate
int previous_seed;
short pos = 16000, neg = -16000;            //scaling noise level

interrupt void c_int11()
{
 previous_seed = noisefunc(previous_seed);      //call ASM function
 if(previous_seed & 0x01)  output_sample(pos);//positive scaling
 else                      output_sample(neg);//negative scaling
}

void main ()
{
 comm_intr();                               //init DSK,codec,McBSP
 previous_seed = noisefunc(0x7E521603);     //call ASM function
 while (1);                                 //infinite loop
}
```

FIGURE 3.12. C program that calls an ASM function to generate a 32-bit noise sequence (noisegen_casm.c).

```
;Noisegen_casmfunc.asm Noise generation C-called function

            .def    _noisefunc  ;ASM function called from C
_noisefunc  ZERO    A2          ;init A2 for seed manipulation
            MV      A4,A1       ;seed in A1
            SHR     A1,17,A1    ;shift right 17->bit 17 to LSB
            ADD     A1,A2,A2    ;add A1 to A2 => A2
            SHR     A1,11,A1    ;shift right 11->bit 28 to LSB
            ADD     A1,A2,A2    ;add again
            SHR     A1,2,A1     ;shift right 2->bit 30 to LSB
            ADD     A1,A2,A2    ;
            SHR     A1,1,A1     ;shift right 1->bit 31 to LSB
            ADD     A1,A2,A2    ;
            AND     A2,1,A2     ;Mask LSB of A2
            SHL     A4,1,A4     ;shift seed left 1
            OR      A2,A4,A4    ;Put A2 into LSB of A4
            B       B3          ;return to calling function
            NOP     5           ;5 delays for branch
```

FIGURE 3.13. ASM function called from C to generate a 32-bit noise sequence (noisegen_casmfunc.asm).

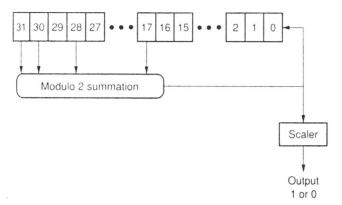

FIGURE 3.14. A 32-bit noise generator diagram.

3. The LSB of the resulting summation is selected. This bit is either a 1 or a 0 and is scaled accordingly to a positive or negative value.
4. The seed value is shifted left by one, and the resulting bit from the previous step is placed in the LSB position and the process repeated with the new (shifted by one) seed value.

The 32-bit noise generator diagram is shown in Figure 3.14. Within the *asm* function, the seed value is moved from A4 to A1. Shifting this seed value right by 17 places bit 17 in the LSB position, where the addition is meaningful. The resulting summation is shifted right by 11 to place bit 28 (already shifted by 17) in the LSB position. This procedure is repeated, adding bits 17, 28, 30, and 31. The LSB, which is a 1 or a 0, is then placed into A4, and returned to the C calling function, where it is scaled as either a positive or a negative value, respectively. On each interrupt, this LSB bit, 1 or 0, represents the noise sample.

Build and run this project as **Noisegen_casm**. Sampling at 48 kHz, verify that the noise spectrum is flat, with a bandwidth of approximately 23 kHz. Connect the output to a speaker to verify the generated noise. Change the scaling values to ±8000 and verify that the level of the generated noise is reduced.

Set a breakpoint in the *asm* function and view the value of A4 before it is returned to the C calling function and verify the noise sequence as 1, 1, 1, 1, 0, 1, 0, 1, 1, 1, 1, 0, 0, 1, 1, 0, . . . This noise sequence will repeat after $(2^N - 1)$ with N as a 32-bit seed.

Example 3.5: Code Detection Using C Calling an ASM Function (`Code_casm`)

This example detects a four-digit code set initially in the main C source program. Figure 3.15 shows the main C source program `code_casm.c` that calls the *asm* function `code_casmfunc.asm`, shown in Figure 3.16. The code is set with *code1*,

```
//Code_casm.c Calls ASM function.If code match slider values

#include <stdio.h>
short digit1=1,digit2=1,digit3=1,digit4=1;//init slider values

main()
{
 short code1=1,code2=2,code3=2,code4=4;   //initialize code
 short result;
 DSK6713_init();                          //init BSL
 DSK6713_DIP_init();                       //init dip switches
 while(DSK6713_DIP_get(3) == 1)           //continue til SW #3 pressed
 {
  if(DSK6713_DIP_get(0) == 0)             //if DIP SW #0 is pressed
  {                                        //call ASM function
  result=codefunc(digit1,digit2,digit3,digit4,code1,code2,code3,code4);
  if(result==0) printf("correct match\n");//result from ASM function
  else          printf("no match\n");     //correct match or no match
  }
 }
}
```

FIGURE 3.15. C program that calls an ASM function to detect a four-digit code (code_casm.c).

```
;Code_casmfunc.asm ASM function->if code matches slider values

              .def   _codefunc    ;ASM function called from C
   _codefunc: MV     A8, A2       ;correct code
              MV     B8, B2
              MV     A10, A7
              MV     B10, B7
              CMPEQ A2,A4,A1       ;compare 1st digit(A1=1 if A2=A4)
              CMPEQ A1,0,A1        ;otherwise A1=0
       [A1]   B      DONE          ;done if A1=0 since no match
              NOP    5
              MV     B2,A2
              CMPEQ A2,B4,A1       ;compare 2nd digit
              CMPEQ A1,0,A1
       [A1]   B      DONE
              NOP    5
              MV     A7,A2
              CMPEQ A2,A6,A1       ;compare 3rd digit
              CMPEQ A1,0,A1
       [A1]   B      DONE
              NOP    5
              MV     B7,A2
              CMPEQ A2,B6,A1       ;compare 4th digit
              CMPEQ A1,0,A1
   DONE:      MV     A1,A4         ;return 1 if complete match
              B      B3            ;return to C program
              NOP    5
              .end
```

FIGURE 3.16. ASM function called from C to detect a four-digit code (code_casmfunc.asm).

..., *code4* as 1, 2, 2, 4, respectively. The initial values of *digit1*, ..., *digit4* set as 1, 1, 1, 1, respectively, are passed to the *asm* function to compare these four digit values with the four code values. Four sliders are used to change the digit values passed to the *asm* function. The C source program, the *asm* function, and the gel file for the sliders are included in the folder **code_casm**.

Build this project example as **code_casm**. Load and run the executable file. Press switch #0 (SW0) slightly and verify that "no match" is continuously being printed (as long as SW0 is pressed). Load the gel file `code_casm.gel` and set the sliders *Digit1*, ..., *Digit4* to positions 1, 2, 2, 4, respectively. Slightly press SW0 and verify that "correct match" is being printed (with SW0 pressed). Change the slider *Digit2* from position 2 to position 3, and again press SW0 to verify that there is no longer a match. The program is in a continuous loop as long as switch #3 (SW3) is *not* pressed. Note that the initial value for the code (code1, ..., code4) can be readily changed.

Example 3.6: Dot Product Using Assembly Program Calling an Assembly Function (dotp4a)

This example takes the sum of products of two arrays, each array with four numbers. See also Example 1.3, which implements it using only C code, and Examples 3.2 through 3.5, which introduced the syntax of assembly code. Figure 3.17 shows a

```
;Dotp4a_init.asm ASM program to init variables.Calls dotp4afunc

                .def        init          ;starting address
                .ref        dotp4afunc    ;called ASM function
                .text                     ;section for code
x_addr          .short      1,2,3,4       ;numbers in x array
y_addr          .short      0,2,4,6       ;numbers in y array
result_addr     .short      0             ;initialize sum of products

init            MVK    result_addr,A4     ;result addr -->A4
                MVK    0,A3               ;A3=0
                STH    A3,*A4             ;init result to 0
                MVK    x_addr,A4          ;A4 = address of x
                MVK    y_addr,B4          ;B4 = address of y
                MVK    4,A6               ;A6 = size of array
                B      dotp4afunc         ;B to function dotp4afunc
                MVK    ret_addr,b3        ;B3=return addr from dotp4a
                NOP    3                  ;3 more delay slots(branch)
ret_addr        MVK    result_addr,A0     ;A0 = result address
                STW    A4,*A0             ;store result
wait            B      wait               ;wait here
                NOP    5                  ;delay slots for branch
```

FIGURE 3.17. ASM program calling an ASM function to find the sum of products (dotp4a_init.asm).

```
;Dotp4afunc.asm Multiply two arrays. Called from dotp4a_init.asm
;A4=x address,B4=y address,A6=count(size of array),B3=return address

                .def        dotp4afunc  ;dot product function
                .text                   ;text section
dotp4afunc      MV          A6,A1       ;move loop count -->A1
                ZERO        A7          ;init A7 for accumulation
loop            LDH         *A4++,A2    ;A2=content of x address
                LDH         *B4++,B2    ;B2=content of y address
                NOP         4           ;4 delay slots for LDH
                MPY         A2,B2,A3    ;A3 = x * y
                NOP                     ;1 delay slot for MPY
                ADD         A3,A7,A7    ;sum of products in A7
                SUB         A1,1,A1     ;decrement loop counter
        [A1]    B           loop        ;branch back to loop till A1=0
                NOP         5           ;5 delay slots for branch
                MV          A7,A4       ;A4=result
                B           B3          ;return from func to addr in B3
                NOP         5           ;5 delay slots for branch
```

FIGURE 3.18. ASM function called from an ASM program to find the sum of products
(dotp4afunc.asm).

listing of the assembly program *dotp4a_init.asm*, which initializes the two
arrays of numbers and calls the assembly function *dotp4afunc.asm*, shown in
Figure 3.18, which takes the sum of products of the two arrays. It also sets a return
address through register B3 and the result address to A0. The addresses of the two
arrays and the size of the array are passed to the function *dotp4afunc.asm*
through registers A4, A6, and B4, respectively. The result from the called function
is "sent back" through A4. The resulting sum of the products is stored in memory
whose address is result_addr. The instruction STW stores the resulting sum of
the products in A4 (in memory pointed by A0). Register A0 serves as a pointer with
the address result_addr.

The instruction MVK moves the 16LSBs (equivalent to MVKL). If a 32-bit
address (or result) is required, then the pair of instructions MVKL and MVKH can
be used to move both the lower and upper 16 bits of the address (or result). The
starting address of the calling ASM program is defined as init. The vector file is
modified and included in the folder **dotp4a** so that the reference to the entry
address is changed from _c_int00 to the entry address init. An alternative
vector file *vectors_dotp4a.asm*, as shown in Figure 3.19, specifies a branch to
that entry address. The called *asm* function *dotp4afunc.asm* calculates the sum
of products. The loop count value was moved to A1 since A6 cannot be used as a
conditional register (only A1, A2, B0, B1, and B2 can be used). The two LDH instructions load (half-word of 16 bits) the addresses of the two arrays starting at *x_addr*
and *y_addr* into registers A2 and B2, respectively. For example, the instruction

LDH *B4++,B2

```
;vectors_dotp4a.asm Alternative vector file for dotp4a project

        .ref    init       ;starting addr in init file
        .sect   "vectors"  ;in section vectors
rst:    mvkl .s2 init,b0    ;init addr 16 LSB -->B0
        mvkh .s2 init,b0    ;init addr 16 MSB -->B0
        b       b0          ;branch to addr init
        nop     5
```

FIGURE 3.19. Alternative vector file that specifies the entry address in the calling ASM program for the sum of products (vectors_dotp4a.asm).

loads the content in memory (the first value in the second array starting at *y_address*) pointed at by B4 (the address of the second array) into B2. Then register B4, used as a pointer, is postincremented to the next higher address in memory that contains the second value in the second array. Register A7 is used to accumulate and move the sum of products to register A4, since the result is passed to the calling function through A4.

Support files for this project include (no library file is necessary):

1. dotp4a_init.asm
2. dotp4afunc.asm
3. vecs_dotp4a.asm

The vector file *vecs_dotp4a.asm* (modified vector file) or the alternative vector file *vectors_dotp4a.asm* shown in Figure 3.19 are both included in the folder **dotp4a**. Build and run this project as **dotp4a**. Modify the Linker Option (Project → Options) to select "No Autoinitialization." Otherwise, the warning "entry point symbol *_c_int00* undefined" is displayed when this project is built (it can be ignored). This is because the "conventional" entry point is not used in this project, since there is no *main* function in C.

Set a breakpoint at the first branch instruction in the program dotp4a_init.asm:

B dotp4afunc

Select View → Memory, set address to *result_addr*, and use the 16-bit signed integer. Right-click on the memory window and deselect "Float in Main Window." This allows you to have a better display of the Memory window while viewing the source file *dotp4a_init.asm*.

Select Run. Execution stops at the set breakpoint. The content in memory at the address *result_addr* is zero (the called function *dotp4afunc.asm* is not yet executed). Run again, then halt, since execution is within the infinite wait loop instruction:

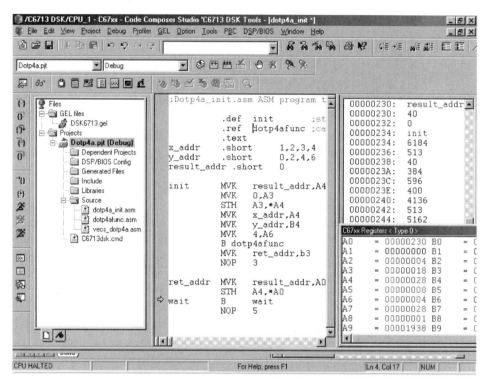

FIGURE 3.20. CCS windows for the sum of products in the project `dotp4a`.

```
wait B wait ;wait here
```

Verify that the resulting sum of products is A4 = `0x28` = 40. Note that A0 contains the result address (`result_addr`). Select View → Registers → Core Registers and verify this address (in hex). Figure 3.20 shows a CCS display of this project. Note from the disassembly file that execution was halted at the infinite wait loop.

Example 3.7: Dot Product Using C Function Calling a Linear Assembly Function (`dotp4clasm`)

Figure 3.21 shows a listing of the C source program `dotp4clasm.c`, which calls the linear assembly function `dotp4clasmfunc.sa`, shown in Figure 3.22. Example 1.3 introduced the dot product implementation using C code only. The previous five examples introduced the syntax of assembly-coded programs.

The section of code invoked by the linear assembler optimizer starts and ends with the linear assembler directives, `.cproc` and `.endproc`, respectively. The name of the linear assembly function called is preceded by an underscore since the calling function is in C. The directive `.def` defines the function.

```
//Dotp4clasm.c Multiplies two arrays using C calling linear ASM func

short dotp4clasmfunc(short *a,short *b,short ncount);  //prototype
#include <stdio.h>                       //for printing statement
#include "dotp4.h"                       //arrays of data values
#define  count 4                         //number of data values
short x[count] = {x_array};              //declare 1st array
short y[count] = {y_array};              //declare 2nd array
volatile int result = 0;                 //result

main()
{
 result = dotp4clasmfunc(x,y,count);     //call linear ASM func
 printf("result = %d decimal \n", result);  //print result
}
```

FIGURE 3.21. C program calling a linear ASM function to find the sum of products (dotp4clasm.c).

```
;Dotp4clasmfunc.sa  Linear assembly function to multiply two arrays
                .ref     _dotp4clasmfunc ;ASM func called from C
_dotp4clasmfunc: .cproc  ap,bp,count     ;start section linear ASM
                .reg     a,b,prod,sum    ;asm optimizer directive
                zero     sum             ;init sum of products
loop:           ldh      *ap++,a         ;pointer to 1st array->a
                ldh      *bp++,b         ;pointer to 2nd array->b
                mpy      a,b,prod        ;product = a*b
                add      prod,sum,sum    ;sum of products -->sum
                sub      count,1,count   ;decrement counter
   [count]      b        loop            ;loop back if count # 0
                .return  sum             ;return sum as result
                .endproc                 ;end linear ASM function
```

FIGURE 3.22. Linear ASM function called from C to find the sum of products (dotp4clasmfunc.sa).

Functional units are optional as in an assembly-coded program. Registers a, b, $prod$, and sum are defined by the linear assembler directive .reg. The addresses of the two arrays x and y and the size of the array (count) are passed to the linear assembly function through the registers ap, bp, and $count$. Both ap and bp are registers used as pointers, as in C code. The instruction field is seen to be as in an assembly-coded program, and the subsequent field uses a syntax as in C programming. For example, the instruction

```
loop:   ldh   *ap++,a
```

(the first time through the loop section of code) loads the content in memory, whose address is specified by register ap, into register a. Then the pointer register ap is postincremented to point to the next higher memory address, pointing at the

memory location containing the second value of x within the x array. The value of
the sum of the products is accumulated in *sum*, which is returned to the C calling
program.

Build and run this project as **dotp4clasm**. Verify that the following is printed:
result = 40. You may wish to profile the linear assembly code function and
compare its execution time with that of the C-coded version in Example 1.3.

Example 3.8: Factorial Using C Calling a Linear Assembly Function (`factclasm`)

Figure 3.23 shows a listing of the C program *factclasm.c*, which calls the linear
asm function *factclasmfunc.sa*, shown in Figure 3.24, to calculate the factorial
of a number less than 8. See also Example 3.3, which finds the factorial of a number
using a C program that calls an asm function. Example 3.7 illustrates a C program

```
//Factclasm.c Factorial of number. Calls linear ASM function

#include <stdio.h>                    //for print statement

void main()
{
  short number = 7;                   //set value
  short result;                       //result of factorial
  result = factclasmfunc(number);     //call ASM function factlasmfunc
  printf("factorial = %d", result);   //result from linear ASM function
}
```

FIGURE 3.23. C program that calls a linear ASM function to find the factorial of a number
(factclasm.c).

```
;Factclasmfunc.sa Linear ASM function called from C to find factorial

                .ref      _factclasmfunc  ;Linear ASM func called from C
_factclasmfunc: .cproc    number          ;start of linear ASM function
                .reg      a,b             ;asm optimizer directive
                mv        number,b        ;setup loop count in b
                mv        number,a        ;move number to a
                sub       b,1,b           ;decrement loop counter
loop:           mpy       a,b,a           ;n(n-1)
                sub       b,1,b           ;decrement loop counter
   [b]          b         loop            ;loop back to loop if count #0
                .return a                 ;result to calling function
                .endproc                  ;end of linear ASM function
```

FIGURE 3.24. Linear ASM function called from C that finds the factorial of a number
(factclasmfunc.sa).

calling a linear ASM function to find the sum of products and is instructive for this project. Examples 3.3 and 3.7 cover the essential background for this example.

Support files for this project include `factclasm.c`, `factclasmfunc.sa`, `rts6700.lib`, and `C6713dsk.cmd`. Build and run this project as **factclasm**. Verify that the result of 7! is printed, or `factorial = 5040`.

3.22 ASSIGNMENTS

1. Write a C program that calls an assembly function that takes input values a and b from the C program to calculate the following: $[a^2 + (a + 1)^2 + (a + 2)^2 + \ldots + (2a - 1)^2] - [b^2 + (b + 1)^2 + (b + 2)^2 + \ldots + (2b - 1)^2]$. Set $a = 3$ and $b = 2$ in the C program and verify that the result is printed as 37.

2. Write a C program that calls an assembly function to obtain the determinant of a 3×3 matrix. Set the matrix values in the C program. The first row values are $\{4, 5, 9\}$; the second row values are $\{8, 6, 5\}$, and the third row values are $\{2, 1, 2\}$. Verify that the resulting determinant is printed within CCS as −38.

3. Write a C program `multi_casm.c` that calls an assembly function `multi_casmfunc.asm` to multiply two numbers using the onboard dip switches. The maximum product is $3 \times 4 = 12$ or $4 \times 3 = 12$. Note that $4 \times 4 = 16$ cannot be represented with the four dip switches. Use delay loops for debouncing the switches. A partial program is included in Figure 3.25. In the main C source program, the values of $m = 100$ and $n = 100$ are to check when the first and second switches are pressed. Sw0 is tested and, if pressed, $m = 1$, representing the first value. Similarly, $m = 2, 3, 4$ if SW1, SW2, or SW3 is pressed, respectively. Then all LEDs are turned off. This process is repeated while $n = 100$ to check for the second value (when the second switch is pressed).

   The function *values* performs the multiplication, adding m (n times) with m and n passed to the asm function through A4 and B4, respectively. Note that `led0` is turned on if led0 = 1 (returned from the function `result0`). Similarly for `led1`, ..., `led3`. Then, m and n are reset to 100 and ii to 1. The *asm* function `multi_casmfunc.asm` includes the functions *values, result0, ...,* *result3.* The functions *result1, result2, result3* are similar to *result0*, but A4 must be shifted first by 1, by 2, and by 3, respectively, in each of these functions. Build and run this project example as **multi_casm**. Slightly press SW2, then SW3 to obtain m = 3 and n = 4, and verify that SW2 and SW3 turn on to represent the result of 12.

4. Write a C program that calls a linear assembly or assembly function to generate a random noise sequence, based on the linear feedback shift register (LFSR) shown in Figure 3.26. In lieu of starting with a 16-bit seed value, 16 integer values are used in an array as the seeds. In this fashion, each 32-bit

```
Partial programs C/ASM function to multiply 2 numbers using switches
  ..
while(m == 100)                        //check for first SW pressed
{
 if(DSK6713_DIP_get(0)== 0)            //true if SW0 is pressed
 {
  m = 1;                               //value if SW0 is pressed
  while(DSK6713_DIP_get(0)==0) DSK6713_LED_on(0);//ON until released
  for(delay=0; delay<5000000; delay++){}        //debounce of SW0
 }
 else if(DSK6713_DIP_get(1)==0)        //true if SW1 is pressed
 {
  m = 2;
  .

  .
  else m = 100;
  .

  .
  while(ii == 0)
    {
    result = values(n, m);             //result from ASM function in A4
    led0 = result0(result);            //returns a 0 or 1 to led0
    if(led0==1)  DSK6713_LED_on(0); //if led0 is 1 turn it on
    .

    .
;ASM function
    ..
_values:    MV    A4,A5        ;setup n as loop counter
            MV    B4,B1
LOOP:       ADD   A5,A4,A4     ;accumulate in A4
            SUB   B1,1,B1      ;decrement loop counter
     [B1]   B     LOOP         ;branch to LOOP if B1#0
            NOP   5            ;five NOPs for delay slots
            SUB   A4,A5,A4     ;answer into A4
            B     B3           ;return to calling routine
            NOP   5            ;five NOPs for delay slots

_result0:   SHL   A4,31,A4     ;shift left 31 bits to keep LSB
            SHRU  A4,31,A4     ;shift right 31 bits to make A4=0 or 1
            B     B3           ;return to calling routine
            NOP   5            ;five NOPs for delay slots
```

FIGURE 3.25. Partial programs (C and ASM function) to multiply two numbers using the dip switches.

seed is treated as a theoretical bit. The "tap points" are chosen as shown (bits 1, 2, 11, 15, and 16) to produce a large string of random numbers [11]. Within the *asm* or *linear asm* function, each integer value is taken as a seed, and you can use instructions such as LDW/STW, repeated 15 times, to move each seed "up." XOR bits 1 and 2, the result of which is XORed with bit 11, and so on, as shown in Figure 3.26. The resulting seed generated is placed at the "bottom"

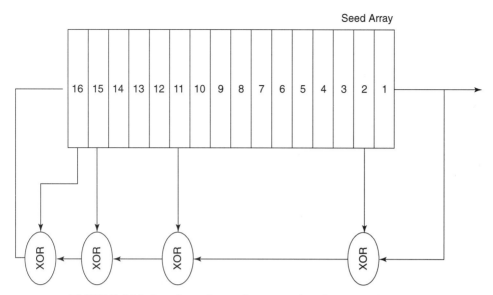

FIGURE 3.26. Pseudorandom noise generation diagram using LFSR.

of the array, and the process is repeated. The output is a 32-bit value. Sampling at 8 kHz, verify that the generated noise spectrum is flat until it rolls off at about 3.8 kHz, which is the cutoff frequency of the antialiasing filter on the codec.

REFERENCES

1. R. Chassaing and D. W. Horning, *Digital Signal Processing with the TMS320C25*, Wiley, New York, 1990.

2. R. Chassaing, *Digital Signal Processing Laboratory Experiments Using C and the TMS320C31 DSK*, Wiley, New York, 1999.

3. R. Chassaing, *Digital Signal Processing with C and the TMS320C30*, Wiley, New York, 1992.

4. R. Chassaing and P. Martin, Parallel processing with the TMS320C40, *Proceedings of the 1995 ASEE Annual Conference*, June 1995.

5. R. Chassaing and R. Ayers, Digital signal processing with the SHARC, *Proceedings of the 1996 ASEE Annual Conference*, June 1996.

6. *TMS320C6000 CPU and Instruction Set*, SPRU189F, Texas Instruments, Dallas, TX, 2000.

7. *TMS320C6000 Peripherals*, SPRU190D, Texas Instruments, Dallas, TX, 2001.

8. *TMS320C6000 Programmer's Guide*, SPRU198G, Texas Instruments, Dallas, TX, 2002.

9. *TMS320C6000 Assembly Language Tools User's Guide*, SPRU186K, Texas Instruments, Dallas, TX, 2002.

10. *TMS320C6000 Optimizing C Compiler User's Guide*, SPRU187K, Texas Instruments, Dallas, TX, 2002.

11. Linear Feedback Shift Registers, New Wave Instruments, 2002, www.newwaveinstruments.com/resources.

4

Finite Impulse Response Filters

- Introduction to the z-transform
- Design and implementation of finite impulse response (FIR) filters
- Programming examples using C and TMS320C6x code

The z-transform is introduced in conjunction with discrete-time signals. Mapping from the s-plane, associated with the Laplace transform, to the z-plane, associated with the z-transform, is illustrated. FIR filters are designed with the Fourier series method and implemented by programming a discrete convolution equation. Effects of window functions on the characteristics of FIR filters are covered.

4.1 INTRODUCTION TO THE z-TRANSFORM

The z-transform is utilized for the analysis of discrete-time signals, similar to the Laplace transform for continuous-time signals. We can use the Laplace transform to solve a differential equation that represents an analog filter or the z-transform to solve a difference equation that represents a digital filter. Consider an analog signal $x(t)$ ideally sampled

$$x_s(t) = \sum_{k=0}^{\infty} x(t)\delta(t - kT) \tag{4.1}$$

Digital Signal Processing and Applications with the C6713 and C6416 DSK   By Rulph Chassaing
ISBN 0-471-69007-4   Copyright © 2005 by John Wiley & Sons, Inc.

where $\delta(t - kT)$ is the impulse (delta) function delayed by kT and $T = 1/F_s$ is the sampling period. The function $x_s(t)$ is zero everywhere except at $t = kT$. The Laplace transform of $x_s(t)$ is

$$X_s(s) = \int_0^\infty x_s(t)e^{-st}\,dt$$
$$= \int_0^\infty \{x(t)\delta(t) + x(t)\delta(t - T) + \cdots\}e^{-st}\,dt \qquad (4.2)$$

From the property of the impulse function

$$\int_0^\infty f(t)\delta(t - kT)\,dt = f(kT)$$

$X_s(s)$ in (4.2) becomes

$$X_s(s) = x(0) + x(T)e^{-sT} + x(2T)e^{-2sT} + \cdots = \sum_{n=0}^\infty x(nT)e^{-nsT} \qquad (4.3)$$

Let $z = e^{sT}$ in (4.3), which becomes

$$X(z) = \sum_{n=0}^\infty x(nT)z^{-n} \qquad (4.4)$$

Let the sampling period T be implied; then $x(nT)$ can be written as $x(n)$, and (4.4) becomes

$$X(z) = \sum_{n=0}^\infty x(n)z^{-n} = ZT\{x(n)\} \qquad (4.5)$$

which represents the z-transform (ZT) of $x(n)$. There is a one-to-one correspondence between $x(n)$ and $X(z)$, making the z-transform a unique transformation.

Exercise 4.1: ZT of Exponential Function $x(n) = e^{nk}$

The ZT of $x(n) = e^{nk}$, $n \geq 0$ and k a constant, is

$$X(z) = \sum_{n=0}^\infty e^{nk}z^{-n} = \sum_{n=0}^\infty \left(e^k z^{-1}\right)^n \qquad (4.6)$$

Using the geometric series, obtained from a Taylor series approximation

$$\sum_{n=0}^{\infty} u^n = \frac{1}{1-u} \qquad |u| < 1$$

(4.6) becomes

$$X(z) = \frac{1}{1 - e^k z^{-1}} = \frac{z}{z - e^k} \tag{4.7}$$

for $|e^k z^{-1}| < 1$ or $|z| > |e^k|$. If $k = 0$, the ZT of $x(n) = 1$ is $X(z) = z/(z-1)$.

Exercise 4.2: ZT of Sinusoid x(n) = sin nωT

A sinusoidal function can be written in terms of complex exponentials. From Euler's formula $e^{ju} = \cos u + j \sin u$,

$$\sin n\omega T = \frac{e^{jn\omega T} - e^{-jn\omega T}}{2j}$$

Then

$$X(z) = \frac{1}{2j} \sum_{n=0}^{\infty} \left(e^{jn\omega T} z^{-n} - e^{-jn\omega T} z^{-n} \right) \tag{4.8}$$

Using the geometric series as in Exercise 4.1, one can solve for $X(z)$; or the results in (4.7) can be used with $k = j\omega T$ in the first summation of (4.8) and $k = -j\omega T$ in the second, to yield

$$\begin{aligned}
X(z) &= \frac{1}{2j} \left(\frac{z}{z - e^{j\omega T}} - \frac{z}{z - e^{-j\omega T}} \right) \\
&= \frac{1}{2j} \frac{z^2 - ze^{-j\omega T} - z^2 + ze^{j\omega T}}{z^2 - z(e^{-j\omega T} + e^{j\omega T}) + 1} \\
&= \frac{z \sin \omega T}{z^2 - 2z \cos \omega T + 1} \\
&= \frac{Cz}{z^2 - Az - B} \qquad |z| > 1
\end{aligned} \tag{4.9}$$

$$\tag{4.10}$$

where $A = 2\cos \omega T$, $B = -1$, and $C = \sin \omega T$. In Chapter 5 we generate a sinusoid based on this result. We can readily generate sinusoidal waveforms of different frequencies by changing the value of ω in (4.9).

Similary, using Euler's formula for $\cos n\omega T$ as a sum of two complex exponentials, one can find the ZT of $x(n) = \cos n\omega T = (e^{jn\omega T} + e^{-jn\omega T})/2$, as

$$X(z) = \frac{z^2 - z\cos\omega T}{z^2 - 2z\cos\omega T + 1} \qquad |z| > 1 \tag{4.11}$$

4.1.1 Mapping from *s*-Plane to *z*-Plane

The Laplace transform can be used to determine the stability of a system. If the poles of a system are on the left side of the $j\omega$ axis on the s-plane, a time-decaying system response will result, yielding a stable system. If the poles are on the right side of the $j\omega$ axis, the response will grow in time, making such a system unstable. Poles located on the $j\omega$ axis, or purely imaginary poles, will yield a sinusoidal response. The sinusoidal frequency is represented by the $j\omega$ axis, and $\omega = 0$ represents dc (direct current).

In a similar fashion, we can determine the stability of a system based on the location of its poles on the z-plane associated with the z-transform, since we can find corresponding regions between the s-plane and the z-plane. Since $z = e^{sT}$ and $s = \sigma + j\omega$,

$$z = e^{\sigma T} e^{j\omega T} \tag{4.12}$$

Hence, the magnitude of z is $|z| = e^{\sigma T}$ with a phase of $\theta = \omega T = 2\pi f/F_s$, where F_s is the sampling frequency. To illustrate the mapping from the s-plane to the z-plane, consider the following regions from Figure 4.1.

σ < 0
Poles on the left side of the $j\omega$ axis (region 2) in the s-plane represent a stable system, and (4.12) yields a magnitude of $|z| < 1$, because $e^{\sigma T} < 1$. As σ varies from

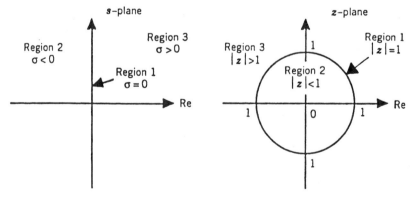

FIGURE 4.1. Mapping from the s-plane to the z-plane.

$-\infty$ to 0^-, $|z|$ will vary from 0 to 1^-. Hence, poles *inside* the unit circle within region 2 in the *z*-plane will yield a stable system. The response of such system will be a decaying exponential if the poles are real or a decaying sinusoid if the poles are complex.

σ > 0

Poles on the right side of the *j*ω axis (region 3) in the *s*-plane represent an unstable system, and (4.12) yields a magnitude of $|z| > 1$, because $e^{\sigma T} > 1$. As σ varies from 0^+ to ∞, $|z|$ will vary from 1^+ to ∞. Hence, poles *outside* the unit circle within region 3 in the *z*-plane will yield an unstable system. The response of such system will be an increasing exponential if the poles are real or a growing sinusoid if the poles are complex.

σ = 0

Poles on the *j*ω axis (region 1) in the *s*-plane represent a marginally stable system, and (4.12) yields a magnitude of $|z| = 1$, which corresponds to region 1. Hence, poles *on* the unit circle in region 1 in the *z*-plane will yield a sinusoid. In Chapter 5 we implement a sinusoidal signal by programming a difference equation with its poles *on* the unit circle. Note that from Exercise 4.2 the poles of $X(s) = \sin n\omega T$ in (4.9) or $X(s) = \cos n\omega T$ in (4.11) are the roots of $z^2 - 2z \cos \omega T + 1$, or

$$p_{1,2} = \frac{2\cos\omega T \pm \sqrt{4\cos^2 \omega T - 4}}{2}$$
$$= \cos\omega T \pm \sqrt{-\sin^2 \omega T} = \cos\omega T \pm j \sin\omega T \qquad (4.13)$$

The magnitude of each pole is

$$|p_1| = |p_2| = \sqrt{\cos^2 \omega T + \sin^2 \omega T} = 1 \qquad (4.14)$$

The phase of z is $\theta = \omega T = 2\pi f/F_s$. As the frequency f varies from zero to $\pm F_s/2$, the phase θ will vary from 0 to π.

4.1.2 Difference Equations

A digital filter is represented by a difference equation in a similar fashion as an analog filter is represented by a differential equation. To solve a difference equation, we need to find the *z*-transform of expressions such as $x(n - k)$, which corresponds to the *k*th derivative $d^k x(t)/dt^k$ of an analog signal $x(t)$. The order of the difference equation is determined by the largest value of *k*. For example, $k = 2$ represents a second-order derivative. From (4.5)

$$X(z) = \sum_{n=0}^{\infty} x(n)z^{-n} = x(0) + x(1)z^{-1} + x(2)z^{-2} + \cdots \qquad (4.15)$$

Then the *z*-transform of $x(n - 1)$, which corresponds to a first-order derivative dx/dt, is

$$ZT[x(n-1)] = \sum_{n=0}^{\infty} x(n-1)z^{-n}$$

$$= x(-1) + x(0)z^{-1} + x(1)z^{-2} + x(2)z^{-3} + \cdots$$

$$= x(-1) + z^{-1}[x(0) + x(1)z^{-1} + x(2)z^{-2} + \cdots]$$

$$= x(-1) + z^{-1}X(z) \tag{4.16}$$

where we used (4.15), and $x(-1)$ represents the initial condition associated with a first-order difference equation. Similarly, the ZT of $x(n-2)$, equivalent to a second derivative $d^2x(t)/dt^2$ is

$$ZT[x(n-2)] = \sum_{n=0}^{\infty} x(n-2)z^{-n}$$

$$= x(-2) + x(-1)z^{-1} + x(0)z^{-2} + x(1)z^{-3} + \cdots$$

$$= x(-2) + x(-1)z^{-1} + z^{-2}[x(0) + x(1)z^{-1} + \cdots]$$

$$= x(-2) + x(-1)z^{-1} + z^{-2}X(z) \tag{4.17}$$

where $x(-2)$ and $x(-1)$ represent the two initial conditions required to solve a second-order difference equation. In general,

$$ZT[x(n-k)] = z^{-k}\sum_{m=1}^{k} x(-m)z^{m} + z^{k}X(z) \tag{4.18}$$

If the initial conditions are all zero, then $x(-m) = 0$ for $m = 1, 2, \ldots, k$, and (4.18) reduces to

$$ZT[x(n-k)] = z^{-k}X(z) \tag{4.19}$$

4.2 DISCRETE SIGNALS

A discrete signal $x(n)$ can be expressed as

$$x(n) = \sum_{m=-\infty}^{\infty} x(m)\delta(n-m) \tag{4.20}$$

where $\delta(n-m)$ is the impulse sequence $\delta(n)$ delayed by m, which is equal to 1 for $n = m$ and is 0 otherwise. It consists of a sequence of values $x(1), x(2), \ldots$, where n is the time, and each sample value of the sequence is taken one sample time apart, determined by the sampling interval or sampling period $T = 1/F_s$.

The signals and systems that we deal with in this book are linear and time-invariant, where both superposition and shift invariance apply. Let an input signal $x(n)$ yield an output response $y(n)$, or $x(n) \rightarrow y(n)$. If $a_1x_1(n) \rightarrow a_1y_1(n)$ and $a_2x_2(n) \rightarrow a_2y_2(n)$, then $a_1x_1(n) + a_2x_2(n) \rightarrow a_1y_1(n) + a_2y_2(n)$, where a_1 and a_2 are constants.

This is the superposition property, where an overall output response is the sum of the individual responses to each input. Shift invariance implies that if the input is delayed by m samples, the output response will also be delayed by m samples, or $x(n-m) \rightarrow y(n-m)$. If the input is a unit impulse $\delta(n)$, the resulting output response is $h(n)$, or $\delta(n) \rightarrow h(n)$, and $h(n)$ is designated as the impulse response. A delayed impulse $\delta(n-m)$ yields the output response $h(n-m)$ by the shift-invariance property.

Furthermore, if this impulse is multiplied by $x(m)$, then $x(m)\delta(n-m) \rightarrow x(m)h(n-m)$. Using (4.20), the response becomes

$$y(n) = \sum_{m=-\infty}^{\infty} x(m)h(n-m) \qquad (4.21)$$

which represents a convolution equation. For a causal system, (4.21) becomes

$$y(n) = \sum_{m=-\infty}^{\infty} x(m)h(n-m) \qquad (4.22)$$

Letting $k = n - m$ in (4.22) yields

$$y(n) = \sum_{k=0}^{\infty} h(k)x(n-k) \qquad (4.23)$$

4.3 FIR FILTERS

Filtering is one of the most useful signal processing operations [1–47]. DSp are now available to implement digital filters in real time. The TMS320C6x instruction set and architecture makes it well suited for such filtering operations. An analog filter operates on continuous signals and is typically realized with discrete components such as operational amplifiers, resistors, and capacitors. However, a digital filter, such as an FIR filter, operates on discrete-time signals and can be implemented with a DSp such as the TMS320C6x. This involves use of an ADC to capture an external input signal, processing the input samples, and sending the resulting output through a DAC.

Within the last few years, the cost of DSp has been reduced significantly, which adds to the numerous advantages that digital filters have over their analog counterparts. These include higher reliability, accuracy, and less sensitivity to temperature and aging. Stringent magnitude and phase characteristics can be achieved with a digital filter. Filter characteristics such as center frequency, bandwidth, and filter type can readily be modified. A number of tools are available to design and implement within a few minutes an FIR filter in real time using the TMS320C6x-based DSK. The filter design consists of the approximation of a transfer function with a resulting set of coefficients.

Different techniques are available for the design of FIR filters, such as a commonly used technique that utilizes the Fourier series, as discussed in Section 4.4. Computer-aided design techniques such as that of Parks and McClellan are also used for the design of FIR filters [5,6].

The convolution equation (4.23) is very useful for the design of FIR filters, since we can approximate it with a finite number of terms, or

$$y(n) = \sum_{k=0}^{N-1} h(k)x(n-k) \tag{4.24}$$

If the input is a unit impulse $x(n) = \delta(0)$, the output impulse response will be $y(n) = h(n)$. We will see in Section 4.4 how to design an FIR filter with N coefficients $h(0), h(1), \ldots, h(N-1)$, and N input samples $x(n), x(n-1), \ldots, x(n-(N-1))$. The input sample at time n is $x(n)$, and the delayed input samples are $x(n-1), \ldots, x(n-(N-1))$. Equation (4.24) shows that an FIR filter can be implemented with knowledge of the input $x(n)$ at time n and of the delayed inputs $x(n-k)$. It is non-recursive, and no feedback or past outputs are required. Filters with feedback (recursive) that require past outputs are discussed in Chapter 5. Other names used for FIR filters are transversal and tapped-delay filters.

The z-transform of (4.24) with zero initial conditions yields

$$Y(z) = h(0)X(z) + h(1)z^{-1}X(z) + h(2)z^{-2}X(z) + \cdots + h(N-1)z^{-(N-1)}X(z) \tag{4.25}$$

Equation (4.24) represents a convolution in time between the coefficients and the input samples, which is equivalent to a multiplication in the frequency domain, or

$$Y(z) = H(z)X(z) \tag{4.26}$$

where $H(z) = ZT[h(k)]$ is the transfer function, or

$$\begin{aligned} H(z) &= \sum_{k=0}^{N-1} h(k)z^{-k} = h(0) + h(1)z^{-1} + h(2)z^{-2} + \cdots + h(N-1)z^{-(N-1)} \\ &= \frac{h(0)z^{(N-1)} + h(1)z^{N-2} + h(2)z^{N-3} + \cdots + h(N-1)}{z^{N-1}} \end{aligned} \tag{4.27}$$

which shows that there are $N-1$ poles, all of which are located at the origin. Hence, this FIR filter is inherently stable, with its poles located only inside the unit circle. We usually describe an FIR filter as a filter with "no poles." Figure 4.2 shows an FIR filter structure representing (4.24) and (4.25).

A very useful feature of an FIR filter is that it can guarantee *linear phase*. The linear phase feature can be very useful in applications such as speech analysis, where phase distortion can be critical. For example, with linear phase, all input sinusoidal

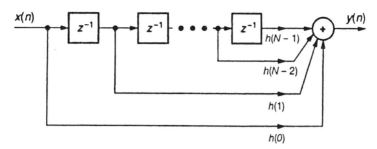

FIGURE 4.2. FIR filter structure showing delays.

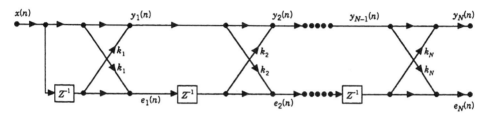

FIGURE 4.3. FIR lattice structure.

components are delayed by the same amount. Otherwise, harmonic distortion can occur. Linear phase filters are FIR filters; however, not all FIR filters have linear phase.

The Fourier transform of a delayed input sample $x(n-k)$ is $e^{-j\omega kT}X(j\omega)$, yielding a phase of $\theta = -\omega kT$, which is a linear function in terms of ω. Note that the group delay function, defined as the derivative of the phase, is a constant, or $d\theta/d\omega = -kT$.

4.4 FIR LATTICE STRUCTURE

The lattice structure is commonly used for applications in adaptive filtering and speech processing [48,49], such as in a linear predictive coding (LPC) application. An Nth-order lattice structure is shown in Figure 4.3. The coefficients $k_1, k_2, \ldots, k_N$ are commonly referred to as *reflection coefficients* (or *k*-parameters). An advantage of this structure is that the frequency response is not as sensitive as the previous structure to small changes in the coefficients. From the first section in Figure 4.3, with $N = 1$, we have

$$y_1(n) = x(n) + k_1 x(n-1) \tag{4.28}$$

$$e_1(n) = k_1 x(n) + x(n-1) \tag{4.29}$$

From the second section (cascaded with the first), using (4.28) and (4.29),

$$
\begin{aligned}
y_2(n) &= y_1(n) + k_2 e_1(n-1) \\
&= x(n) + k_1 x(n-1) + k_2 k_1 x(n-1) + k_2 x(n-2) \\
&= x(n) + (k_1 + k_1 k_2)x(n-1) + k_2 x(n-2)
\end{aligned}
\tag{4.30}
$$

and

$$
\begin{aligned}
e_2(n) &= k_2 y_1(n) + e_1(n-1) \\
&= k_2 x(n) + k_2 k_1 x(n-1) + k_1 x(n-1) + x(n-2) \\
&= k_2 x(n) + (k_1 + k_1 k_2)x(n-1) + x(n-2)
\end{aligned}
\tag{4.31}
$$

For a specific section i,

$$
y_i(n) = y_{i-1}(n) + k_i e_{i-1}(n-1)
\tag{4.32}
$$

$$
e_i(n) = k_i y_{i-1}(n) + e_{i-1}(n-1)
\tag{4.33}
$$

It is instructive to see that (4.30) and (4.31) have the same coefficients but in reversed order. It can be shown that this property also holds true for a higher-order structure. In general, for an Nth-order FIR lattice system, (4.30) and (4.31) become

$$
y_N(n) = \sum_{i=0}^{N} a_i x(n-i)
\tag{4.34}
$$

and

$$
e_N(n) = \sum_{i=0}^{N} a_{N-i} x(n-i)
\tag{4.35}
$$

with $a_0 = 1$. If we take the ZT of (4.34) and (4.35) and find their impulse responses,

$$
Y_N(z) = \sum_{i=0}^{N} a_i z^{-i}
\tag{4.36}
$$

$$
E_N(z) = \sum_{i=0}^{N} a_{N-i} z^{-i}
\tag{4.37}
$$

It is interesting to note that

$$
E_N(z) = z^{-N} Y_N(1/z)
\tag{4.38}
$$

Equations (4.36) and (4.37) are referred to as image polynomials. For two sections, $k_2 = a_2$; in general,

$$k_N = a_N \qquad (4.39)$$

For this structure to be useful, it is necessary to find the relationship between the k-parameters and the impulse response coefficients. The lattice network is highly structured, as seen in Figure 4.3 and as demonstrated through the previous difference equations. Starting with k_N in (4.39), we can recursively (with reverse recursion) compute the preceding k-parameters, $k_{N-1}, \ldots, k_1$.

Consider an intermediate section r and, using (4.36) and (4.37),

$$Y_r(z) = Y_{r-1}(z) + k_r z^{-1} E_{r-1}(z) \qquad (4.40)$$

$$E_r(z) = k_r Y_{r-1}(z) + z^{-1} E_{r-1}(z) \qquad (4.41)$$

Solving for $E_{r-1}(z)$ in (4.41) and substituting it into (4.40), $Y_r(z)$ becomes

$$Y_r(z) = Y_{r-1}(z) + k_r z^{-1} \frac{E_r(z) - k_r Y_{r-1}(z)}{z^{-1}} \qquad (4.42)$$

Equation (4.42) now can be solved for $Y_{r-1}(z)$ in terms of $Y_r(z)$, or

$$Y_{r-1}(z) = \frac{Y_r(z) - k_r E_r(z)}{1 - k_r^2}, \quad |k_r| = 1 \qquad (4.43)$$

Using (4.38) with $N = r$, (4.43) becomes

$$Y_{r-1}(z) = \frac{Y_r(z) - k_r z^{-r} Y_r(1/z)}{1 - k_r^2} \qquad (4.44)$$

Equation (4.44) is an important relationship that shows that by using a reversed recursion procedure, we can find Y_{r-1} from Y_r, where $1 \leq r \leq N$. Consequently, we can also find the k-parameters starting with k_r and proceeding to k_1. For r sections, (4.36) can be written as

$$Y_r(z) = \sum_{i=0}^{r} a_{ri} z^{-i} \qquad (4.45)$$

Replacing i by $r - i$, and z by $1/z$, (4.45) becomes

$$Y_r\left(\frac{1}{z}\right) = \sum_{i=0}^{r} a_{r(r-i)} z^{r-i} \qquad (4.46)$$

Using (4.45) and (4.46), equation (4.44) becomes

$$\sum_{i=0}^{r} a_{(r-1)i} z^{-i} = \frac{\sum_{i=0}^{r} a_{ri} z^{-i} - k_r z^{-r} \sum_{i=0}^{r} a_{r(r-i)} z^{r-i}}{1 - k_r^2} \tag{4.47}$$

$$= \frac{\sum_{i=0}^{r} a_{ri} z^{-i} - k_r \sum_{i=0}^{r} a_{r(r-i)} z^{-i}}{1 - k_r^2} \tag{4.48}$$

from which

$$a_{(r-1)i} = \frac{a_{ri} - k_r a_{r(r-i)}}{1 - k_r^2}, \quad i = 0, 1, \ldots, r-1 \tag{4.49}$$

with $r = N, N - 1, \ldots, 1$, $|k_r| \neq 1$, $i = 0, 1, \ldots, r - 1$, and

$$k_r = a_{rr}, \quad r = N, N-1, \ldots, 1 \tag{4.50}$$

Exercise 4.3: FIR Lattice Structure

This example illustrates the use of (4.49) and (4.50) to compute the k-parameters. Given that the impulse response of an FIR filter in the frequency domain is

$$Y_2(z) = 1 + 0.2 z^{-1} - 0.5 z^{-2}$$

Then, from (4.45), with $r = 2$,

$$Y_2(z) = a_{20} + a_{21} z^{-1} + a_{22} z^{-2}$$

where $a_{20} = 1$, $a_{21} = 0.2$, and $a_{22} = -0.5$. Starting with $r = 2$ in (4.50),

$$k_2 = a_{22} = -0.5$$

Using (4.49), for $i = 0$,

$$a_{10} = \frac{a_{20} - k_2 a_{22}}{1 - k_2^2} = \frac{1 - (-0.5)(-0.5)}{1 - (-0.5)^2} = 1$$

and, for $i = 1$,

$$a_{11} = \frac{a_{21} - k_2 a_{21}}{1 - k_2^2} = \frac{0.2 - (-0.5)(0.2)}{1 - (-0.5)^2} = 0.4$$

From (4.50),

$$k_1 = a_{11} = 0.4$$

Note that the values for the k-parameters $k_2 = -0.5$ and $k_1 = 0.4$ can be verified using (4.30). In the next chapter, we will continue our discussions on lattice structures in conjunction with IIR filters.

4.5 FIR IMPLEMENTATION USING FOURIER SERIES

The design of an FIR filter using a Fourier series method is such that the magnitude response of its transfer function $H(z)$ approximates a desired magnitude response. The transfer function desired is

$$H_d(\omega) = \sum_{n=-\infty}^{\infty} C_n e^{jn\omega T} \qquad |n| < \infty \qquad (4.51)$$

where C_n are the Fourier series coefficients. Using a normalized frequency variable v such that $v = f/F_N$, where F_N is the Nyquist frequency, or $F_N = F_s/2$, the desired transfer function in (4.51) can be written as

$$H_d(v) = \sum_{n=-\infty}^{\infty} C_n e^{jn\pi v} \qquad (4.52)$$

where $\omega T = 2\pi f/F_s = \pi v$ and $|v| < 1$. The coefficients C_n are defined as

$$C_n = \tfrac{1}{2} \int_{-1}^{1} H_d(v) e^{-jn\pi v} dv$$

$$= \tfrac{1}{2} \int_{-1}^{1} H_d(v)(\cos n\pi v - j\sin n\pi v) dv \qquad (4.53)$$

Assume that $H_d(v)$ is an even function (frequency selective filter); then (4.53) reduces to

$$C_n = \int_{0}^{1} H_d(v) \cos n\pi v dv \qquad n \geq 0 \qquad (4.54)$$

since $H_d(v) \sin n\pi v$ is an odd function and

$$\int_{-1}^{1} H_d(v) \sin n\pi v dv = 0$$

with $C_n = C_{-n}$. The desired transfer function $H_d(v)$ in (4.52) is expressed in terms of an infinite number of coefficients, and to obtain a realizable filter, we must truncate (4.52), which yields the approximated transfer function

$$H_a(v) = \sum_{n=-Q}^{Q} C_n e^{jn\pi v} \tag{4.55}$$

where Q is positive and finite and determines the order of the filter. The larger the value of Q, the higher the order of the FIR filter and the better the approximation in (4.55) of the desired transfer function. The truncation of the infinite series with a finite number of terms results in ignoring the contribution of the terms outside a rectangular window function between $-Q$ and $+Q$. In Section 4.6 we see how the characteristics of a filter can be improved by using window functions other than rectangular.

Let $z = e^{j\pi v}$; then (4.55) becomes

$$H_a(z) = \sum_{n=-Q}^{Q} C_n z^n \tag{4.56}$$

with the impulse response coefficients $C_{-Q}, C_{-Q+1}, \ldots, C_{-1}, C_0, C_1, \ldots, C_{Q-1}, C_Q$. The approximated transfer function in (4.56), with positive powers of z, implies a non-causal or not realizable filter that would produce an output before an input is applied. To remedy this situation, we introduce a delay of Q samples in (4.56) to yield

$$H(z) = z^{-Q} H_a(z) = \sum_{n=-Q}^{Q} C_n z^{n-Q} \tag{4.57}$$

Let $n - Q = -i$; then $H(z)$ in (4.57) becomes

$$H(z) = \sum_{i=0}^{2Q} C_{Q-i} z^{-i} \tag{4.58}$$

Let $h_i = C_{Q-i}$ and $N - 1 = 2Q$; then $H(z)$ becomes

$$H(z) = \sum_{i=0}^{N-1} h_i z^{-i} \tag{4.59}$$

where $H(z)$ is expressed in terms of the impulse response coefficients h_i, and $h_0 = C_Q, h_1 = C_{Q-1}, \ldots, h_Q = C_0, h_{Q+1} = C_{-1} = C_1, \ldots, h_{2Q} = C_{-Q}$. The impulse response coefficients are symmetric about h_Q, with $C_n = C_{-n}$.

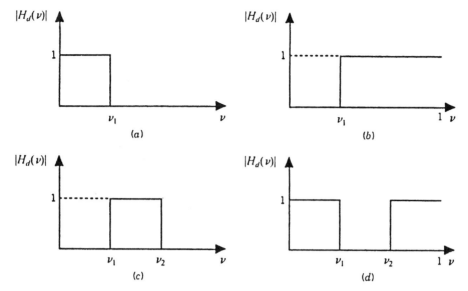

FIGURE 4.4. Desired transfer function: (*a*) lowpass; (*b*) highpass; (*c*) bandpass; (*d*) bandstop.

The order of the filter is $N = 2Q + 1$. For example, if $Q = 5$, the filter will have 11 coefficients $h_0, h_1, \ldots, h_{10}$, or

$$h_0 = h_{10} = C_5$$
$$h_1 = h_9 = C_4$$
$$h_2 = h_8 = C_3$$
$$h_3 = h_7 = C_2$$
$$h_4 = h_6 = C_1$$
$$h_5 = C_0$$

Figure 4.4 shows the desired transfer functions $H_d(v)$ ideally represented for the frequency-selective filters: lowpass, highpass, bandpass, and bandstop for which the coefficients $C_n = C_{-n}$ can be found.

1. *Lowpass:* $C_0 = v_1$

$$C_n = \int_0^{v_1} H_d(v) \cos n\pi v \, dv = \frac{\sin n\pi v_1}{n\pi} \tag{4.60}$$

2. *Highpass:* $C_0 = 1 - v_1$

$$C_n = \sum_{v_1}^{1} H_d(v) \cos n\pi v \, dv = -\frac{\sin n\pi v_1}{n\pi} \tag{4.61}$$

3. *Bandpass:* $C_0 = v_2 - v_1$

$$C_n = \int_{v_1}^{v_2} H_d(v)\cos n\pi v \, dv = \frac{\sin n\pi v_2 - \sin n\pi v_1}{n\pi} \tag{4.62}$$

4. *Bandstop:* $C_0 = 1 - (v_2 - v_1)$

$$C_n = \int_0^{v_1} H_d(v)\cos n\pi v \, dv + \int_{v_2}^1 H_d(v)\cos n\pi v \, dv = \frac{\sin n\pi v_1 - \sin n\pi v_2}{n\pi} \tag{4.63}$$

where v_1 and v_2 are the normalized cutoff frequencies shown in Figure 4.4. Several filter-design packages are currently available for the design of FIR filters, as discussed later. When we implement an FIR filter, we develop a generic program such that the specific coefficients will determine the filter type (e.g., whether lowpass or bandpass).

Exercise 4.4: Lowpass FIR Filter

We will find the impulse response coefficients of an FIR filter with $N = 11$, a sampling frequency of $10\,\text{kHz}$, and a cutoff frequency $f_c = 1\,\text{kHz}$. From (4.60),

$$C_0 = v_1 = \frac{f_c}{F_N} = 0.2$$

where $F_N = F_s/2$ is the Nyquist frequency and

$$C_n = \frac{\sin 0.2n\pi}{n\pi} \qquad n = \pm 1, \pm 2, \ldots, \pm 5 \tag{4.64}$$

Since the impulse response coefficients $h_i = C_{Q-i}$, $C_n = C_{-n}$, and $Q = 5$, the impulse response coefficients are

$$\begin{aligned}
h_0 &= h_{10} = 0 & h_3 &= h_7 = 0.1514 \\
h_1 &= h_9 = 0.0468 & h_4 &= h_6 = 0.1872 \\
h_2 &= h_8 = 0.1009 & h_5 &= 0.2
\end{aligned} \tag{4.65}$$

These coefficients can be calculated with a utility program (on the accompanying CD) and inserted within a generic filter program, as described later. Note the symmetry of these coefficients about $Q = 5$. While $N = 11$ for an FIR filter is low for a practical design, doubling this number can yield an FIR filter with much better characteristics, such as selectivity. For an FIR filter to have linear phase, the coefficients must be symmetric, as in (4.65).

The program *Amplit. cpp* (on the CD), described in Appendix D, plots the magnitude and phase of a transfer function. It can be used to show that the coefficients in (4.65) yield a lowpass filter (use 1 as the coefficient for the denominator).

4.6 WINDOW FUNCTIONS

We truncated the infinite series in the transfer function equation (4.52) to arrive at (4.55). We essentially put a rectangular window function with an amplitude of 1 between $-Q$ and $+Q$ and ignored the coefficients outside that window. The wider this rectangular window, the larger Q is and the more terms we use in (4.55) to get a better approximation of (4.52). The rectangular window function can therefore be defined as

$$w_R(n) = \begin{cases} 1 & \text{for } |n| \leq Q \\ 0 & \text{otherwise} \end{cases} \tag{4.66}$$

The transform of the rectangular window function $\omega_R(n)$ yields a sinc function in the frequency domain. It can be shown that

$$W_R(v) = \sum_{n=-Q}^{Q} e^{jn\pi v} = e^{-jQ\pi v} \left(\sum_{n=0}^{2Q} e^{jn\pi v} \right) = \frac{\sin\left[\left(\frac{2Q+1}{2} \right) \pi v \right]}{\sin(\pi v/2)} \tag{4.67}$$

which is a sinc function that exhibits high sidelobes or oscillations caused by the abrupt truncation, specifically, near discontinuities.

A number of window functions are currently available to reduce these high-amplitude oscillations; they provide a more gradual truncation to the infinite series expansion. However, while these alternative window functions reduce the amplitude of the sidelobes, they also have a wider mainlobe, which results in a filter with lower selectivity. A measure of a filter's performance is a ripple factor that compares the peak of the first sidelobe to the peak of the mainlobe (their ratio). A compromise or trade-off is to select a window function that can reduce the sidelobes while approaching the selectivity that can be achieved with the rectangular window function. The width of the mainlobe can be reduced by increasing the width of the window (order of the filter). Later, we will plot the magnitude response of an FIR filter that shows the undesirable sidelobes.

In general, the Fourier series coefficients can be written as

$$C'_n = C_n w(n) \tag{4.68}$$

where $w(n)$ is the window function. In the case of the rectangular window function, $C_n' = C_n$. The transfer function in (4.59) can then be written as

$$H'(z) = \sum_{i=0}^{N-1} h_i' z^{-i} \tag{4.69}$$

where

$$h_i' = C_{Q-i}' \qquad 0 \le i \le 2Q \tag{4.70}$$

The rectangular window has its highest sidelobe level, down by only $-13\,\mathrm{dB}$ from the peak of its mainlobe, resulting in oscillations with an amplitude of considerable size. On the other hand, it has the narrowest mainlobe that can provide high selectivity. The following window functions are commonly used in the design of FIR filters [12].

4.6.1 Hamming Window

The Hamming window function [12,25] is

$$w_H(n) = \begin{cases} 0.54 + 0.46\cos(n\pi/Q) & \text{for } |n| \le Q \\ 0 & \text{otherwise} \end{cases} \tag{4.71}$$

which has the highest or first sidelobe level at approximately $-43\,\mathrm{dB}$ from the peak of the main lobe.

4.6.2 Hanning Window

The Hanning or raised cosine window function is

$$w_{HA}(n) = \begin{cases} 0.5 + 0.5\cos(n\pi/Q) & \text{for } |n| \le Q \\ 0 & \text{otherwise} \end{cases} \tag{4.72}$$

which has the highest or first sidelobe level at approximately $-31\,\mathrm{dB}$ from the peak of the mainlobe.

4.6.3 Blackman Window

The Blackman window function is

$$w_B(n) = \begin{cases} 0.42 + 0.5\cos(n\pi/Q) + 0.08\cos(2n\pi/Q) & |n| \le Q \\ 0 & \text{otherwise} \end{cases} \tag{4.73}$$

which has the highest sidelobe level down to approximately −58 dB from the peak of the mainlobe. While the Blackman window produces the largest reduction in the sidelobe compared with the previous window functions, it has the widest mainlobe. As with the previous windows, the width of the mainlobe can be decreased by increasing the width of the window.

4.6.4 Kaiser Window

The design of FIR filters with the Kaiser window has become very popular in recent years. It has a variable parameter to control the size of the sidelobe with respect to the mainlobe. The Kaiser window function is

$$w_K(n) = \begin{cases} I_0(b)/I_0(a) & |n| \leq Q \\ 0 & \text{otherwise} \end{cases} \tag{4.74}$$

where a is an empirically determined variable, and $b = a[1 - (n/Q)^2]^{1/2}$. $I_0(x)$ is the modified Bessel function of the first kind defined by

$$I_0(x) = 1 + \frac{0.25x^2}{(1!)^2} + \frac{(0.25x^2)^2}{(2!)^2} + \cdots = 1 + \sum_{n=1}^{\infty} \left[\frac{(x/2)^n}{n!} \right]^2 \tag{4.75}$$

which converges rapidly. A trade-off between the size of the sidelobe and the width of the mainlobe can be achieved by changing the length of the window and the parameter a.

4.6.5 Computer-Aided Approximation

An efficient technique is the computer-aided iterative design based on the Remez exchange algorithm, which produces equiripple approximation of FIR filters [5,6]. The order of the filter and the edges of both passbands and stopbands are fixed, and the coefficients are varied to provide this equiripple approximation. This minimizes the ripple in both the passbands and the stopbands. The transition regions are left unconstrained and are considered "don't care" regions, where the solution may fail. Several commercial filter design packages include the Parks–McClellan algorithm for the design of an FIR filter.

4.7 PROGRAMMING EXAMPLES USING C AND ASM CODE

Within minutes, an FIR filter can be designed and implemented in real time. Several filter design packages are available for the design of FIR filters. They are described in Appendix D using MATLAB [50] and in Appendix E using DigiFilter and a homemade package (on the accompanying CD).

TABLE 4.1   Memory Organization for Coefficients and Samples (Initially)

i	Coefficients	Samples
0	h(0)	x(n)
1	h(1)	x(n - 1)
2	h(2)	x(n - 2)
.	.	.
.	.	.
.	.	.
N - 1	h(N - 1)	x(n - (N - 1))

Several examples illustrate the implementation of FIR filters. Most of the programs are in C. A few examples using mixed C and ASM code illustrate the use of a circular buffer as a more efficient way to update delay samples, with the circular buffer in internal or external memory. These examples illustrate modulation, up-sampling, down-sampling, aliasing, and so on. The convolution equation (4.24) is used to program and implement these filters, or

$$y(n) = \sum_{i=0}^{N-1} h(i)x(n-i)$$

We can arrange the impulse response coefficients within a buffer (array) so that the first coefficient, $h(0)$, is at the beginning (first location) of the buffer (lower-memory address). The last coefficient, $h(N-1)$, can reside at the end (last location) of the coefficients buffer (higher-memory address). The delay samples are organized in memory so that the newest sample, $x(n)$, is at the beginning of the samples buffer, while the oldest sample, $x(n-(N-1))$, is at the end of the buffer. The coefficients and the samples can be arranged in memory as shown in Table 4.1. Initially, all the samples are set to zero.

Time n

The newest sample, $x(n)$, at time n is acquired from an ADC and stored at the beginning of the sample buffer. The filter's output at time n is computed from the convolution equation (4.24), or

$$y(n) = h(0)x(n) + h(1)x(n-1) + \cdots + h(N-2)x(n-(N-2)) + h(N-1)x(n-(N-1))$$

The delay samples are then updated so that $x(n-k) = x(n+1-k)$ can be used to calculate $y(n+1)$, the output for the next unit of time, or sample period T_s. All the samples are updated except the newest sample. For example, $x(n-1) = x(n)$, and $x(n-(N-1)) = x(n-(N-2))$. This updating process has the effect of "moving the data" (down) in memory (see Table 4.2, associated with time $n+1$).

TABLE 4.2   Memory Organization to Illustrate Update of Samples

	Samples		
Coefficients	Time n	Time n + 1	Time n + 2
h(0)	x(n)	x(n + 1)	x(n + 2)
h(1)	x(n - 1)	x(n)	x(n + 1)
h(2)	x(n - 2)	x(n - 1)	x(n)
.	.	.	.
.	.	.	.
.	.	.	.
h(N - 3)	x(n - (N - 3))	x(n - (N - 4))	x(n - (N - 5))
h(N - 2)	x(n - (N - 2))	x(n - (N - 3))	x(n - (N - 4))
h(N - 1)	x(n - (N - 1))	x(n - (N - 2))	x(n - (N - 3))

Time n + 1

At time $n + 1$, a new input sample $x(n + 1)$ is acquired and stored at the top of the sample buffer, as shown in Table 4.2. The output $y(n + 1)$ can now be calculated as

$$y(n+1) = h(0)x(n+1) + h(1)x(n) + \cdots + h(N-2)x(n-(N-3)) \\ + h(N-1)x(n-(N-2))$$

The samples are then updated for the next unit of time.

Time n + 2

At time $n + 2$, a new input sample, $x(n + 2)$, is acquired. The output becomes

$$y(n+2) = h(0)x(n+2) + h(1)x(n+1) + \cdots + h(N-1)x(n-(N-3))$$

This process continues to calculate the filter's output and updating the delay samples at each unit of time (sample period).

Example 4.7 illustrates four different ways of arranging the coefficients and samples in memory and of calculating the convolution equation (e.g., the newest sample at the end of the buffer and the oldest sample at the beginning).

Example 4.1: FIR Filter Implementation: Bandstop and Bandpass (FIR)

Figure 4.5 shows a listing of the C source program *FIR.c*, which implements an FIR filter. It is a generic FIR program, since the coefficient file included, *bs2700.cof* (Figure 4.6), specifies the filter's characteristics. This coefficient file, which contains 89 coefficients, represents an FIR bandstop (notch) filter centered at 2700 Hz. The number of coefficients N is defined in the coefficient file. This filter was designed using MATLAB's graphical user interface (GUI) filter designer SPTool, described in Appendix D [50]. Figure 4.7 shows the filter's characteristics (MATLAB's order of 88 corresponds to 89 coefficients). MATLAB's FDATool can be used in the place of SPTool (see Appendix D).

```
//Fir.c  FIR filter. Include coefficient file with length N

#include "bs2700.cof"              //coefficient file
#include "dsk6713_aic23.h"         //codec-dsk support file
Uint32 fs=DSK6713_AIC23_FREQ_8KHZ; //set sampling rate
int yn = 0;                        //initialize filter's output
short dly[N];                      //delay samples

interrupt void c_int11()           //ISR
{
     short i;

     dly[0]=input_sample();        //input newest sample
     yn = 0;                       //initialize filter's output
     for (i = 0; i< N; i++)
         yn += (h[i] * dly[i]);    //y(n) += h(i)* x(n-i)
     for (i = N-1; i > 0; i--)     //starting @ end of buffer
         dly[i] = dly[i-1];        //update delays with data move
     output_sample(yn >> 15);      //scale output filter sample
     return;
}

void main()
   {
     comm_intr();                  //init DSK, codec, McBSP
     while(1);                     //infinite loop
   }
```

FIGURE 4.5. Generic FIR filter program (FIR.c).

```
//bs2700.cof FIR bandstop coefficients designed with MATLAB

#define N 89                     //number of coefficients

short h[N]= {-14,23,-9,-6,0,8,16,-58,50,44,-147,119,67,-245,
            200,72,-312,257,53,-299,239,20,-165,88,0,105,
            -236,33,490,-740,158,932,-1380,392,1348,-2070,
            724,1650,-2690,1104,1776,-3122,1458,1704,29491,
            1704,1458,-3122,1776,1104,-2690,1650,724,-2070,
            1348,392,-1380,932,158,-740,490,33,-236,105,0,
            88,-165,20,239,-299,53,257,-312,72,200,-245,67,
            119,-147,44,50,-58,16,8,0,-6,-9,23,-14};
```

FIGURE 4.6. Coefficients for a FIR bandstop filter (bs2700.cof).

A buffer dly[N] is created for the delay samples. The newest input sample, $x(n)$, is acquired through dly[0] and stored at the beginning of the buffer. The coefficients are stored in another buffer, $h[N]$, with $h[0]$ at the beginning of the coefficients' buffer. The samples and coefficients are then arranged in their respective buffer, as shown in Table 4.1.

Two "for" loops are used within the interrupt service routine (we will also implement an FIR filter using one loop). The first loop implements the convolution equa-

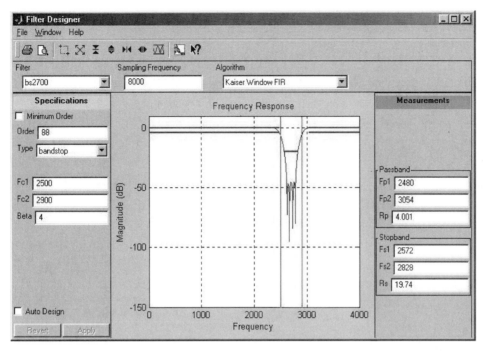

FIGURE 4.7. MATLAB's filter designer SPTool, showing the characteristics of a FIR band-stop filter centered at 2700 Hz.

tion with N coefficients and N delay samples for a specific time n. At time n the output is

$$y(n) = h(0)x(n) + h(1)x(n-1) + \cdots + h(N-1)x(n-(N-1))$$

The delay samples are then updated within the second loop to be used for calculating $y(n)$ at time $n + 1$, or $y(n + 1)$. The newly acquired input sample always resides at the beginning of the samples buffer (in this example). The memory location that contained the sample $x(n)$ now contains the newly acquired sample $x(n + 1)$. The output $y(n + 1)$ at time $n + 1$ is then calculated. This scheme uses a data move to update the delay samples.

Example 4.7 illustrates how various memory organizations can be used for both the delay samples and the filter coefficients, as well as for updating the delay samples within the same loop as the convolution equation. We also illustrate the use of a circular buffer with a pointer to update the delay samples in lieu of moving the data in memory. The output is scaled (right-shifted by 15) before it is sent to the codec's DAC. This allows for a fixed-point implementation as well.

Bandstop, Centered at 2700 Hz (bs2700.cof)

Build and run this project as **FIR**. Input a sinusoidal signal and vary the input frequency slightly below and above 2700 Hz. Verify that the output is a minimum at 2700 Hz.

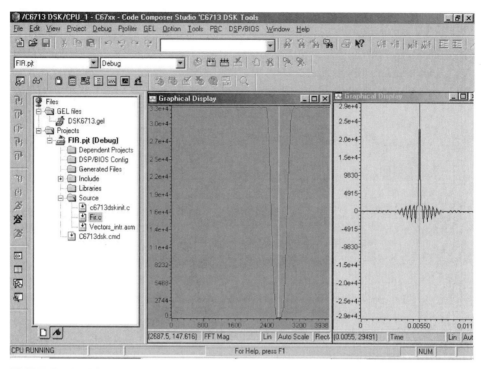

FIGURE 4.8. CCS plots displaying the FFT magnitude of the bandstop filter's coefficients and its impulse response.

Figure 4.8 shows a plot of CCS project windows. It shows the FFT magnitude of the filter's coefficients h (see Example 1.2, with a starting address of h) using a 128-point FFT. The characteristics of the FIR bandstop filter, centered at 2700 Hz, are displayed. Figure 4.8 also shows a CCS time-domain plot, or the impulse response of the filter.

With noise as input, the output frequency response of the bandstop filter can also be verified. The pseudorandom noise sequence developed in Chapter 2, or another noise source, can be used as input to the FIR filter, as illustrated later. Figure 4.9 shows a plot of the frequency response of the filter with a notch at 2700 Hz implemented in real time. This plot is obtained using an Hewlett-Packard (HP) 3561A dynamic signal analyzer with an input noise source from the analyzer. The roll-off at approximately 3850 Hz is due to the antialiasing lowpass filter on the codec.

Bandpass, Centered at 1750 Hz (*bp1750.cof*)

Within CCS, edit the program *FIR.c* to include the coefficient file *bp1750.cof* in lieu of *bs2700.cof*. The file *bp1750.cof* represents an FIR bandpass filter (81 coefficients) centered at 1750 Hz, as shown in Figure 4.10. This filter was designed

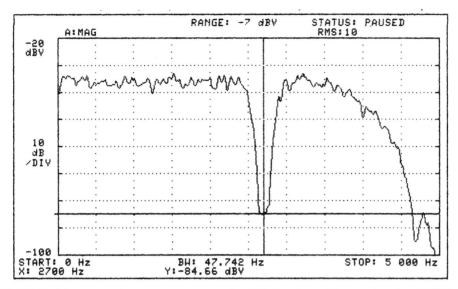

FIGURE 4.9. Output frequency response of a FIR bandstop filter centered at 2700 Hz, obtained with a signal analyzer.

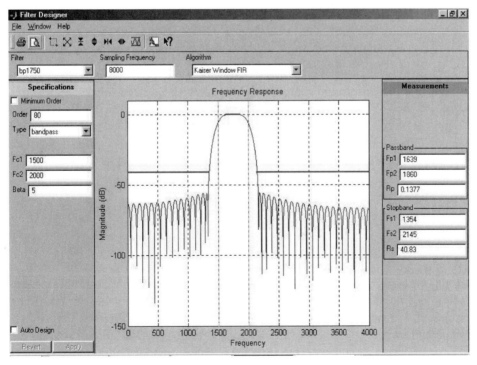

FIGURE 4.10. MATLAB's filter designer SPTool, showing the characteristics of a FIR bandpass filter centered at 1750 Hz.

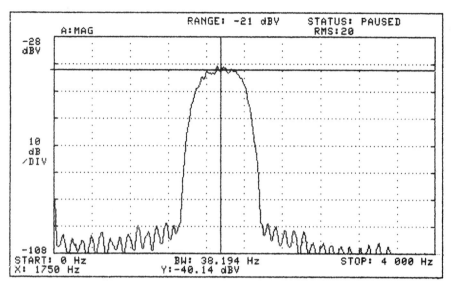

FIGURE 4.11. Output frequency response of a FIR bandpass filter centered at 1750 Hz, obtained with a signal analyzer.

with MATLAB's SPTool (Appendix D). Select the incremental Build, and the new coefficient file *bp1750.cof* will automatically be included in the project. Run again and verify an FIR bandpass filter centered at 1750 Hz. Figure 4.11 shows a real-time plot of the output frequency response obtained with the HP signal analyzer.

Example 4.2: Effects on Voice Using Three FIR Lowpass Filters (*FIR3LP*)

Figure 4.12 shows a listing of the program *FIR3lp.c*, which implements three FIR lowpass filters with cutoff frequencies at 600, 1500, and 3000 Hz, respectively. The three lowpass filters were designed with MATLAB's SPTool to yield the corresponding three sets of coefficients. This example expands on the generic FIR program in Example 4.1.

LP_number selects the desired lowpass filter to be implemented. For example, if LP_number is set to 0, h[0][i] is equal to hlp600[i] (within the "for" loop in the function *main*), which is the address of the first set of coefficients. The coefficients file LP600.cof represents an 81-coefficient FIR lowpass filter with a 600-Hz cutoff frequency, using the Kaiser window function. Figure 4.13 shows a listing of this coefficient file (the other two sets are on the CD). That filter is then implemented. LP_number can be changed to 1 or 2 to implement the 1500- or 3000-Hz lowpass filter, respectively. With the GEL file FIR3LP.gel (Figure 4.14), one can vary LP_number from 0 to 2 and slide through the three different filters.

Build this project as **FIR3LP.** Use the .wav file *TheForce.wav* (on the CD) as input and observe the effects of the three lowpass filters on the input voice. With

```
//Fir3LP.c FIR using 3 lowpass coefficients with three different BW

#include "lp600.cof"                  //coeff file LP @ 600 Hz
#include "lp1500.cof"                 //coeff file LP @ 1500 Hz
#include "lp3000.cof"                 //coeff file LP @ 3000 Hz
#include "dsk6713_aic23.h"            //codec-dsk support file
Uint32 fs=DSK6713_AIC23_FREQ_8KHZ;   //set sampling rate
short LP_number = 0;                 //start with 1st LP filter
int yn = 0;                          //initialize filter's output
short dly[N];                        //delay samples
short h[3][N];                       //filter characteristics 3xN

interrupt void c_int11()             //ISR
{
 short i;
 dly[0] = input_sample();            //newest input @ top of buffer
 yn = 0;                             //initialize filter output
 for (i = 0; i< N; i++)
 yn +=(h[LP_number][i]*dly[i]);      //y(n) += h(LP#,i)*x(n-i)
 for (i = N-1; i > 0; i--)           //starting @ bottom of buffer
     dly[i] = dly[i-1];              //update delays with data move
 output_sample(yn >> 15);            //output filter
 return;                             //return from interrupt
}

void main()
{
 short i;
 for (i=0; i<N; i++)
  {
     dly[i] = 0;                     //init buffer
     h[0][i] = hlp600[i];            //start addr of LP600 coeff
     h[1][i] = hlp1500[i];           //start addr of LP1500 coeff
     h[2][i] = hlp3000[i];           //start addr of LP3000 coeff
  }
 comm_intr();                        //init DSK, codec, McBSP
 while(1);                           //infinite loop
}
```

FIGURE 4.12. FIR program to implement three different lowpass filters using a slider for selection (FIR3LP.c).

```
//LP600.cof FIR lowpass filter coefficients using Kaiser window

#define N 81                    //length of filter

short hlp600[N]={0,-6,-14,-22,-26,-24,-13,8,34,61,80,83,63,19,-43,-113,
-171,-201,-185,-117,0,146,292,398,428,355,174,-99,-416,-712,-905,-921,
-700,-218,511,1424,2425,3391,4196,4729,4915,4729,4196,3391,2425,1424,
511,-218,-700,-921,-905,-712,-416,-99,174,355,428,398,292,146,0,-117,
-185,-201,-171,-113,-43,19,63,83,80,61,34,8,-13,-24,-26,-22,-14,-6,0};
```

FIGURE 4.13. Coefficient file for a FIR lowpass filter with a 600-Hz cutoff frequency (LP600.cof).

```
/*FIR3LP.gel Gel file to step through three different LP filters*/

menuitem "Filter Characteristics"

slider Filter(0,2,1,1,filterparameter) /*from 0 to 2,incr by 1*/
{
      LP_number = filterparameter;      /*for 3 LP filters*/
}
```

FIGURE 4.14. GEL file for selecting one of three FIR lowpass filter coefficients (FIR3LP.gel).

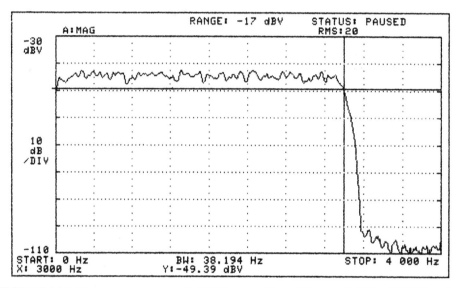

FIGURE 4.15. Frequency response of a FIR lowpass filter with a bandwidth of 3000 Hz using LP3000.cof, obtained with a signal analyzer.

the lower bandwidth of 600 Hz, using the first set of coefficients, the frequency components of the speech signal above 600 Hz are suppressed. Connect the output to a speaker or a spectrum analyzer to verify such results, and observe the different bandwidths of the three FIR lowpass filters. The shareware utility Goldwave generates different signals, including noise, using a sound card (see Appendix E). The output from the sound card with the noise generated by Goldwave can be used as the input to the DSK. Connecting the output from the DSK as the input to the sound card, Goldwave can also be used as a virtual spectrum analyzer. The frequency responses of these three lowpass filters can be obtained readily in real time. Figure 4.15 shows the frequency response of the 3000-Hz lowpass FIR filter, obtained with an HP signal analyzer.

Example 4.3: Implementation of Four Different Filters: Lowpass, Highpass, Bandpass, and Bandstop (FIR4types)

This example is similar to Example 4.2 and illustrates the GEL (slider) file to step through four different types of FIR filters. Each filter has 81 coefficients, designed with MATLAB's SPTool. The four coefficient files (on the accompanying CD) are:

1. *lp1500.cof*: lowpass with bandwidth of 1500 Hz
2. *hp2200.cof*: highpass with bandwidth of 2200 Hz
3. *bp1750.cof*: bandpass with center frequency at 1750 Hz
4. *bs790.cof*: bandstop with center frequency at 790 Hz

The program *FIR4types.c* (on the CD) implements this project. The program FIR3LP.c (Example 4.2) is modified slightly to incorporate the implementation of a fourth filter.

Build and run this project as **FIR4types**. Load the GEL file *FIR4types.gel* (on the CD) and verify the implementation of the four different FIR filters. This example can readily be expanded to implement more FIR filters.

Figure 4.16 shows the frequency response of the FIR bandstop filter centered at 790 Hz, using the coefficient file bs790.cof.

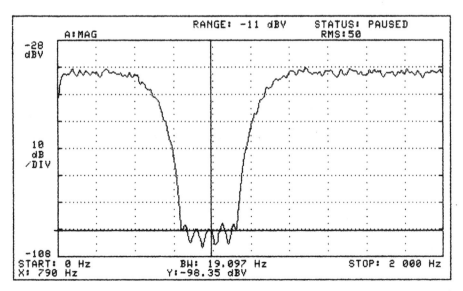

FIGURE 4.16. Frequency response of a FIR bandstop filter centered at 790 Hz using bs790.cof, obtained with a signal analyzer.

Example 4.4: FIR Implementation with a Pseudorandom Noise Sequence as Input to a Filter (FIRPRN)

The program *FIRPRN.c* (Figure 4.17) implements an FIR filter using an internally generated pseudorandom noise sequence as input to the filter. This input is the pseudorandom noise sequence generated in Example 2.16. The coefficient file *BP55.cof* uses a float data format and is shown in Figure 4.18. It represents a

```
//FIRPRN.c FIR with internally generated input noise sequence

#include "DSK6713_AIC23.h"          //codec-dsk support file
Uint32 fs=DSK6713_AIC23_FREQ_8KHZ; //set sampling rate
#include "bp55.cof"                 //BP @ Fs/4 coeff file in float
#include "noise_gen.h"              //header file for noise sequence
short dly[N], fb;                   //delay samples,feedback variable
shift_reg sreg;

short prn(void)                     //pseudorandom noise generation
{
 short prnseq;                      //for pseudorandom sequence
 if(sreg.bt.b0) prnseq = -16000;    //scaled negative noise level
 else           prnseq = 16000;     //scaled positive noise level
 fb  =(sreg.bt.b0)^(sreg.bt.b1);    //XOR bits 0,1
 fb ^=(sreg.bt.b11)^(sreg.bt.b13);  //with bits 11,13 ->fb
 sreg.regval<<=1;                   //shift register 1 bit to left
 sreg.bt.b0 = fb;                   //close feedback path
 return prnseq;                     //return generated sequence
}

interrupt void c_int11()            //ISR
{
    int i, yn = 0;                  //initialize filter's output
    dly[0] = prn();                 //input noise sequence
    for (i = 0; i< N; i++)
       yn +=(h[i]*dly[i]);          //y(n)+= h(i)*x(n-i)
    for (i = N-1; i > 0; i--)       //start @ bottom of buffer
       dly[i] = dly[i-1];           //data move to update delays
    output_sample((short)yn);       //output filter
    return;                         //return from interrupt
}

void main()
{
    short i;
    sreg.regval = 0xFFFF;           //shift register to nominal values
    fb = 1;                         //initial feedback value
    for (i = 0; i<N; i++)
       dly[i] = 0;                  //init buffer
    comm_intr();                    //init DSK, codec, McBSP
    while(1);                       //infinite loop
}
```

FIGURE 4.17. FIR program with a pseudorandom noise sequence as input (FIRPRN.c).

```
//bp55.cof Coefficients for FIR bandpass filter centered @ Fs/4

#define N 55                    //number of coefficients

float h[N]=
{1.7619E-017, 7.0567E-003, 2.2150E-018,-1.0962E-002, 4.0310E-017,
 1.3946E-002, 7.1787E-018,-1.4588E-002, 3.9928E-017, 1.1474E-002,
 5.9881E-018,-3.5159E-003,-6.6174E-018,-9.7476E-003,-1.7919E-017,
 2.7932E-002,-9.4329E-017,-4.9740E-002, 3.3834E-017, 7.3066E-002,
-3.6228E-017,-9.5284E-002, 3.2194E-017, 1.1365E-001,-2.2165E-017,
-1.2576E-001, 7.8980E-018, 1.3000E-001, 7.8980E-018,-1.2576E-001,
-2.2165E-017, 1.1365E-001, 3.2194E-017,-9.5284E-002,-3.6228E-017,
 7.3066E-002, 3.3834E-017,-4.9740E-002,-9.4329E-017, 2.7932E-002,
-1.7919E-017,-9.7476E-003,-6.6174E-018,-3.5159E-003, 5.9881E-018,
 1.1474E-002, 3.9928E-017,-1.4588E-002, 7.1787E-018, 1.3946E-002,
 4.0310E-017,-1.0962E-002, 2.2150E-018, 7.0567E-003, 1.7619E-017};
```

FIGURE 4.18. Coefficient file in float format for a FIR bandpass filter centered at $F_s/4$ (BP55.cof).

55-coefficient FIR bandpass filter with a center frequency at $F_s/4$. A filter development package (on the CD) that generates filter coefficients in float or hexadecimal format is described in Appendix E.

Build this project as **FIRPRN**. Run this project (with the coefficient file *BP55.cof*) and verify that the output is an FIR bandpass filter centered at 2 kHz. To verify the output as the noise sequence, output dly[0] in lieu of yn when calling the function output_sample.

Testing Different FIR Filters

Halt the program. Edit the C source program to include and test different coefficient files (on the CD) that represent different FIR filters, all using float format values. Each coefficient file contains 55 coefficients (except comb14.cof).

1. BP55.cof: bandpass with center frequency $F_s/4$
2. BS55.cof: bandstop with center frequency $F_s/4$
3. LP55.cof: lowpass with cutoff frequency $F_s/4$
4. HP55.cof: highpass with bandwidth $F_s/4$
5. Pass2b.cof: with two passbands
6. Pass3b.cof: with three passbands
7. Pass4b.cof: with four passbands
8. Comb14.cof: with multiple notches (comb filter)

These filters were designed with MATLAB (see Appendix D). Figure 4.19*a* shows the real-time output frequency response of an FIR filter with two passbands, using

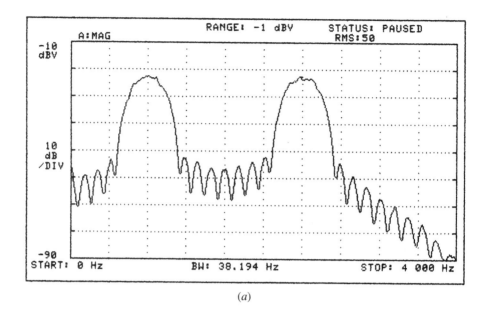

(a)

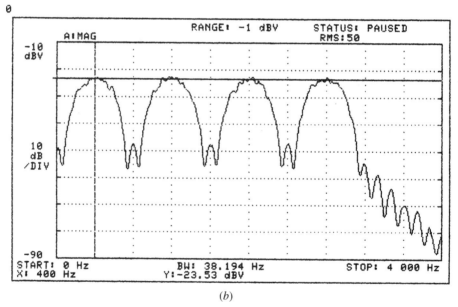

(b)

FIGURE 4.19. Output frequency responses obtained with an HP analyzer using program FIRPRN.c: (a) FIR filter with two passbands; (b) FIR filter with four passbands.

the coefficient file *pass2b.cof.* Figure 4.19b shows the frequency response of an FIR filter with four passbands using the coefficients file *pass4b.cof.* These plots were obtained with the HP 3561A signal analyzer. An example of a comb filter is described by $y(n) = 1/8[x(n)-x(n-8)]$ having eight zeros, and a frequency response with four notches at $\pi/4$, $\pi/2$, $3\pi/4$, and $\pi/8$.

Example 4.5: FIR Filter with Internally Generated Pseudorandom Noise as Input to a Filter and Output Stored in Memory (FIRPRNbuf)

This example builds on the previous one that generates a pseudorandom noise sequence as the input to an FIR filter, with the filter's output also stored in a memory buffer. Figure 4.20 shows a listing of the program FIRPRNbuf.c, which implements this project example.

The input to the filter is a software-generated noise sequence using dly[0] as the newest noise sequence. The coefficient file BP41.cof represents a 41-coefficient FIR bandpass filter centered at $F_s/8$.

Build and run this project as **FIRPRNbuf**. Verify the output frequency response of a 1-kHz FIR bandpass filter. Use CCS to verify the FFT magnitude plot as shown in Figure 4.21. Select/set for the plot:

1. *Display type:* FFT magnitude
2. *Start address:* yn_buffer
3. *Acquisition buffer size:* 1024
4. *FFT frame size:* 1024
5. *FFT order:* 10
6. *DSP data type:* 16-bit signed integer
7. *Sampling rate:* 8000 Hz

Use the default settings for the other fields. The FFT order is M, where 2^M = FFT frame size.

Figure 4.22 shows the real-time frequency response of the FIR bandpass filter, centered at $F_s/8$, displayed using an HP analyzer. Change the output buffer so that the noise sequence (in lieu of *yn*) is stored in memory using

```
yn_buffer[buffercount] = dly[0];
```

Run the program again and plot the FFT magnitude of the noise sequence. It does not appear quite flat since the resulting plot is not averaged. You can also output the noise sequence using

```
output_sample(dly[0]);
```

in the program. With the output to a spectrum analyzer with averaging capability, verify that the noise spectrum is quite flat until about 3800 Hz, the bandwidth of the antialiasing filter on the codec (it looks like a lowpass filter with a bandwidth of 3800 Hz). Figure 4.23 shows the spectrum of this noise sequence using the HP analyzer (averaged with the analyzer). Use a GEL file to develop a slider so that the DSK output is either the noise sequence generated internally, dly[0], or the filter's output, y(n).

```
//FIRPRNbuf.c FIR filter with input noise sequence & output in buffer

#include "DSK6713_AIC23.h"          //codec-DSK support file
Uint32 fs=DSK6713_AIC23_FREQ_8KHZ;  //set sampling rate
#include "bp41.cof"                 //BP @ 1 kHz coefficient file
#include "noise_gen.h"              //header file for noise sequence
int yn = 0;                         //initialize filter's output
short dly[N];                       //delay samples
short buffercount = 0;              //init buffer count
const short bufferlength = 1024;    //buffer size
short yn_buffer[1024];              //output buffer
short fb;                           //feedback variable
shift_reg sreg;

short prn(void)                     //pseudorandom noise generation
{
 short prnseq;                      //for pseudorandom sequence
 if(sreg.bt.b0) prnseq =-16000;     //scaled negative noise level
 else           prnseq = 16000;     //scaled positive noise level
 fb  =(sreg.bt.b0)^(sreg.bt.b1);    //XOR bits 0,1
 fb ^=(sreg.bt.b11)^(sreg.bt.b13);  //with bits 11,13 ->fb
 sreg.regval<<=1;                   //shift register 1 bit to left
 sreg.bt.b0 = fb;                   //close feedback path
 return prnseq;                     //return sequence
}

interrupt void c_int11()            //ISR
{
 short i;
 dly[0] = prn();                    //input noise sequence
 yn = 0;                            //initialize filter's output
 for (i = 0; i< N; i++)
   yn +=(h[i]*dly[i]) >>15;         //y(n)+=h(i)*x(n-i)
 for (i = N-1; i > 0; i--)          //start @ bottom of buffer
   dly[i] = dly[i-1];               //data move to update delays
 output_sample((short)yn);          //output filter
 yn_buffer[buffercount] = yn;       //filter's output into buffer
 buffercount++;                     //increment buffer count
 if(buffercount==bufferlength) buffercount=0;  //reinit buffer count
 return;                            //return from interrupt
}

void main()
{
     sreg.regval = 0xFFFF;          //shift register to nominal values
     fb = 1;                        //initial feedback value
     comm_intr();                   //init DSK, codec, McBSP
     while(1);                      //infinite loop
}
```

FIGURE 4.20. FIR program using an internally generated input pseudorandom noise sequence and output stored in memory (FIRPRNbuf.c).

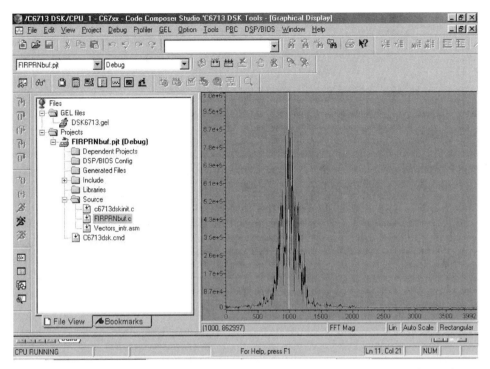

FIGURE 4.21. CCS output frequency response of a 1-kHz FIR bandpass filter using an internally generated noise sequence as input to the filter for project `FIRPRNbuf`.

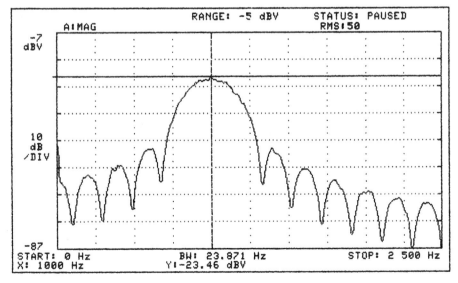

FIGURE 4.22. Frequency response of a 1-kHz FIR bandpass filter with a signal analyzer, using program `FIRPRNbuf.c`

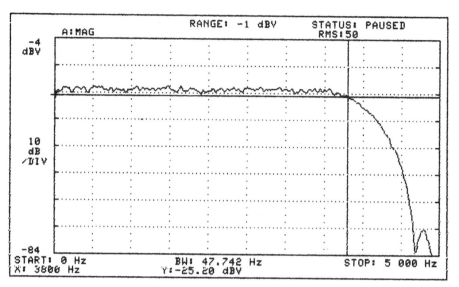

FIGURE 4.23. Spectrum of an internally generated pseudorandom noise sequence with a signal analyzer, using program FIRPRNbuf.c.

Example 4.6: Two Notch Filters to Recover Corrupted Input Voice (Notch2)

This example illustrates the implementation of two notch (bandstop) FIR filters to remove two undesired sinusoidal signals corrupting an input voice signal. The voice signal (*TheForce.wav* on the CD) was "added" (using Goldwave) with the two undesired sinusoidal signals at frequencies of 900 and 2700 Hz, to produce the corrupted input signal *corruptvoice.wav* (on the CD).

Figure 4.24 shows a listing of the program *NOTCH2.c*, which implements the two notch filters in cascade (series). Two coefficient files, BS900.cof and BS2700.cof (on the CD), each containing 89 coefficients and designed with MATLAB, are included in the filter program *NOTCH2.c*. They represent two FIR notch filters, centered at 900 and 2700 Hz, respectively. A buffer is used for the delay samples of each filter. The output of the first notch filter, centered at 900 Hz, becomes the input to the second notch filter, centered at 2700 Hz.

Build this project as **notch2**. Input (play) the corrupted voice signal *corruptvoice.wav*. Verify that the slider in position 1 (as set initially) outputs the corrupted voice signal, as shown in Figure 4.25. This plot is obtained with Goldwave using the DSK output as the input to a sound card (see Appendix E). The plot is shown on only one side (left channel) since a mono signal is used. Observe the two spikes (representing the two sinusoidal signals) at 900 and 2700 Hz, respectively. Change the slider to position 2 and verify that the two undesirable sinusoidal signals are removed. Output *y1out* in lieu of *y2out* and verify that only the 2700-Hz spike corrupts the input voice signal. Note that *y1out* is the output of the first notch filter.

```
//Notch2.c Two FIR notch filters to remove sinusoidal noise signals

#include "DSK6713_AIC23.h"          //codec-DSK support file
Uint32 fs=DSK6713_AIC23_FREQ_8KHZ;  //set sampling rate
#include "bs900.cof"                //BS @ 900 Hz coefficient file
#include "bs2700.cof"               //BS @ 2700 Hz coefficient file
short dly1[N]={0};                  //delay samples for 1st filter
short dly2[N]={0};                  //delay samples for 2nd filter
int y1out = 0, y2out = 0;           //init output of each filter
short out_type = 1;                 //slider for output type

interrupt void c_int11()            //ISR
{
 short i;
 dly1[0] = input_sample();          //newest input @ top of buffer
 y1out = 0;                         //init output of 1st filter
 y2out = 0;                         //init output of 2nd filter
 for (i = 0; i< N; i++)
   y1out += h900[i]*dly1[i];        //y1(n)+=h900(i)*x(n-i)
 dly2[0]=(y1out>>15);               //out of 1st filter->in 2nd filter
 for (i = 0; i< N; i++)
   y2out += h2700[i]*dly2[i];       //y2(n)+=h2700(i)*x(n-i)
 for (i = N-1; i > 0; i--)          //from bottom of buffer
  {
   dly1[i] = dly1[i-1];             //update samples of 1st buffer
   dly2[i] = dly2[i-1];             //update samples of 2nd buffer
  }
 if(out_type==1) output_sample(dly1[0]); //corrupted input(voice+sines)
 if(out_type==2) output_sample((short)(y2out>>15)); //out of 2nd filter
 return;                            //return from ISR
}

void main()
{
     comm_intr();                   //init DSK, codec, McBSP
     while(1);                      //infinite loop
}
```

FIGURE 4.24. Program implementing two FIR notch filters in cascade to remove two undesired sinusoidal signals (NOTCH2.c).

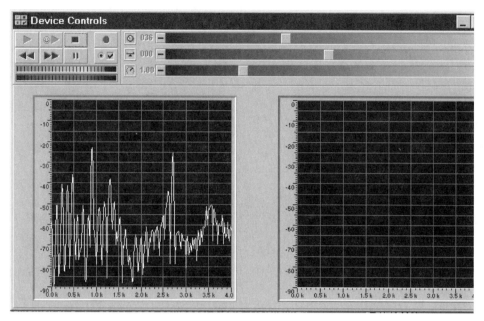

FIGURE 4.25. Spectrum of voice signal corrupted by two sinusoidal signals at frequencies of 900 and 2700 Hz (obtained with Goldwave).

Example 4.7: FIR Implementation Using Four Different Methods (FIR4ways)

Figure 4.26 shows a listing of the program *FIR4ways.c*, which implements an FIR filter using four alternative methods for convolving/updating the delay samples. This example extends Example 4.1, where the first method (method A) is used. In this first method with two "for" loops, the delay samples are arranged in memory with the newest sample at the beginning of the buffer and the oldest sample at the end of the buffer. The convolution starts with the newest sample and the first coefficient using

$$y(n) = h(0)x(n) + h(1)x(n-1) + \cdots + h(N-1)x(n-(N-1))$$

Each data value is "moved down" in memory to update the delay samples, with the newest sample being the newly acquired input sample. The size of the array for the delay samples is now set at $N + 1$, not at N, to illustrate the third method (method C). The other three methods use a buffer size of N for the delay samples. The bottom (end) of the buffer in this example refers to memory location N, not $N + 1$. Note that in this case the unused data $x(n - N)$ in memory location $(N + 1)$ is not updated by using the index $i < N$.

The second method (method B) performs the convolution and updates the delay samples using one loop. The convolution starts with the oldest coefficient and the

```
//FIR4ways.c FIR with alternative ways of storing/updating samples

#include "DSK6713_AIC23.h"      //codec-DSK file support
Uint32 fs=DSK6713_AIC23_FREQ_8KHZ;//set sampling rate
#include "bp41.cof"             //BP coeff centered at Fs/8
#define METHOD 'A'              //change to B or C or D
int yn = 0;                     //initialize filter's output
short dly[N+1];                 //delay samples array(one extra)

interrupt void c_int11()        //ISR
{
   short i;
   yn = 0;                      //initialize filter's output

#if METHOD == 'A'               //if 1st method
   dly[0] = input_sample();     //newest sample @ top of buffer
   for (i = 0; i< N; i++)
      yn += (h[i] * dly[i]);    //y(n)=h[0]*x[n]+..+h[N-1]x[n-(N-1)]
   for (i = N-1; i > 0; i--)    //from bottom of buffer
     dly[i] = dly[i-1];         //update sample data move "down"

#elif METHOD == 'B'             //if 2nd method
   dly[0] = input_sample();     //newest sample @ top of buffer
   for (i = N-1; i >= 0; i--)   //start @ bottom to convolve
     {
      yn += (h[i] * dly[i]);    //y=h[N-1]x[n-(N-1)]+...+h[0]x[n]
      dly[i] = dly[i-1];        //update sample data move "down"
     }

#elif METHOD == 'C'             //use xtra memory location
   dly[0] = input_sample();     //newest sample @ top of buffer
   for (i = N-1; i>=0; i--)     //start @ bottom of buffer
     {
      yn += (h[i] * dly[i]);    //y=h[N-1]x[n-(N-1)]+...+h[0]x[n]
      dly[i+1] = dly[i];        //update sample data move "down"
     }

#elif METHOD == 'D'             //1st convolve before loop
   dly[N-1] = input_sample();   //newest sample @ bottom of buffer
   yn = h[N-1] * dly[0];        //y=h[N-1]x[n-(N-1)] (only one)
   for (i = 1; i<N; i++)        //convolve the rest
     {
      yn +=(h[N-(i+1)]*dly[i]); //h[N-2]x[n-(N-2)]+...+h[0]x[n]
      dly[i-1] = dly[i];        //update sample data move "up"
     }
#endif
output_sample((short)(yn>>15)); //output filter
return;                         //return from ISR
}

void main()
{
    comm_intr();                //init DSK, codec, McBSP
    while(1);                   //infinite loop
}
```

FIGURE 4.26. FIR program using four alternative methods for convolution and updating of delay samples (FIR4ways.c).

oldest sample, "moving up" through the buffers using

$$y(n) = h(N-1)x(n-(N-1)) + h(N-2)x(n-(N-2)) + \cdots + h(0)x(n)$$

The updating scheme is similar to that of the first method. In method B, when $i = 0$, the newest sample is updated by an invalid data value residing at the memory location preceding the start of the sample buffer. But this invalid data item is then replaced by a newly acquired input sample with `dly[0]` before $y(n)$ is calculated for the next unit of time. Or, one could use an "if" statement to update the delay samples for all values of i except for $i = 0$.

The third method (method C) uses $N + 1$ memory locations to update the delay samples. The unused data at memory location $N + 1$ is also updated. This extra memory location is used so that a valid data item in that location is not overwritten during the update of the delay samples. The fourth method (method D) performs the first convolution expression "outside" the loop. The delay samples in the previous methods were arranged in memory so that the newest sample, $x(n)$, is at the beginning of the buffer and the oldest sample, $x(n - (N - 1))$, is at the end. However, in this method, the newest input sample is acquired through `dly[N - 1]` so that the newest sample is now at the end of the buffer and the updating process moves the data "up in memory."

Build and run this project as **FIR4ways**. Verify that the output is an FIR bandpass filter centered at 1 kHz, as in the example *FIRPRNbuf*. Change the method to test (define) the other three methods and verify that the resulting output is the same.

Example 4.8: Voice Scrambling Using Filtering and Modulation (Scrambler)

This example illustrates a voice scrambling/descrambling scheme. The approach makes use of basic algorithms for filtering and modulation. Modulation was introduced in the AM example in Chapter 2. With voice as input, the resulting output is scrambled voice. The original unscrambled voice is recovered when the output of the DSK is used as the input to a second DSK running the same program.

The scrambling method used is commonly referred to as *frequency inversion*. It takes an audio range, represented by the band 0.3 to 3 kHz, and "folds" it about a carrier signal. The frequency inversion is achieved by multiplying (modulating) the audio input by a carrier signal, causing a shift in the frequency spectrum with upper and lower sidebands. On the lower sideband that represents the audible speech range, the low tones are high tones, and vice versa.

Figure 4.27 is a block diagram of the scrambling scheme. At point A we have a bandlimited signal 0 to 3 kHz. At point B we have a double-sideband signal with suppressed carrier. At point C the upper sideband is filtered out. Its attractiveness comes from its simplicity, since only simple DSP algorithms are utilized: filtering, and sine generation and modulation.

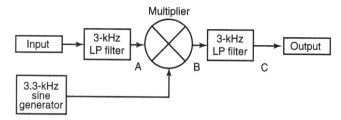

FIGURE 4.27. Block diagram of a scrambler/descrambler scheme.

Figure 4.28 shows a listing of the program `Scrambler.c`, which implements this project. The input signal is first lowpass-filtered and the resulting output (at point A in Figure 4.27) is multiplied (modulated) by a 3.3-kHz sine function with data values in a buffer (lookup table). The modulated signal (at point B) is filtered again, and the overall output is a scrambled signal (at point C).

There are three functions in Figure 4.28 in addition to the function *main*. One of the functions, *filtmodfilt*, calls a filter function to implement the first lowpass filter as an antialiasing filter. The resulting output (filtered input) becomes the input to a multiplier/modulator. The function *sinemod* modulates (multiplies) the filtered input with the 3.3-kHz sine data values. This produces higher and lower sideband components. The modulated output is again filtered, so that only the lower sideband components are kept.

A buffer is used to store the 114 coefficients that represent the lowpass filter. The coefficient file `lp114.cof` is on the CD. Two other buffers are used for the delay samples, one for each filter. The samples are arranged in memory as

$$x(n-(N-1)), x(n-(N-2)), \ldots, x(n-1), x(n)$$

with the oldest sample at the beginning of the buffer and the newest sample at the end (bottom) of the buffer. The file `sine160.h` with 160 data values over 33 cycles is on the CD. The frequency generated is $f = F_s$ (number of cycles)/(number of points) = 16,000(33)/160 = 3.3 kHz.

Using the resulting output as the input to a second DSK running the same algorithm, the original unscrambled input is recovered as the output of the second DSK. Note that the program can still run on the first DSK when the USB connector cable is removed from the DSK.

An optional up-sampling (by a factor of 2) scheme is used to obtain a 16-kHz sampling rate. This scheme is achieved by "processing" the input data twice while retaining only the second result. This allows for a wider input signal bandwidth to be scrambled, resulting in a better performance.

Build and run this project as **Scrambler**. First, test this project using a 2-kHz input sine wave. The resulting output is a lower sideband signal of 1.3 kHz, obtained as (3.3 kHz − 2 kHz). The upper sideband signal of (3.3 + 2 kHz) is filtered out by the second lowpass filter (actually by the antialiasing filter on the codec).

```
//Scrambler.c Voice scrambler/de-scrambler program
#include "dsk6713_aic23.h"              //codec-dsk support file
Uint32 fs=DSK6713_AIC23_FREQ_8KHZ;     //set sampling rate
#include "sine160.h"                    //sine data values
#include "LP114.cof"                    //filter coefficient file
short filtmodfilt(short data);
short filter(short inp,short *dly);
short sinemod(short input);
static short filter1[N],filter2[N];
short input, output;

void main()
{
 short i;
 comm_poll();                          //init DSK using polling
 for (i=0; i< N; i++)
    {
     filter1[i] = 0;                    //init 1st filter buffer
     filter2[i] = 0;                    //init 2nd filter buffer
    }
 while(1)
    {
     input=input_sample();             //input new sample data
     filtmodfilt(input);               //process sample twice(upsample)
     output=filtmodfilt(input);        //and throw away 1st result
     output_sample(output);            //then output
    }
}

short filtmodfilt(short data)          //filtering & modulating
{
 data = filter(data,filter1);          //newest in ->1st filter
 data = sinemod(data);                 //modulate with 1st filter out
 data = filter(data,filter2);          //2nd LP filter
 return data;
}

short filter(short inp,short *dly) //implements FIR
{
 short i;
 int yn;
 dly[N-1] = inp;                       //newest sample @bottom buffer
 yn = dly[0] * h[N-1];                 //y(0)=x(n-(N-1))*h(N-1)
 for (i = 1; i < N; i++)               //loop for the rest
    {
     yn += dly[i] * h[N-(i+1)];        //y(n)=x[n-(N-1-i)]*h[N-1-i]
     dly[i-1] = dly[i];                //data up to update delays
    }
 yn = (yn >>15);                       //filter's output
 return yn;                            //return y(n) at time n
}

short sinemod(short input)             //sine generation/modulation
{
 static short i=0;
 input=(input*sine160[i++])>>11; //(input)*(sine data)
 if(i>= NSINE) i = 0;                  //if end of sine table
 return input;                         //return modulated signal
}
```

FIGURE 4.28. Voice scrambler program (`Scrambler.c`).

A second DSK is used to recover/unscramble the original signal (simulating the receiving end). Use the output of the first DSK as the input to the second DSK. Run the same program on the second DSK. This produces the reverse procedure, yielding the original unscrambled signal. If the same 2-kHz original input is considered, the 1.3 kHz as the scrambled signal becomes the input to the second DSK. The resulting output is the original signal of 2 kHz (3.3 − 1.3 kHz), the lower sideband signal.

With a sweeping input sinusoidal signal increasing in frequency, the resulting output is the sweeping signal "decreasing" in frequency. Use as input the .wav file *TheForce.wav* and verify the scrambling/descrambling scheme.

The up-sampling scheme is optional since a 16-kHz sampling rate can be set directly in the program and commenting the line of code

```
filtmodfilt(input);
```

Verify the up-sampling scheme. Are the results the same as before, with an 8-kHz sampling rate and processing the input twice?

Interception of the speech signal can be made more difficult by changing the modulation frequency dynamically and including (or omitting) the carrier frequency according to a predefined sequence: for example, a code for no modulation, another for modulating at frequency f_{c1}, and a third code for modulating at frequency f_{c2}. This project was first implemented using the TMS320C25 [51] and also on the TMS320C31 DSK without the need for up-sampling.

Example 4.9: Illustration of Aliasing Effects with Down-Sampling (aliasing)

Figure 4.29 shows a listing of the program aliasing.c, which implements this project. To illustrate the effects of aliasing, the processing rate is down-sampled by a factor of 2 to an equivalent 4-kHz rate. Note that the antialiasing and reconstruction filters on the AIC23 codec are fixed and cannot be bypassed or altered. Up-sampling and lowpass filtering are then needed to output the 4-kHz rate samples to the AIC23 codec sampling at 8 kHz.

Build this project as **aliasing**. Load the slider file aliasing.gel (on the CD). With antialiasing initially set to zero in the program, aliasing will occur.

1. Input a sinusoidal signal and verify that for an input signal frequency up to 2 kHz, the output is essentially a loop program (delayed input). Increase the input signal frequency to 2.5 kHz and verify that the output is an aliased 1.5-kHz signal. Similarly, a 3- and a 3.5-kHz input signal yield an aliased output signal of 1 and 0.5 kHz, respectively. Input signals with frequencies beyond 3.9 kHz are supressed due to the AIC23 codec's antialiasing filter.

2. Change the slider position to 1, so that antialiasing at the down-sampled rate of 4 kHz is desired. For an input signal frequency up to about 1.9 kHz, the output is

```
//Aliasing.c illustration of downsampling, aliasing, upsampling

#include "DSK6713_AIC23.h"       //codec-DSK support file
Uint32 fs=DSK6713_AIC23_FREQ_8KHZ;//set sampling rate
#include "lp33.cof"              //lowpass at 1.9 kHz
short flag = 0;                  //toggles for 2x down-sampling
short indly[N],outdly[N];        //antialias and reconst delay lines
float yn; int i;                 //filter output, index
short antialiasing = 0;          //init for no antialiasing filter

interrupt void c_int11()         //ISR
{
 indly[0]=(float)(input_sample());//new sample to antialias filter
 yn = 0.0;                       //initialize downsampled value
 if (flag == 0) flag = 1;        //don't discard at next sampling
 else
   {
    if (antialiasing == 1)       //if antialiasing filter desired
      {                          //compute downsampled value
       for (i = 0 ; i < N ; i++) //using LP @ 1.9 kHz filter coeffs
          yn += (h[i]*indly[i]); //filter is implemented using float
      }
    else                         //if filter is bypassed
         yn = indly[0];          //downsampled value is input value
    flag = 0;                    //next input value will be discarded
   }
 for (i = N-1; i > 0; i--)
       indly[i] = indly[i-1];    //update input buffer

 outdly[0] = (yn);               //input to reconst filter
 yn = 0.0;                       //4 kHz sample values and zeros
 for (i = 0 ; i < N ; i++)       //are filtered at 8 kHz rate
       yn += (h[i]*outdly[i]);   //by reconstruction lowpass filter

 for (i = N-1; i > 0; i--)
       outdly[i] = outdly[i-1];  //update delays

 output_sample((short)yn);       //8kHz rate sample
 return;                         //return from interrupt
}

void main()
{
 comm_intr();                    //init DSK, codec, McBSP
 while(1);                       //infinite loop
}
```

FIGURE 4.29. Program to illustrate aliasing and antialiasing down-sampling to a rate of 4kHz (aliasing.c).

a delayed version of the input. Increase the input signal frequency beyond 1.9kHz and verify that the output reduces to zero. This is due to the 1.9-kHz (at the down-sampling rate of 4kHz) antialiasing lowpass filter, implemented using the coefficient file lp33.cof (on the CD). In lieu of a sinusoidal signal as input, you can use a swept sinusoidal input signal.

Example 4.10: Implementation of an Inverse FIR Filter (`FIRinverse`)

Figure 4.30 shows a listing of the program `FIRinverse.c`, which implements an inverse FIR filter. An original input sequence to an FIR filter can be recovered using an inverse FIR filter. A slider is used to select among the input noise, the output of an FIR filter, or the inverse of the FIR filter that is the original input noise. The transfer function of an FIR filter of order N is

$$H(z) = \sum_{i=0}^{N-1} h_i z^{-i}$$

where h_i represents the impulse response coefficients. The output sequence of the FIR filter is

```
//FIRinverse.c Implementation of inverse FIR Filter

#include "DSK6713_AIC23.h"              //codec-DSK support file
Uint32 fs=DSK6713_AIC23_FREQ_8KHZ;      //set sampling rate
#include "bp41.cof"                      //coefficient file BP @ Fs/8
int yn;                                  //filter's output
short dly[N];                            //delay samples
int out_type = 1;                        //select output with slider

interrupt void c_int11()                 //ISR
{
 short i;
 dly[0] = input_sample();                //newest input sample data
 yn = 0;                                 //initialize filter's output
 for (i = 0; i<N; i++)
    yn += (h[i]*dly[i]);                 //y(n) += h(i)*x(n-i)
 if(out_type==1) output_sample(dly[0]);  //original input
 if(out_type==2) output_sample((short)(yn>>15));//output->FIR filter
 if(out_type==3)                         //calculate inverse FIR
   {
    for (i = N-1; i>1; i--)
       yn -= (h[i]*dly[i]);              //calculate inverse FIR filter
    yn = yn/h[0];                        //scale output of inverse filter
    output_sample((short)(yn>>8));       //output of inverse filter
   }
 for (i = N-1; i>0; i--)                 //from bottom of buffer
    dly[i] = dly[i-1];                   //update delay samples
 return;                                 //return from ISR
}

void main()
{
 comm_intr();                            //init DSK, codec, McBSP
 while(1);                               //infinite loop
}
```

FIGURE 4.30. Program to implement an inverse FIR filter (FIRinverse.c).

$$y(n) = \sum_{i=0}^{N-1} h_i x(n-i) = h_0 x(n) + h_1 x(n-1) + \cdots + h_{N-1} x(n-(N-1))$$

where $x(n-i)$ represents the input sequence. The original input sequence, x, can then be recovered, using $\hat{x}(n)$ as an estimate of $x(n)$, or

$$\hat{x}(n) = \frac{y(n) - \sum_{i=1}^{N-1} h_i \hat{x}(n-i)}{h_0}$$

Build this project as **FIRinverse**. Use noise as input (from Goldwave or from a noise generator, or modify the program to use the pseudorandom noise sequence, etc.). Verify that the output is the input noise sequence, with the slider in position 1 (default). Change the slider to position 2 and verify the output as an FIR bandpass filter centered at 1 kHz. With the slider in position 3, the inverse of the FIR filter is calculated, so that the output is the original input noise sequence.

Example 4.11: FIR Implementation Using C Calling an ASM Function (*FIRcasm*)

The C program FIRcasm.c (Figure 4.31) calls the ASM function *FIRcasm-func.asm* (Figure 4.32), which implements an FIR filter.

Build and run this project as **FIRcasm**. Verify that the output is a 1-kHz FIR bandpass filter. Two buffers are created: dly for the data samples and h for the filter's coefficients. On each interrupt, a new data sample is acquired and stored at the end (higher-memory address) of the buffer dly. The delay samples and the filter coefficients are arranged in memory as shown in Table 4.3. The delay samples are stored in memory starting with the oldest sample. The newest sample is at the end of the buffer. The coefficients are arranged in memory with h(0) at the beginning of the coefficient buffer and $h(N-1)$ at the end.

The addresses of the delay sample buffer, the filter coefficient buffer, and the size of each buffer are passed to the ASM function through registers A4, B4, and A6, respectively. The size of each buffer through register A6 is doubled since data in each memory location are stored as bytes. The pointers A4 and B4 are incremented or decremented every two bytes (two memory locations). The end address of the coefficients' buffer is in B4, which is at $2N-1$.

The two 16-bit load (LDH) instructions load the content in memory pointed by (whose address is specified by) A4 and the content in memory at the address specified by B4. This loads the oldest sample and last coefficient, $x(n-(N-1))$ and $h(N-1)$, respectively. A4 is then postincremented to point at $x(n-(N-2))$, and B4 is postdecremented to point at $h(N-2)$. After the first accumulation, the oldest sample is updated. The content in memory at the address specified by A4 is loaded into A7, then stored at the preceding memory location. This is because A4 is

```
//FIRcasm.c FIR C program calling ASM function fircasmfunc.asm

#include "DSK6713_AIC23.h"            //codec-DSK support file
Uint32 fs=DSK6713_AIC23_FREQ_8KHZ;   //set sampling rate
#include "bp41.cof"                   //BP @ Fs/8 coefficient file
int yn = 0;                          //initialize filter's output
short dly[N];                        //delay samples

interrupt void c_int11()             //ISR
{
 dly[N-1] = input_sample();          //newest sample @bottom buffer
 yn = fircasmfunc(dly,h,N);          //to ASM func through A4,B4,A6
 output_sample((short)(yn>>15));     //filter's output
 return;                             //return from ISR
}

void main()
{
 short i;
 for (i = 0; i<N; i++)
     dly[i] = 0;                     //init buffer for delays
 comm_intr();                        //init DSK, codec, McBSP
 while(1);                           //infinite loop
}
```

FIGURE 4.31. C program calling an ASM function for FIR implementation (`FIRcasm.c`).

```
;FIRcasmfunc.asm ASM function called from C to implement FIR
;A4 = Samples address, B4 = coeff address, A6 = filter order
;Delays organized as:x(n-(N-1))...x(n);coeff as h[0]...h[N-1]

            .def    _fircasmfunc
_fircasmfunc:                    ;ASM function called from C
            MV    A6,A1          ;setup loop count
            MPY   A6,2,A6        ;since dly buffer data as byte
            ZERO  A8             ;init A8 for accumulation
            ADD   A6,B4,B4       ;since coeff buffer data as byte
            SUB   B4,1,B4        ;B4=bottom coeff array h[N-1]
  loop:                         ;start of FIR loop
            LDH   *A4++,A2       ;A2 = x[n-(N-1)+i] i=0,1,...,N-1
            LDH   *B4--,B2       ;B2 = h[N-1-i] i=0,1,...,N-1
            NOP   4
            MPY   A2,B2,A6       ;A6 = x[n-(N-1)+i]*h[N-1-i]
            NOP
            ADD   A6,A8,A8       ;accumlate in A8
            LDH   *A4,A7         ;A7=x[(n-(N-1)+i+1]update delays
            NOP   4             ;using data move "up"
            STH   A7,*-A4[1]     ;-->x[(n-(N-1)+i] update sample
            SUB   A1,1,A1        ;decrement loop count
     [A1]   B     loop          ;branch to loop if count # 0
            NOP   5
            MV    A8,A4          ;result returned in A4
            B     B3            ;return addr to calling routine
            NOP   4
```

FIGURE 4.32. FIR ASM function called from C (`FIRcasmfunc.asm`).

TABLE 4.3 Memory Organization of Coefficients and Samples for FIRcasm

Coefficients	Samples		
	Time n		Time n + 1
h(0)	A4 → x(n - (N - 1))		A4 → x(n - (N - 2))
h(1)	x(n - (N - 2))		x(n - (N - 3))
h(2)	x(n - (N - 3))		x(n - (N - 4))
.	.		.
.	.		.
.	.		.
h(N - 2)	x(n - 1)		x(n)
B4 → h(N - 1)	x(n)	← newest →	x(n + 1)

postdecremented *without* modification to point at the memory location containing the oldest sample. As a result, the oldest sample, $x(n - (N - 1))$, is replaced (updated) by $x(n - (N - 2))$. The updating of the delay samples is for the next unit of time. As the output at time n is being calculated, the samples are updated or "primed" for time $(n + 1)$. At time n the filter's output is

$$y(n) = h(N - 1)x(n - (N - 1)) + h(N - 2)x(n - (N - 2)) + \cdots + h(1)x(n - 1) + h(0)x(n)$$

The loop is processed 41 times. For each time $n, n + 1$, and $n + 2, \ldots$, an output value is calculated, with each sample updated for the next unit of time. The newest sample is also updated in this process, with an invalid data value residing at the memory location beyond the end of the buffer. But this is remedied since for each unit of time, the newest sample, acquired through the ADC of the codec, overwrites it.

Accumulation is in A8 and the result, for each unit of time, is moved to A4 to be returned to the calling function. The address of the calling function is in B3.

Viewing Update of Samples in Memory

1. Select → *View* → *Memory* using a 16-bit hex format and a starting address of `dly`. The delay samples are within 82 (not 41) memory locations, each location specified with a byte. The coefficients also occupy 82 memory locations in the buffer h. You can verify the content in the coefficient buffer stored as a 16-bit or half-word value. Right-click on the memory window and deselect "Float in Main Window" for a better display with both source program and memory.

2. Select → *View* → *Mixed C/ASM*. Place a breakpoint within the function *FIRcasmfunc.asm* at the move instruction

```
MV  A8,A4
```

You can either double-click on that line of code or right-mouse-click to Toggle Breakpoint).

3. Select → *Debug* → *Animate* (introduced in Chapter 1). Execution halts at the set breakpoint for each unit of time. Observe the end (bottom) memory location of the delay samples' buffer. Verify that the newest sample data value is placed at the end of the buffer. This value is then moved up the buffer to a lower address. Observe after a while that the samples are being updated, with each value in the buffer moving up in memory. You can also observe the register (pointer) A4 incrementing by 2 (two bytes) and B4 decrementing by 2.

Example 4.12: FIR Implementation Using C Calling a Faster ASM Function (FIRcasmfast)

The same C calling program, *FIRcasm.c*, is used in this example as in Example 4.11. It calls the ASM function *Fircasmfunc* within the file *FIRcasmfunc-fast.asm,* as shown in Figure 4.33. This ASM function executes faster than the function in the previous example by having parallel instructions and rearranging the sequence of instructions. There are two parallel instructions: LDH/LDH and SUB/LDH.

1. The number of NOPs is reduced from 19 to 11.

2. The SUB instruction to decrement the loop count is moved up the program.

3. The sequence of some instructions is changed to "fill" some of the NOP slots.

```
;FIRCASMfuncfast.asm C-called faster function to implement FIR
            .def    _fircasmfunc
_fircasmfunc:                   ;ASM function called from C
            MV      A6,A1       ;setup loop count
            MPY     A6,2,A6     ;since dly buffer data as byte
            ZERO    A8          ;init A8 for accumulation
            ADD     A6,B4,B4    ;since coeff buffer data as byte
            SUB     B4,1,B4     ;B4 = bottom coeff array h[N-1]
loop:                           ;start of FIR loop
            LDH     *A4++,A2    ;A2 = x[n-(N-1)+i] i=0,1,...,N-1
   ||       LDH     *B4--,B2    ;B2 = h[N-1-i] i=0,1,...,N-1
            SUB     A1,1,A1     ;decrement loop count
   ||       LDH     *A4,A7      ;A7=x[(n-(N-1)+i+1]update delays
            NOP     4
            STH     A7,*-A4[1]  ;-->x[(n-(N-1)+i] update sample
   [A1]     B       loop        ;branch to loop if count # 0
            NOP     2
            MPY     A2,B2,A6    ;A6=x[n-(N-1)+i]*h[N-1-i]
            NOP
            ADD     A6,A8,A8    ;accumlate in A8
            B       B3          ;return addr to calling routine
            MV      A8,A4       ;result returned in A4
            NOP     4
```

FIGURE 4.33. FIR ASM function with parallel instructions for faster execution (FIRcasmfuncfast.asm).

For example, the conditional branch instruction executes *after* the ADD instruction to accumulate in A8, since branching has five delay slots. Additional changes to make it faster would also make it less comprehensible due to further resequencing of the instructions.

Build this project as **FIRcasmfast**, so that the linker option names the output executable file FIRcasmfast.out. The resulting output is the same 1-kHz bandpass filter as in the previous example.

Example 4.13: FIR Implementation Using C Calling an ASM Function with a Circular Buffer (FIRcirc)

The C program *FIRcirc.c* (Figure 4.34) calls the ASM function *FIRcircfunc.asm* (Figure 4.35). This example expands Example 4.12 to implement an FIR filter using a circular buffer. The coefficients within the file *bp1750.cof* were designed with MATLAB using the Kaiser window and represent a 128-coefficient FIR bandpass filter with a center frequency of 1750 Hz. Figure 4.36 displays the characteristics of this filter, obtained from MATLAB's filter designer SPTool (described in Appendix D).

In lieu of moving the data to update the delay samples, a pointer is used. The 16 LSBs of the address mode register are set with a value of

```
0x0040 = 0000 0000 0100 0000
```

```
//FIRcirc.c  C program calling ASM function using circular buffer

#include "DSK6713_AIC23.h"              //codec-DSK support file
Uint32 fs=DSK6713_AIC23_FREQ_8KHZ;     //set sampling rate
#include "bp1750.cof"                   //BP at 1750 Hz coeff file
int yn = 0;                             //init filter's output

interrupt void c_int11()               //ISR
{
 short sample_data;
 sample_data = (input_sample());       //newest input sample data
 yn = fircircfunc(sample_data,h,N);    //ASM func passing to A4,B4,A6
 output_sample((short)(yn>>15));       //filter's output
 return;                               //return to calling function
}

void main()
{
 comm_intr();                          //init DSK, codec, McBSP
 while(1);                             //infinite loop
}
```

FIGURE 4.34. C program calling an ASM function using a circular buffer (FIRcirc.c).

```
;FIRcircfunc.asm ASM function called from C using circular addressing
;A4=newest sample, B4=coefficient address, A6=filter order
;Delay samples organized: x[n-(N-1)]...x[n]; coeff as h(0)...h[N-1]

              .def   _fircircfunc
              .def    last_addr
              .def    delays
              .sect "circdata"      ;circular data section
              .align 256            ;align delay buffer 256-byte boundary
delays        .space 256            ;init 256-byte buffer with 0's
last_addr     .int   last_addr-1    ;point to bottom of delays buffer
              .text                 ;code section
_fircircfunc:                       ;FIR function using circ addr
              MV    A6,A1           ;setup loop count
              MPY   A6,2,A6         ;since dly buffer data as byte
              ZERO  A8              ;init A8 for accumulation
              ADD   A6,B4,B4        ;since coeff buffer data as bytes
              SUB   B4,1,B4         ;B4=bottom coeff array h[N-1]
              MVKL  0x00070040,B6   ;select A7 as pointer and BK0
              MVKH  0x00070040,B6   ;BK0 for 256 bytes (128 shorts)
              MVC   B6,AMR          ;set address mode register AMR
              MVKL  last_addr,A9    ;A9=last circ addr(lower 16 bits)
              MVKH  last_addr,A9    ;last circ addr (higher 16 bits)
              LDW   *A9,A7          ;A7=last circ addr
              NOP   4
              STH   A4,*A7++        ;newest sample-->last address
loop:                               ;begin FIR loop
              LDH   *A7++,A2        ;A2=x[n-(N-1)+i] i=0,1,...,N-1
       ||     LDH   *B4--,B2        ;B2=h[N-1-i] i=0,1,...,N-1
              SUB   A1,1,A1         ;decrement count
       [A1]   B     loop            ;branch to loop if count # 0
              NOP   2
              MPY   A2,B2,A6        ;A6=x[n-(N-1)+i]*h[N-1+i]
              NOP
              ADD   A6,A8,A8        ;accumulate in A8
              STW   A7,*A9          ;store last circ addr to last_addr
              B     B3              ;return addr to calling routine
              MV    A8,A4           ;result returned in A4
              NOP   4
```

FIGURE 4.35. FIR ASM function using a circular buffer for updating samples (FIRcircfunc.asm).

This selects A7 mode as the circular buffer pointer register. The 16 MSBs of AMR are set with N = 0x0007 to select the block BK0 as a circular buffer. The buffer size is $2^{N+1} = 256$. A circular buffer is used in this example only for the delay samples. It is also possible to use a second circular buffer for the coefficients. For example, using

```
0x0140 = 0000 0001 0100 0000
```

would select two pointers, B4 and A7.

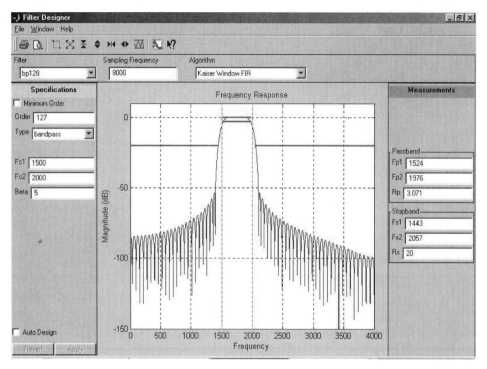

FIGURE 4.36. Frequency characteristics of a 128-coefficient FIR bandpass filter centered at 1750 Hz using MATLAB's filter designer SPTool described in Appendix D.

Within a C program, an inline assembly code can be used with the `asm` statement. For example,

```
asm(" MVK   0x0040,B6")
```

Note the blank space after the first quotation mark so that the instruction does not start in column 1. The circular mode of addressing eliminates the data move to update the delay samples, since a *pointer* can be moved to achieve the same results and much faster. Initially, the register pointer A7 points to the last address in the sample buffer. Consider for now the sample buffer only, since it is circular. (Note that the coefficient's buffer is not made to be circular.)

1. *Time n.* At time n, A7 points to the end of the buffer, where the newest sample is stored. It is then postincremented to point to the beginning of the buffer, as shown in Table 4.4. Then the section of code within the loop starts and calculates

$$y(n) = h(N-1)x(n-(N-1)) + h(N-2)x(n-(N-2)) + \cdots + h(1)x(n-1) + h(0)x(n)$$

TABLE 4.4   Memory Organization of Coefficients and Samples Using a Circular Buffer

Coefficients	Samples		
	Time n	Time n + 1	Time n + 2
h(0)	A7 → x(n - (N - 1))	newest → x(n + 1)	x(n + 1)
h(1)	x(n - (N - 2))	A7 → x(n - (N - 2))	newest → x(n + 2)
h(2)	x(n - (N - 3))	x(n - (N - 3))	A7 → x(n - (N - 3))
.	.	.	.
.	.	.	.
.	.	.	.
h(N - 2)	x(n - 1)	x(n - 1)	x(n - 1)
h(N - 1)	newest → x(n)	x(n)	x(n)

After the last multiplication, $h(0)x(n)$, A7 is postincremented to point to the beginning address of the buffer. The resulting filter's output at time n is then returned to the calling function. Before the loop starts for each unit of time, A7 always contains the address where the newest sample is to be stored. While the newly acquired sample is passed to the ASM function through A4 at each unit of time $n, n + 1, n + 2, \ldots$, A4 is stored in A7, which always contains the "last" address where the subsequent new sample is to be stored.

2. *Time $n + 1$.* At time $(n + 1)$, the newest sample, $x(n + 1)$, is passed to the ASM function through A4. The 16-bit store (STH) instruction stores that sample into memory whose address is in A7, which is at the beginning of the buffer. It is then postincremented to point at the address containing $x(n - (N - 2))$, as shown in Table 4.4. The output is now

$$y(n+1) = h(N-1)x(n-(N-2)) + h(N-2)x(n-(N-3)) + \cdots + h(1)x(n) \\ + h(0)x(n+1)$$

The last multiplication always involves $h(0)$ and the newest sample.

3. *Time $n + 2$.* At time $(n + 2)$, the filter's output is

$$y(n+2) = h(N-1)x(n-(N-3)) + h(N-2)x(n-(N-4)) + \cdots + h(1)x(n+1) \\ + h(0)x(n+2)$$

Note that for each unit of time, the newly acquired sample overwrites the oldest sample at the previous unit of time. At each time $n, n + 1, \ldots$, the filter's output is calculated within the ASM function and the result is sent to the calling C function, where a new sample is acquired at each sample period.

The conditional branch instruction was moved up, as in Example 4.12. Branching to loop takes effect (due to five delay slots) after the ADD instruction to accumulate in A8. One can save the content of AMR at the end of processing one buffer and restore it before using it again with a pair of MVC instructions: MVC AMR,Bx and MVC Bx,AMR using a B register.

Build and run this project as **FIRcirc**. Verify an FIR bandpass filter centered at 1750 Hz. Halt, and Restart the program.

Place a breakpoint within the ASM function FIRcircfunc.asm at the branch instruction to return to the calling C function (B   B3). View memory at the address delays and verify that this buffer of size 256 is initialized to zero. Right-click on the memory window to toggle "Float in Main Window" (for a better display). Run the program. Execution stops at the breakpoint. Verify that the newest sample (16 bits) is stored at the end (higher address) of the buffer (at 0x3FE and 0x3FF). Memory location 0x400 (in A9) contains the address 0x301, where the subsequent new sample is to be stored. This address represents the starting address of the buffer. View the core registers and verify that A7 contains this address.

Run the project again and observe the new sample stored at the beginning of the buffer. This 16-bit data sample is stored at 0x300 and 0x301. Animate now and observe where each new sample is being stored in memory. Note that A7 is incremented to 0x303, 0x305, ... The circular method of updating the delays is more efficient. It is important that the buffer is aligned on a boundary with a power of 2. While a buffer may be "naturally aligned," one must make sure that it is (an address with LSBs as zeros) if such buffer is to be used as circular.

Example 4.14: FIR Implementation Using C Calling an ASM Function Using a Circular Buffer in External Memory (FIRcirc_ext)

This example implements an FIR filter using a circular buffer in external memory. The same C source program FIRcirc.c and ASM function FIRcircfunc.asm as in the previous example are used, but with a modified linker command file.

This linker command file FIRcirc_ext.cmd is listed in Figure 4.37. The section circdata designates the memory section buffer_ext, which starts in external memory at 0x80000000.

Build this project as **FIRcirc_ext**. Load the executable file and view the memory at the address delays. This should display the external memory section that starts at 0x80000000. Verify that the circular buffer is in external memory, where all the delay samples are initialized to zero. Place a breakpoint as in Example 4.13, run the program up to the breakpoint, and verify that the newest input sample is stored at the end of the circular buffer at 0x800000FE and 0x800000FF. Register A9 contains the last address, and register A7 contains the address where the subsequent 16-bit input sample is to be stored (0x80000001). Run the program again (to the set breakpoint) and verify that the subsequent acquired sample is stored at the beginning of the buffer at the address 0x80000001. Remove the breakpoint, Restart/run, and verify that the output is the same FIR bandpass filter centered at 1750 Hz, as in Example 4.13.

```
/*FIRcirc_ext.cmd Linker command file for external memory*/

MEMORY
{
  IVECS:      org =            0h,  len =       0x220
  IRAM:       org = 0x00000220,  len = 0x0002FFFF
  SRAM_EXT1:  org = 0x80000000,  len = 0x00000110
  SRAM_EXT2:  org = 0x80000110,  len = 0x00100000
  FLASH:      org = 0x90000000,  len = 0x00020000
}

SECTIONS
{
  circdata :> SRAM_EXT1 /*buffer in external mem*/
  .vecs    :> IVECS     /*Created in vectors file*/
  .text    :> IRAM      /*Created by C Compiler*/
  .bss     :> IRAM
  .cinit   :> IRAM
  .stack   :> IRAM
  .sysmem  :> IRAM
  .const   :> IRAM
  .switch  :> IRAM
  .far     :> IRAM
  .cio     :> IRAM
  .csldata :> IRAM
}
```

FIGURE 4.37. Linker command file for a circular buffer in external memory (FIRcirc_ext.cmd).

4.8 ASSIGNMENTS

1. **(a)** Design a 65-coefficient FIR lowpass filter with a cutoff frequency of 2500 Hz and a sampling frequency of 8 kHz. Implement it in real time using the Hamming window function.

   (b) Compare the filter's characteristics between the Hamming, Hanning, and Kaiser windows.

2. The coefficient file LP1500_256.cof (in the folder **FIR**) represents 256 coefficients of an FIR lowpass filter, with a bandwidth of 1500 Hz, when sampling at 48 kHz. Implement this filter to achieve this 1500-Hz bandwidth. Hint: the C-coded examples in this chapter may not be efficient enough to implement this filter at a sampling rate of 48 kHz (what about an ASM-coded FIR function with a circular buffer to update the delays?).

3. Design and implement a multiband FIR filter with two passbands, one centered at 2500 and the other at 3500 Hz. Select a sampling frequency of 16 kHz.

4. In lieu of using an internal noise generator coded in C as input to a C-coded FIR function (see *FIRPRN*), use the input noise generated in ASM code (see *noisegen_casm*).

5. In lieu of using an internal noise generator coded in C as input to a C-coded FIR function (see *FIRPRN*), use the input noise generated in ASM code (see *noisegen_casm*) to an ASM-coded FIR function.

REFERENCES

1. W. J. Gomes III and R. Chassaing, Filter design and implementation using the TMS320C6x interfaced with MATLAB, *Proceedings of the 2000 ASEE Annual Conference*, 2000.

2. A. V. Oppenheim and R. Schafer, *Discrete-Time Signal Processing*, Prentice Hall, Upper Saddle River, NJ, 1989.

3. B. Gold and C. M. Rader, *Digital Signal Processing of Signals*, McGraw-Hill, New York, 1969.

4. L. R. Rabiner and B. Gold, *Theory and Application of Digital Signal Processing*, Prentice Hall, Upper Saddle River, NJ, 1975.

5. T. W. Parks and J. H. McClellan, Chebychev approximation for nonrecursive digital filter with linear phase, *IEEE Transactions on Circuit Theory*, Vol. CT-19, 1972, pp. 189–194.

6. J. H. McClellan and T. W. Parks, A unified approach to the design of optimum linear phase digital filters, *IEEE Transactions on Circuit Theory*, Vol. CT-20, 1973, pp. 697–701.

7. J. F. Kaiser, Nonrecursive digital filter design using the I0-sinh window function, *Proceedings of the IEEE International Symposium on Circuits and Systems*, 1974.

8. J. F. Kaiser, Some practical considerations in the realization of linear digital filters, *Proceedings of the 3rd Allerton Conference on Circuit System Theory*, Oct. 1965, pp. 621–633.

9. L. B. Jackson, *Digital Filters and Signal Processing*, Kluwer Academic, Norwell, MA, 1996.

10. J. G. Proakis and D. G. Manolakis, *Digital Signal Processing: Principles, Algorithms, and Applications*, Prentice Hall, Upper Saddle River, NJ, 1996.

11. R. G. Lyons, *Understanding Digital Signal Processing*, Addison-Wesley, Reading, MA, 1997.

12. F. J. Harris, On the use of windows for harmonic analysis with the discrete Fourier transform, *Proceedings of the IEEE*, Vol. 66, 1978, pp. 51–83.

13. I. F. Progri, W. R. Michalson, and R. Chassaing, Fast and efficient filter design and implementation on the TMS320C6711 digital signal processor, *International Conference on Acoustics, Speech, and Signal Processing Student Forum*, May 2001.

14. B. Porat, *A Course in Digital Signal Processing*, Wiley, New York, 1997.

15. T. W. Parks and C. S. Burrus, *Digital Filter Design*, Wiley, New York, 1987.

16. S. D. Stearns and R. A. David, *Signal Processing in Fortran and C*, Prentice Hall, Upper Saddle River, NJ, 1993.

17. N. Ahmed and T. Natarajan, *Discrete-Time Signals and Systems*, Reston Publishing, Reston, VA, 1983.

18. S. J. Orfanidis, *Introduction to Signal Processing*, Prentice Hall, Upper Saddle River, NJ, 1996.

19. A. Antoniou, *Digital Filters: Analysis, Design, and Applications*, McGraw-Hill, New York, 1993.

20. E. C. Ifeachor and B. W. Jervis, *Digital Signal Processing: A Practical Approach*, Addison-Wesley, Reading, MA, 1993.

21. P. A. Lynn and W. Fuerst, *Introductory Digital Signal Processing with Computer Applications*, Wiley, New York, 1994.

22. R. D. Strum and D. E. Kirk, *First Principles of Discrete Systems and Digital Signal Processing*, Addison-Wesley, Reading, MA, 1988.

23. D. J. DeFatta, J. G. Lucas, and W. S. Hodgkiss, *Digital Signal Processing: A System Approach*, Wiley, New York, 1988.

24. C. S. Williams, *Designing Digital Filters*, Prentice Hall, Upper Saddle River, NJ, 1986.

25. R. W. Hamming, *Digital Filters*, Prentice Hall, Upper Saddle River, NJ, 1983.

26. S. K. Mitra and J. F. Kaiser, eds., *Handbook for Digital Signal Processing*, Wiley, New York, 1993.

27. S. K. Mitra, *Digital Signal Processing: A Computer-Based Approach*, McGraw-Hill, New York, 2001.

28. R. Chassaing, B. Bitler, and D. W. Horning, Real-time digital filters in C, *Proceedings of the 1991 ASEE Annual Conference*, June 1991.

29. R. Chassaing and P. Martin, Digital filtering with the floating-point TMS320C30 digital signal processor, *Proceedings of the 21st Annual Pittsburgh Conference on Modeling and Simulation*, May 1990.

30. S. D. Stearns and R. A. David, *Signal Processing in Fortran and C*, Prentice Hall, Upper Saddle River, NJ, 1993.

31. R. A. Roberts and C. T. Mullis, *Digital Signal Processing*, Addison-Wesley, Reading, MA, 1987.

32. E. P. Cunningham, *Digital Filtering: An Introduction*, Houghton Mifflin, Boston, 1992.

33. N. J. Loy, *An Engineer's Guide to FIR Digital Filters*, Prentice Hall, Upper Saddle River, NJ, 1988.

34. H. Nuttall, Some windows with very good sidelobe behavior, *IEEE Transactions on Acoustics, Speech, and Signal Processing*, Vol. ASSP-29, No. 1, Feb. 1981.

35. L. C. Ludemen, *Fundamentals of Digital Signal Processing*, Harper & Row, New York, 1986.

36. M. Bellanger, *Digital Processing of Signals: Theory and Practice*, Wiley, New York, 1989.

37. M. G. Bellanger, *Digital Filters and Signal Analysis*, Prentice Hall, Upper Saddle River, NJ, 1986.

38. F. J. Taylor, *Principles of Signals and Systems*, McGraw-Hill, New York, 1994.

39. F. J. Taylor, *Digital Filter Design Handbook*, Marcel Dekker, New York, 1983.

40. W. D. Stanley, G. R. Dougherty, and R. Dougherty, *Digital Signal Processing*, Reston Publishing, Reston, VA, 1984.

41. R. Kuc, *Introduction to Digital Signal Processing*, McGraw-Hill, New York, 1988.

42. H. Baher, *Analog and Digital Signal Processing*, Wiley, New York, 1990.

43. J. R. Johnson, *Introduction to Digital Signal Processing*, Prentice Hall, Upper Saddle River, NJ, 1989.

44. S. Haykin, *Modern Filters*, Macmillan, New York, 1989.

45. T. Young, *Linear Systems and Digital Signal Processing*, Prentice Hall, Upper Saddle River, NJ, 1985.

46. A. Ambardar, *Analog and Digital Signal Processing*, PWS, Boston, MA, 1995.

47. A. W. M. van den Enden and N. A. M. Verhoeckx, *Discrete-Time Signal Processing*, Prentice-Hall International, Hemel Hempstead, Hertfordshire, England, 1989.

48. A. H. Gray and J. D. Markel, Digital lattice and ladder filter synthesis, *IEEE Transactions on Acoustics, Speech, and Signal Processing*, Vol. ASSP-21, Dec. 1973, pp. 491–500.

49. A. H. Gray and J. D. Markel, A normalized digital filter structure, *IEEE Transactions on Acoustics, Speech, and Signal Processing*, Vol. ASSP-23, June 1975, pp. 258–277.

50. MATLAB, MathWorks, Natick, MA, 2003.

51. R. Chassaing and D. W. Horning, *Digital Signal Processing with the TMS320C25*, Wiley, New York, 1990.

5

Infinite Impulse Response Filters

- Infinite impulse response filter structures: direct form I, direct form II, cascade, parallel, and lattice
- Bilinear transformation for filter design
- Sinusoidal waveform generation using difference equation
- Filter design and utility packages
- Programming examples using TMS320C6x and C code

The FIR filter discussed in Chapter 4 has no analog counterpart. In this chapter we discuss the infinite impulse response (IIR) filter that makes use of the vast knowledge already acquired with analog filters. The design procedure involves the conversion of an analog filter to an equivalent discrete filter using the bilinear transformation (BLT) technique. As such, the BLT procedure converts a transfer function of an analog filter in the s-domain into an equivalent discrete-time transfer function in the z-domain.

5.1 INTRODUCTION

Consider a general input–output equation of the form

$$y(n) = \sum_{k=0}^{N} a_k x(n-k) - \sum_{j=1}^{M} b_j y(n-j) \tag{5.1}$$

$$= a_0 x(n) + a_1 x(n-1) + a_2 x(n-2) + \cdots + a_N x(n-N)$$
$$- b_1 y(n-1) - b_2 y(n-2) - \cdots - b_M y(n-M) \tag{5.2}$$

Digital Signal Processing and Applications with the C6713 and C6416 DSK   By Rulph Chassaing
ISBN 0-471-69007-4   Copyright © 2005 by John Wiley & Sons, Inc.

This recursive type of equation represents an IIR filter. The output depends on the inputs as well as past outputs (with feedback). The output $y(n)$, at time n, depends not only on the current input $x(n)$, at time n, and on past inputs $x(n-1)$, $x(n-2)$, ..., $x(n-N)$, but also on past outputs $y(n-1)$, $y(n-2)$, ..., $y(n-M)$.

If we assume all initial conditions to be zero in (5.2), the z-transform of (5.2) becomes

$$Y(z) = a_0 X(z) + a_1 z^{-1} X(z) + a_2 z^{-2} X(z) + \cdots + a_N z^{-N} X(z)$$
$$- b_1 z^{-1} Y(z) - b_2 z^{-2} Y(z) - \cdots - b_M z^{-M} Y(z) \tag{5.3}$$

Let $N = M$ in (5.3); then the transfer function $H(z)$ is

$$H(z) = \frac{Y(z)}{X(z)} = \frac{a_0 + a_1 z^{-1} + a_2 z^{-2} + \cdots + a_N z^{-N}}{1 + b_1 z^{-1} + b_2 z^{-2} + \cdots + b_N z^{-N}} = \frac{N(z)}{D(z)} \tag{5.4}$$

where $N(z)$ and $D(z)$ represent the numerator and denominator polynomial, respectively. Multiplying and dividing by z^N, $H(z)$ becomes

$$H(z) = \frac{a_0 z^N + a_1 z^{N-1} + a_2 z^{N-2} + \cdots + a_N}{z^N + b_1 z^{N-1} + b_2 z^{N-2} + \cdots + b_N} = C \prod_{i=1}^{N} \frac{z - z_i}{z - p_i} \tag{5.5}$$

which is a transfer function with N zeros and N poles. If all the coefficients b_j in (5.5) are zero, this transfer function reduces to the transfer function with N poles at the origin in the z-plane representing the FIR filter discussed in Chapter 4. For a system to be stable, all the poles must reside inside the unit circle, as discussed in Chapter 4. Hence, for an IIR filter to be stable, the magnitude of each of its poles must be less than 1, or:

1. If $|P_i| < 1$, then $h(n) \to 0$, as $n \to \infty$, yielding a stable system.
2. If $|P_i| > 1$, then $h(n) \to \infty$, as $n \to \infty$, yielding an unstable system.

If $|P_i| = 1$, the system is marginally stable, yielding an oscillatory response. Furthermore, multiple-order poles on the unit circle yield an unstable system. Note again that with all the coefficients $b_j = 0$, the system reduces to a nonrecursive and stable FIR filter.

5.2 IIR FILTER STRUCTURES

There are several structures that can represent an IIR filter, as discussed next.

5.2.1 Direct Form I Structure

With the direct form I structure shown in Figure 5.1, the filter in (5.2) can be realized. There is an implied summer (not shown) in Figure 5.1. For an Nth-order filter,

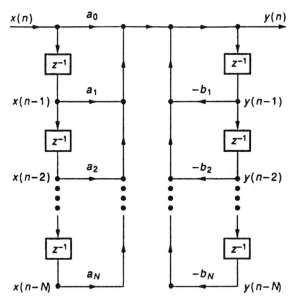

FIGURE 5.1. Direct form I IIR filter structure.

this structure has $2N$ delay elements, represented by z^{-1}. For example, a second-order filter with $N = 2$ will have four delay elements.

5.2.2 Direct Form II Structure

The direct form II structure shown in Figure 5.2 is one of the most commonly used structures. It requires half as many delay elements as the direct form I. For example, a second-order filter requires two delay elements z^{-1}, as opposed to four with the direct form I. To show that (5.2) can be realized with the direct form II, let a delay variable $U(z)$ be defined as

$$U(z) = \frac{X(z)}{D(z)} \tag{5.6}$$

where $D(z)$ is the denominator polynomial of the transfer function in (5.4). From (5.4) and (5.6), $Y(z)$ becomes

$$Y(z) = \frac{N(z)X(z)}{D(z)} = N(z)U(z)$$
$$= U(z)(a_0 + a_1 z^{-1} + a_2 z^{-2} + \cdots + a_N z^{-N}) \tag{5.7}$$

where $N(z)$ is the numerator polynomial of the transfer function in (5.4). From (5.6)

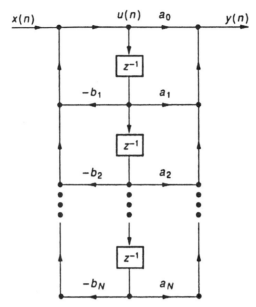

FIGURE 5.2. Direct form II IIR filter structure.

$$X(z) = U(z)D(z) = U(z)\left(1 + b_1 z^{-1} + b_2 z^{-2} + \cdots + b_N z^{-N}\right) \tag{5.8}$$

Taking the inverse z-transform of (5.8) yields

$$x(n) = u(n) + b_1 u(n-1) + b_2 u(n-2) + \cdots + b_N u(n-N) \tag{5.9}$$

Solving for $u(n)$ in (5.9) gives us

$$u(n) = x(n) - b_1 u(n-1) - b_2 u(n-2) - \cdots - b_N u(n-N) \tag{5.10}$$

Taking the inverse z-transform of (5.7) yields

$$y(n) = a_0 u(n) + a_1 u(n-1) + a_2 u(n-2) + \cdots + a_N u(n-N) \tag{5.11}$$

The direct form II structure can be represented by (5.10) and (5.11). The delay variable $u(n)$ at the middle top of Figure 5.2 satisfies (5.10), and the output $y(n)$ in Figure 5.2 satisfies (5.11).

Equations (5.10) and (5.11) are used to program an IIR filter. Initially, $u(n-1)$, $u(n-2), \ldots$ are set to zero. At time n, a new sample $x(n)$ is acquired, and (5.10) is used to solve for $u(n)$. The filter's output at time n then becomes

$$y(n) = a_0 u(n) + 0$$

At time $n + 1$, a newer sample $x(n + 1)$ is acquired and the delay variables in (5.10) are updated, or

$$u(n+1) = x(n+1) - b_1 u(n) - 0$$

where $u(n - 1)$ is updated to $u(n)$. From (5.11), the output at time $n + 1$ is

$$y(n+1) = a_0 u(n+1) + a_1 u(n) + 0$$

and so on, for time $n + 2, n + 3, \ldots$, when, for each specific time, a new input sample is acquired and the delay variables and then the output are calculated using (5.10) and (5.11), respectively.

5.2.3 Direct Form II Transpose

The direct form II transpose structure is a modified version of the direct form II and requires the same number of delay elements. The following steps yield a transpose structure from a direct form II version:

1. Reverse the directions of all the branches.
2. Reverse the roles of the input and output (input $\leftrightarrow$ output).
3. Redraw the structure such that the input node is on the left and the output node is on the right (as is typically done).

The direct form II transpose structure is shown in Figure 5.3. To verify this, let $u_0(n)$ and $u_1(n)$ be as shown in Figure 5.3. Then, from the transpose structure,

$$u_0(n) = a_2 x(n) - b_2 y(n) \tag{5.12}$$

$$u_1(n) = a_1 x(n) - b_1 y(n) + u_0(n-1) \tag{5.13}$$

$$y(n) = a_0 x(n) + u_1(n-1) \tag{5.14}$$

Equation (5.13) becomes, using (5.12) to find $u_0(n - 1)$,

$$u_1(n) = a_1 x(n) - b_1 y(n) + [a_2 x(n-1) - b_2 y(n-1)] \tag{5.15}$$

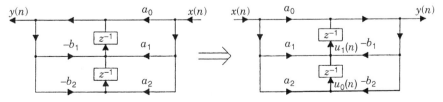

FIGURE 5.3. Direct form II transpose IIR filter structure.

Equation (5.14) becomes, using (5.15) to solve for $u_1(n-1)$,

$$y(n) = a_0 x(n) + [a_1 x(n-1) - b_1 y(n-1) + a_2 x(n-2) - b_2 y(n-2)] \qquad (5.16)$$

which is the same general I/O equation (5.2) for a second-order system. This transposed structure implements first the zeros and then the poles, whereas the direct form II structure implements the poles first.

5.2.4 Cascade Structure

The transfer function in (5.5) can be factored as

$$H(z) = CH_1(z)H_2(z) \cdots H_r(z) \qquad (5.17)$$

in terms of first- or second-order transfer functions. The cascade (or series) structure is shown in Figure 5.4. An overall transfer function can be represented with cascaded transfer functions. For each section, the direct form II structure or its transpose version can be used. Figure 5.5 shows a fourth-order IIR structure in terms of two direct form II second-order sections in cascade. The transfer function $H(z)$, in terms of cascaded second-order transfer functions, can be written as

$$H(z) = \prod_{i=1}^{N/2} \frac{a_{0i} + a_{1i} z^{-1} + a_{2i} z^{-2}}{1 + b_{1i} z^{-1} + b_{2i} z^{-2}} \qquad (5.18)$$

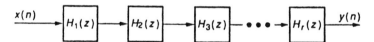

FIGURE 5.4. Cascade form IIR filter structure.

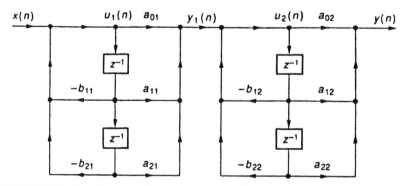

FIGURE 5.5. Fourth-order IIR filter with two direct form II sections in cascade.

where the constant C in (5.17) is incorporated into the coefficients, and each section is represented by i. For example, $N = 4$ for a fourth-order transfer function, and (5.18) becomes

$$H(z) = \frac{\left(a_{01} + a_{11}z^{-1} + a_{21}z^{-2}\right)\left(a_{02} + a_{12}z^{-1} + a_{22}z^{-2}\right)}{\left(1 + b_{11}z^{-1} + b_{21}z^{-2}\right)\left(1 + b_{12}z^{-1} + b_{22}z^{-2}\right)} \qquad (5.19)$$

as can be verified in Figure 5.5. From a mathematical standpoint, proper ordering of the numerator and denominator factors does not affect the output result. However, from a practical standpoint, proper ordering of each second-order section can minimize quantization noise [1–5]. Note that the output of the first section, $y_1(n)$, becomes the input to the second section. With an intermediate output result stored in one of the registers, a premature truncation of the intermediate output becomes negligible. A programming example will illustrate the implementation of an IIR filter cascaded into second-order direct form II sections.

5.2.5 Parallel Form Structure

The transfer function in (5.5) can be represented as

$$H(z) = C + H_1(z) + H_2(z) + \cdots + H_r(z) \qquad (5.20)$$

which can be obtained using a partial fraction expansion (PFE) on (5.5). This parallel form structure is shown in Figure 5.6. Each of the transfer functions $H_1(z)$,

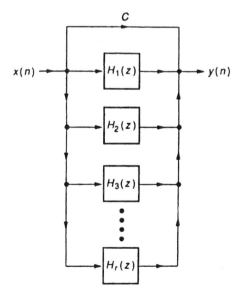

FIGURE 5.6. Parallel form IIR filter structure.

$H_2(z), \ldots$ can be either first- or second-order functions. As with the cascade structure, the parallel form can be efficiently represented in terms of second-order direct form II structure sections. $H(z)$ can be expressed as

$$H(z) = C + \sum_{i=1}^{N/2} \frac{a_{0i} + a_{1i}z^{-1} + a_{2i}z^{-2}}{1 + b_{1i}z^{-1} + b_{2i}z^{-2}} \tag{5.21}$$

For example, for a fourth-order transfer function, $H(z)$ in (5.21) becomes

$$H(z) = C + \frac{a_{01} + a_{11}z^{-1} + a_{21}z^{-2}}{1 + b_{11}z^{-1} + b_{21}z^{-2}} + \frac{a_{02} + a_{12}z^{-1} + a_{22}z^{-2}}{1 + b_{12}z^{-1} + b_{22}z^{-2}} \tag{5.22}$$

This fourth-order parallel structure is represented in terms of two direct form II sections as shown in Figure 5.7. From Figure 5.7, the output $y(n)$ can be expressed in terms of the output of each section, or

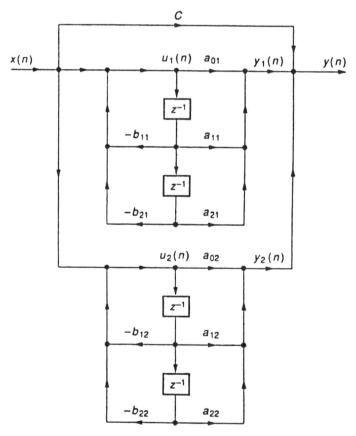

FIGURE 5.7. Fourth-order IIR filter with two direct form II sections in parallel.

$$y(n) = Cx(n) + \sum_{i=1}^{N/2} y_i(n) \tag{5.23}$$

The quantization error associated with the coefficients of an IIR filter depends on the amount of shift in the position of its transfer function's poles and zeros in the complex plane. This implies that the shift in the position of a particular pole depends on the positions of all the other poles. To minimize this dependency of poles, an Nth-order IIR filter is typically implemented as cascaded second-order sections.

5.2.6 Lattice Structure

The lattice structure is used in applications such as adaptive filtering and speech processing.

All-Pole Lattice Structure

We discussed the lattice structure in the previous chapter, where we derived the k-parameters for an FIR or "all-zero" filter (except for poles at $z = 0$). Consider now an all-pole lattice structure associated with an IIR filter. This system is the inverse of the all-zero FIR lattice of Figure 4.3, with N poles (except for zeros at $z = 0$). A solution for this system can be developed from the results obtained with the FIR lattice structure. We can solve (4.52) and (4.53) backwards, computing $y_{i-1}(n)$ in terms of $y_i(n)$, and so on. For example, (4.52) becomes

$$y_{i-1}(n) = y_i(n) - k_i e_{i-1}(n-1) \tag{5.24}$$

and (4.53) is repeated here as

$$e_i(n) = k_i y_{i-1}(n) + e_{i-1}(n-1) \tag{5.25}$$

Equations (5.24) and (5.25) are represented by the ith section lattice structure in Figure 5.8, which can be extended for a higher-order all-pole IIR lattice structure. For example, given the transfer function of an IIR filter with all poles, the reciprocal would be the transfer function of an FIR filter with all zeros. We also want to make sure that this IIR system is stable by having all the poles inside the unit circle. It can be shown that this is so if $|k_i| < 1$, $i = 1, 2, \ldots, N$. Therefore, we can test the stability of a system by using the recursive equation (4.49) to find the k-parameters and check that each $|k_i| < 1$.

Exercise 5.1: All-Pole Lattice Structure

The lattice structure for an all-pole system can be found. Let the transfer function be

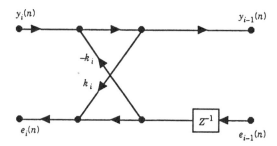

FIGURE 5.8. All-pole IIR lattice filter structure for the ith section.

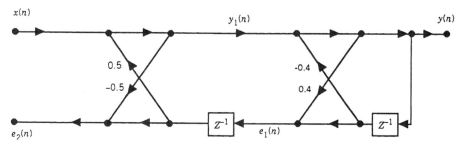

FIGURE 5.9. All-pole IIR lattice filter structure with two sections.

$$H(z) = \frac{1}{1 + 0.2z^{-1} - 0.5z^{-2}} \qquad (5.26)$$

This transfer function is the inverse of the transfer function for the all-zero FIR lattice structure in Exercise 4.3, where the k-parameters were found to be

$$k_1 = 0.4$$
$$k_2 = -0.5$$

Figure 5.9 shows the IIR lattice structure for this example, extending Figure 5.8 to a two-stage structure.

IIR Lattice Structure with Poles and Zeros

For an IIR lattice structure with poles and zeros, the previous results for all-zero and all-pole structures can be used. The notation used for the coefficients must be changed to reflect both the numerator and denominator polynomials in an IIR system. Figure 5.10 shows the IIR lattice structure with both poles and zeros. It shows a ladder (bottom half) portion added to the all-pole structure. A set of coefficients c_i, expressed in terms of both the numerator (a_i) and denominator (b_i)

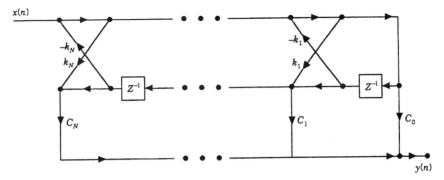

FIGURE 5.10. Nth-order IIR lattice filter structure with both poles and zeros.

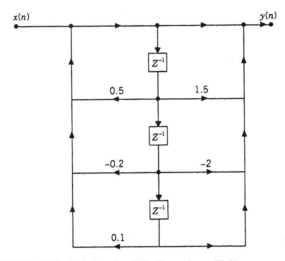

FIGURE 5.11. Third-order IIR direct form II filter structure.

coefficients, can be computed recursively,

$$c_i = a_i - \sum_{r=i+1}^{N} c_r b_{r(r-i)}, \qquad i = 0, 1, \ldots, N \qquad (5.27)$$

A more thorough discussion can be found in [6] and [7].

Exercise 5.2: Lattice Structure with Poles and Zeros

This exercise converts a third-order IIR direct form II structure into a lattice structure. Figure 5.11 shows a third-order IIR filter using the direct form II, and Figure

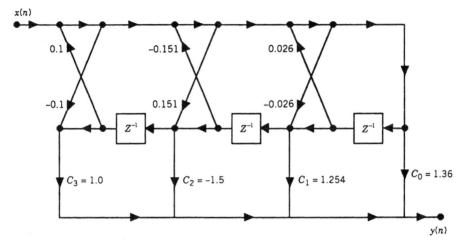

FIGURE 5.12. Third-order IIR lattice filter structure with poles and zeros.

5.12 shows the equivalent IIR lattice structure. The transfer function from Figure 5.11 is

$$H(z) = \frac{1 + 1.5z^{-1} - 2z^{-2} + z^{-3}}{1 - 0.5z^{-1} + 0.2z^{-2} - 0.1z^{-3}} \tag{5.28}$$

Using the results associated with an all-pole structure, and changing the coefficients a_i into b_i to reflect the denominator polynomial, (4.45) becomes

$$Y_3(z) = 1 + b_{31}z^{-1} + b_{32}z^{-2} + b_{33}z^{-3} = 1 - 0.5z^{-1} + 0.2z^{-2} - 0.1z^{-3}$$

Starting with $r = 3$, we have

$$k_3 = b_{33} = -0.1$$

Using (4.49), with $r = 3$ and $i = 0$, we have

$$b_{20} = \frac{b_{30} - k_3 b_{33}}{1 - k_3^2} = \frac{1 - (-0.1)(-0.1)}{1 - (-0.1)^2} = 1$$

For $r = 3$ and $i = 1$, we have

$$b_{21} = \frac{b_{31} - k_3 b_{32}}{1 - k_3^2} = \frac{(-0.5) - (-0.1)(0.2)}{1 - (-0.1)^2} = -0.0303$$

and, for $i = 2$,

$$b_{22} = \frac{b_{32} - k_3 b_{31}}{1 - k_3^2} = \frac{(0.2) - (-0.1)(-0.5)}{1 - (-0.1)^2} = 0.1515$$

from which

$$k_2 = b_{22} = 0.1515$$

From (4.45), with $r = 2$ and $i = 1$,

$$Y_2(z) = 1 + b_{21}z^{-1} + b_{22}z^{-2} = 1 + (-0.0303)z^{-1} + (0.1515)z^{-2}$$

From (4.49), with $r = 2$ and $i = 1$,

$$b_{11} = \frac{b_{21} - k_2 b_{21}}{1 - k_2^2} = \frac{(-0.0303) - (0.1515)(-0.0303)}{1 - (0.1515)^2} = -0.0263$$

from which

$$k_1 = b_{11} = -0.0263$$

The k-parameters $k_1, k_2,$ and k_3 provide the solution for the top half of the IIR lattice structure in Figure 5.12. We can now use the recursive relationship in (5.27) to compute the c_i coefficients that will give us the bottom part of the structure in Figure 5.12. We will now use both a's and b's in applying (4.49). Here, from the numerator polynomial (with a_{ri} replaced by a_i) in (5.28),

$$a_0 = 1$$
$$a_1 = 1.5$$
$$a_2 = -2$$
$$a_3 = 1$$

and, from the denominator polynomial in (5.28),

$$b_{31} = -0.5$$
$$b_{32} = 0.2$$
$$b_{33} = -0.1$$

Starting with c_3 and working backwards using (5.27), the coefficients c_i can be found, or

$$c_3 = a_3 = 1$$
$$c_2 = a_2 - \{c_3 b_{31}\} = -2 - 1(-0.5) = -1.5$$
$$c_1 = a_1 - \{c_2 b_{21} + c_3 b_{32}\}$$
$$= 1.5 - \{(-1.5)(-0.0303) + (1)(0.2)\}$$
$$= 1.2545$$
$$c_0 = a_0 - \{c_1 b_{11} + c_2 b_{22} + c_3 b_{33}\}$$
$$= 1 - \{(1.2545)(-0.0263) + (-1.5)(0.1515) + (1)(-0.1)\}$$
$$= 1.3602$$

The lattice structure can be quite useful for applications in adaptive filtering and speech processing. Although this structure is not as computationally efficient as the direct or cascade forms, requiring more multiplication operations, it is less sensitive to quantization effects [6–8].

5.3 BILINEAR TRANSFORMATION

The BLT is the most commonly used technique for transforming an analog filter into a discrete filter. It provides one-to-one mapping from the analog s-plane to the digital z-plane, using

$$s = K\frac{z-1}{z+1} \tag{5.29}$$

The constant K in (5.29) is commonly chosen as $K = 2/T$, where T represents a sampling variable. Other values for K can be selected, since it has no consequence in the design procedure. We choose $T = 2$ or $K = 1$ for convenience to illustrate the BLT procedure. Solving for z in (5.29) gives us

$$z = \frac{1+s}{1-s} \tag{5.30}$$

This transformation allows the following:

1. The left region in the s-plane, corresponding to $\sigma < 0$, maps *inside* the unit circle in the z-plane.
2. The right region in the s-plane, corresponding to $\sigma > 0$, maps *outside* the unit circle in the z-plane.
3. The imaginary $j\omega$ axis in the s-plane maps *on* the unit circle in the z-plane.

Let ω_A and ω_D represent the analog and digital frequencies, respectively. With $s = j\omega_A$ and $z = e^{j\omega_D T}$, (5.29) becomes

$$j\omega_A = \frac{e^{j\omega_D T} - 1}{e^{j\omega_D T} + 1} = \frac{e^{j\omega_D T/2}\left(e^{j\omega_D T/2} - e^{-j\omega_D T/2}\right)}{e^{j\omega_D T/2}\left(e^{j\omega_D T/2} + e^{-j\omega_D T/2}\right)} \tag{5.31}$$

Using Euler's expressions for sine and cosine in terms of complex exponential functions, ω_A from (5.31) becomes

$$\omega_A = \tan\frac{\omega_D T}{2} \tag{5.32}$$

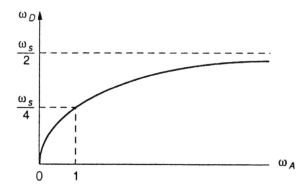

FIGURE 5.13. Relationship between analog and digital frequencies.

which relates the analog frequency ω_A to the digital frequency ω_D. This relationship is plotted in Figure 5.13 for positive values of ω_A. The region corresponding to ω_A between 0 and 1 is mapped into the region corresponding to ω_D between 0 and $\omega_s/4$ in a fairly linear fashion, where ω_s is the sampling frequency in radians. However, the entire region of $\omega_A > 1$ is quite nonlinear, mapping into the region corresponding to ω_D between $\omega_s/4$ and $\omega_s/2$. This compression within this region is referred to as *frequency warping*. As a result, prewarping is done to compensate for this frequency warping. The frequencies ω_A and ω_D are such that

$$H(s)\big|_{s=j\omega_A} = H(z)\big|_{z=e^{j\omega_D T}} \tag{5.33}$$

5.3.1 BLT Design Procedure

The BLT design procedure makes use of a known analog transfer function for the design of a discrete-time filter. It can be applied using well-documented analog filter functions (Butterworth, Chebychev, etc.). Several types of filter design are available with MATLAB, described in Appendix D. Butterworth filters are maximally flat in the passband and in the stopband. Chebyshev types I and II provide equiripple responses in the passbands and stopbands, respectively. For a given specification, these filters are of lower order than Butterworth-type filters, which have monotonic responses in both passbands and stopbands. Chebyshev filters have sharper cutoff frequencies than Butterworth filters, but at the expense of ripples in the passband (type I) or in the stopband (type II). They are useful in applications requiring sharp transitions while tolerating the ripples. An elliptic design has equiripple in both bands and achieves a lower order than a Chebyshev-type design; however, it is more difficult to design, with a highly non-linear phase response in the passbands. Although a Butterworth design requires a higher order, it has a linear phase in the passbands.

Perform the following steps in order to use the BLT technique and find $H(z)$.

1. Obtain a known analog transfer function $H(s)$.

2. Prewarp the desired digital frequency ω_D to obtain the analog frequency ω_A in (5.32).

3. Scale the frequency of the analog transfer function $H(s)$ selected, using

$$H(s)|_{s=s/\omega_A} \qquad (5.34)$$

4. Obtain $H(z)$ using the BLT equation (5.29), or

$$H(z) = H(s/\omega_A)|_{s=(z-1)/(z+1)} \qquad (5.35)$$

In the case of bandpass and bandstop filters with lower and upper cutoff frequencies ω_{D1} and ω_{D2}, the two analog frequencies ω_{A1} and ω_{A2} need to be solved. The exercises in Appendix D further illustrate the BLT procedure.

5.4 PROGRAMMING EXAMPLES USING C AND ASM CODE

Several examples are introduced to illustrate the implementation of an IIR filter using the cascaded direct form II structure and the generation of a tone using a difference equation. An example illustrates the generation of a tone with an assembly-coded function.

Example 5.1: IIR Filter Implementation Using Second-Order Stages in Cascade (IIR)

Figure 5.14 shows a listing of the program IIR.c that implements a generic IIR filter using cascaded second-order stages (sections). The program uses the following two equations associated with each stage (see equations 5.10 and 5.11 for a second-order):

```
u(n)  =  x(n)  -  b₁u(n - 1)  -  b₂u(n - 2)
y(n)  =  a₀u(n)  +  a₁u(n - 1)  +  a₂u(n - 2)
```

The loop section of code within the program is processed five times (the number of stages) for each value of n, or sample period. For the first stage, x(n) is the newly acquired input sample. However, for the other stages, the input x(n) is the output y(n) of the preceding stage.

The coefficients b[i][0] and b[i][1] correspond to b_1 and b_2, respectively; where i represents each stage. The delays dly[i][0] and dly[i][1] correspond to $u(n - 1)$ and $u(n - 2)$, respectively.

```
//IIR.c  IIR filter using cascaded Direct Form II
//Coefficients a's and b's correspond to b's and a's from MATLAB

#include "DSK6713_AIC23.h"         //codec-DSK support file
Uint32 fs=DSK6713_AIC23_FREQ_8KHZ; //set sampling rate
#include "bs1750.cof"              //BS @ 1750 Hz coefficient file
short dly[stages][2] = {0};        //delay samples per stage

interrupt void c_int11()           //ISR
{
 short  i, input;
 int  un, yn;
 input = input_sample();           //input to 1st stage
 for (i = 0; i < stages; i++)      //repeat for each stage
 {
  un=input-((b[i][0]*dly[i][0])>>15) - ((b[i][1]*dly[i][1])>>15);
  yn=((a[i][0]*un)>>15)+((a[i][1]*dly[i][0])>>15)+((a[i][2]*dly[i][1])>>15);
  dly[i][1] = dly[i][0];           //update delays
  dly[i][0] = un;                  //update delays
  input = yn;                      //intermed out->in to next stage
 }
  output_sample((short)yn);        //output final result for time n
  return;                          //return from ISR
}

void main()
{
  comm_intr();                     //init DSK, codec, McBSP
  while(1);                        //infinite loop
}
```

FIGURE 5.14. IIR filter program using second-order sections in cascade (IIR.c).

IIR Bandstop

The coefficient file *bs1750.cof* (Figure 5.15) is obtained from Appendix D. It represents a tenth-order IIR bandstop filter designed with MATLAB's filter designer SPTool, as shown in Figure D.2 in Appendix D. Note that MATLAB's filter designer shows the order as 5, which represents the number of second-order stages. The coefficient file contains the numerator coefficients, a's (three per stage), and the denominator coefficients, b's (two per stage). The a's and b's used in this book correspond to the b's and a's used in MATLAB.

Build and run this project as **IIR**. Verify that the output is an IIR bandstop filter centered at 1750 Hz. Figure 5.16 shows the output frequency response of this IIR bandstop filter obtained with an HP signal analyzer (with noise as the input).

IIR Bandpass and Lowpass

1. Rebuild this project using the coefficient file *bp2000.cof* (on the accompanying CD), which represents a 36th-order (18 stages) Chebyshev type 2 IIR bandpass filter centered at 2 kHz. This filter was designed with MATLAB, as

```
//bs1750.cof IIR bandstop coefficient file, centered at 1,750 Hz

#define stages 5                //number of 2nd-order stages

int a[stages][3]=        {      //numerator coefficients
{27940, -10910, 27940},         //a10, a11, a12 for 1st stage
{32768, -11841, 32768},         //a20, a21, a22 for 2nd stage
{32768, -13744, 32768},         //a30, a31, a32 for 3rd stage
{32768, -11338, 32768},         //a40, a41, a42 for 4th stage
{32768, -14239, 32768}  };

int b[stages][2]=        {      //denominator coefficients
{-11417, 25710},                //b11, b12 for 1st stage
{-9204, 31581},                 //b21, b22 for 2nd stage
{-15860, 31605},                //b31, b32 for 3rd stage
{-10221, 32581},                //b41, b42 for 4th stage
{-15258, 32584}          };     //b51, b52 for 5th stage
```

FIGURE 5.15. Coefficient file for a tenth-order IIR bandstop filter designed with MATLAB in Appendix D (bs1750.cof).

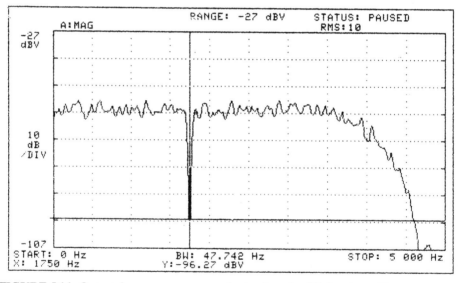

FIGURE 5.16. Output frequency response of a tenth-order IIR bandstop filter centered at 1750 Hz, obtained with an HP signal analyzer.

shown in Figure 5.17. Verify that the filter's output is an IIR bandpass filter centered at 2 kHz. Figure 5.18 shows the output frequency response of this 36th-order IIR bandpass filter, obtained with an HP analyzer.

2. Rebuild this project using the coefficient file *lp2000.cof* (on the CD), which represents an eighth-order IIR lowpass filter with a 2-kHz cutoff frequency (also designed with MATLAB). Verify the output of this IIR lowpass filter.

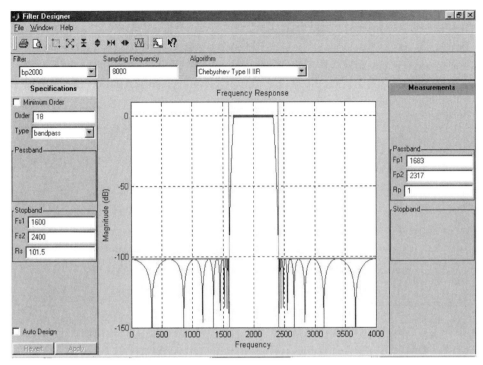

FIGURE 5.17. MATLAB's filter designer (SPTool) displaying frequency characteristics of a 36th-order IIR bandpass filter.

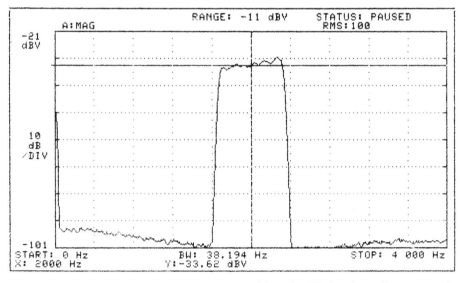

FIGURE 5.18. Output frequency response of a 36th-order IIR bandpass filter centered at 2000 Hz, obtained with an HP signal analyzer.

Example 5.2: Generation of Two Tones Using Two Second-Order Difference Equations (two_tones)

This example generates and adds two tones using a difference equation scheme. The output is also stored in memory and plotted within CCS. The difference equation to generate a sine wave is

$$y(n) = Ay(n-1) - y(n-2)$$

where

$$A = 2\cos(\omega T)$$
$$y(-1) = -\sin(\omega T)$$
$$y(-2) = -\sin(2\omega T)$$

with two initial conditions, $y(-1)$ and $y(-2)$, $\omega = 2\pi f$, and $T = 1/F_s = 1/(8\,\text{kHz}) = 0.125\,\text{ms}$, the sampling period. The z-transform of $y(n)$ is

$$Y(z) = A\{z^{-1}Y(z) + y(-1)\} - \{z^{-2}Y(z) + z^{-1}y(-1) + y(-2)\}$$

which can be written as

$$\begin{aligned} Y(z)\{1 - Az^{-1} + z^{-2}\} &= Ay(-1) - z^{-1}y(-1) - y(-2) \\ &= -2\cos(\omega T)\sin(\omega T) + z^{-1}\sin(\omega T) + \sin(2\omega T) \\ &= z^{-1}\sin(\omega T) \end{aligned}$$

Solving for $Y(z)$ yields

$$Y(z) = z\sin(\omega T)/(z^2 - Az + 1)$$

The inverse z-transform of $Y(z)$ is

$$y(n) = ZT^{-1}\{Y(z)\} = \sin(n\omega T)$$

$f = 1.5\,kHz$

$$\begin{aligned} A = 2\cos(\omega T) &= 0.765 \rightarrow A \times 2^{14} &= 12{,}540 \\ y(-1) = -\sin(\omega T) &= -0.924 \rightarrow y(-1) \times 2^{14} &= -15{,}137 \\ y(-2) = -\sin(2\omega T) &= -0.707 \rightarrow y(-2) \times 2^{14} &= -11{,}585 \end{aligned}$$

$f = 2\,kHz$

$$A = 0$$
$$y(-1) = -1 \rightarrow y(-1) \times 2^{14} = -16{,}384$$
$$y(-2) = 0$$

The coefficient of the second-order difference equation A, along with the two initial conditions, determine the frequency generated. They are scaled for a fixed-point implementation. Using the difference equation

$$y(n) = Ay(n-1) - y(n-2)$$

the output at time $n = 0$ is

$$y(0) = Ay(-1) - y(-2) = -2\cos(\omega T)\sin(\omega T) + \sin(2\omega T) = 0$$

```
//two_tones.c Generates/adds two tones using difference equations

#include "DSK6713_AIC23.h"              //codec-DSK support file
Uint32 fs=DSK6713_AIC23_FREQ_8KHZ;      //set sampling rate
short sinegen(void);                    //for generating tone
short output;                           //for output
short sinegen_buffer[256];              //buffer for output data
const short bufferlength = 256;         //buffer size for plot with CCS
short i = 0;                            //buffer count index
short y1[3] = {0,-15137,-11585};        //y1(0),y1(-1),y1(-2) for 1.5kHz
const short A1 = 12540;                 //A1 = 2coswT scaled by 2^14
short y2[3] = {0,-16384,0};             //y2(0),y2(-1),y2(-2) for 2kHz
const short A2 = 0;                     //A2 = 2coswT scaled by 2^14

interrupt void c_int11()                //ISR
{
 output = sinegen();                    //out from tone generation function
 sinegen_buffer[i] = output;            //output into buffer
 output_sample(output);                 //output result
 i++;                                   //increment buffer count
 if (i == bufferlength) i = 0;          //if buffer count=size of buffer
 return;                                //return to main
}

short sinegen()                         //function to generate tone
{
 y1[0] =((y1[1]*A1)>>14)-y1[2];         //y1(n)= A1*y1(n-1)-y1(n-2)
 y1[2] = y1[1];                         //update y1(n-2)
 y1[1] = y1[0];                         //update y1(n-1)
 y2[0] =((y2[1]*A2)>>14)-y2[2];         //y2(n)= A2*y2(n-1)-y2(n-2)
 y2[2] = y2[1];                         //update y2(n-2)
 y2[1] = y2[0];                         //update y2(n-1)

 return (y1[0] + y2[0]);                //add the two tones
}

void main()
{
 comm_intr();                           //init DSK, codec, McBSP
 while(1);                              //infinite loop
}
```

FIGURE 5.19. Program to generate and add two tones (two_tones.c).

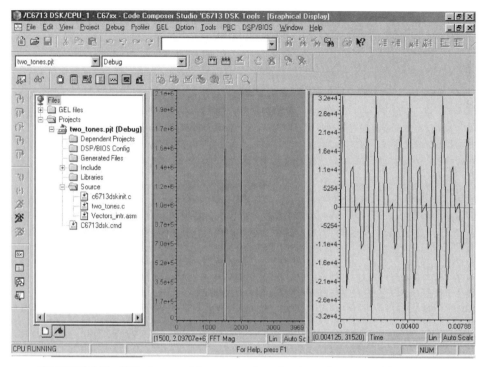

FIGURE 5.20. CCS time- and frequency-domain plots of output with two tones.

Figure 5.19 shows a listing of the program *two_tones.c* that implements a tone generation using a difference equation. The array y1[3] contains the values for y1(0), y1(−1), and y1(−2) to generate a 1.5-kHz tone, and the array y2[3] contains the values for y2(0), y2(−1), and y2(−2) to generate a 2-kHz tone. The function *sinegen* uses the second-order difference equation to generate each tone, then adds the two tones. Scaling by 2^{14} allows for a fixed-point implementation.

Build and run this project as **two_tones**. Verify that the output is the sum of the 1.5- and 2-kHz tones. The output is also stored in a memory buffer. Use CCS to plot the FFT magnitude of the two sinusoids, as shown in Figure 5.20. The starting address of the buffer is sinegen_buffer (see also Example 1.2). Figure 5.20 also shows the time-domain plot of the two sinusoids.

The technique above can be used to generate dual-tone multifrequency: for example, generating and adding the two tones with frequencies of 697 and 1209 Hz, which correspond to the key "3" in a phone.

Example 5.3: Sine Generation Using a Difference Equation (sinegenDE)

This example also generates a sinusoidal tone using an alternative difference equation. See also Example 5.2, which generates/adds two tones. Consider the second-order difference equation obtained in Chapter 4:

$$y(n) = Ay(n-1) + By(n-2) + Cx(n-1)$$

where $B = -1$. Apply an impulse at $n = 1$, so that $x(n - 1) = x(0) = 1$, and 0 otherwise. For $n = 1$,

$$y(1) = Ay(0) + By(-1) + Cx(0) = C$$

with $y(0) = 0$ and $y(-1) = 0$. For $n \geqslant 2$,

$$y(n) = Ay(n-1) - y(n-2).$$

The coefficients $A = 2\cos(\omega T)$ and $C = \sin(\omega T)$ are calculated for a given sampling period $T = 1/F_s$ and a desired frequency ω.

$f = 1.5\,kHz$

$$A = 2\cos(\omega T) = 0.765 \rightarrow A \times 2^{14} = 12{,}540$$
$$y(1) = C = 0.924 \rightarrow C \times 2^{14} = 15{,}137$$
$$y(2) = Ay(1) = 0.707 \rightarrow y(2) \times 2^{14} = 11{,}585$$

$f = 2\,kHz$

$$A = 2\cos(\omega T) = 0$$
$$y(1) = C = \sin(\omega T) = 1 \rightarrow C \times 2^{14} = 16{,}384$$
$$y(2) = Ay(1) - y(0) = AC = 0$$

Figure 5.21 shows a listing of the program $sinegenDE.c$, which generates a sine wave using this alternative difference equation. This difference equation is calculated within the ISR using an alternative scheme to the implementation in Example 5.2. The coefficient A = 0, and the array y[3], which contains $y(0)$, $y(1)$, and $y(2)$, generate a 2-kHz sine wave.

Build and run this project as **sinegenDE**. Verify that the output is a 2-kHz tone. Change the array to y[3]={0,15,137,11,585} and A=12,540. Rebuild/run the program and verify a 1.5-kHz tone generated at the output. A 3-kHz tone can be generated using $A = -23{,}170$ and y[3]={0,11,585,0}.

```
//SinegenDE.c Generates a sinewave using a difference equation

#include "dsk6713_aic23.h"             //codec-DSK file support
Uint32 fs=DSK6713_AIC23_FREQ_8KHZ;    //set sampling rate
short y[3] = {0,16384,0};              //y(0),y(1),y(2)
const short A = 0;                     //A = 2*coswT * 2^14
int n = 2;

interrupt void c_int11()               //ISR
{
 y[n] = ((A*y[n-1])>>14) - y[n-2];     //y(n)=Ay(n-1)-y(n-2)
 y[n-2] = y[n-1];                      //update y(n-2)
 y[n-1] = y[n];                        //update y(n-1)
 output_sample(y[n]);                  //output result
 return;                               //return to main
}

void main()
{
 comm_intr();                          //init DSK, codec, McBSP
 while(1);                             //infinite loop
}
```

FIGURE 5.21. Program to generate a sine wave using a difference equation (sinegenDE.c).

Example 5.4: Generation of a Swept Sinusoid Using a Difference Equation (sweepDE)

Figure 5.22 shows a listing of the program sweepDE. c, which generates a sinusoidal signal, sweeping in frequency. The program implements the difference equation

$$y(n) = Ay(n-1) - y(n-2)$$

where $A = 2\cos(\omega T)$ and the two initial conditions are $y(-1) = \sin(\omega T)$ and $y(-2) = -\sin(2\omega T)$. Example 5.2 illustrates the generation of a sine wave using this difference equation, and Example 2.15 implements a swept sinusoid using an 8000-point lookup table.

An initial signal frequency is set in the program at 500 Hz. The signal's frequency is incremented by 10 Hz until a set maximum frequency of 3500 Hz is reached. The duration of the sinusoidal signal at each frequency generated is set with 200 and can be reduced for a faster sweep.

With an initial frequency of 500 Hz, the constants A $= 30,274$, $y(0) = 0$, $y(-1) = -6270$ and $y(-2) = -11,585$ (see Example 5.2). For each frequency (510, 520, ...) the function coeff_gen is called to calculate a new set of constants A, $y(n-1)$, $y(n-2)$ to implement the difference equation. A slider can be used to control the swept signal, such as the step or incremental frequency and the duration of the sinusoidal signal at each incremental frequency.

```
//SweepDE.c   Generates a sweeping sinusoid using a difference equation

#include "DSK6713_AIC23.h"          //codec-DSK support file
Uint32 fs=DSK6713_AIC23_FREQ_8KHZ;  //set sampling rate
#include <math.h>
#define   two_pi    (2*3.1415926)    //2*pi
#define   two_14          16384      //2^14
#define   T            0.000125      //sample period = 1/Fs
#define   MIN_FREQ         500       //initial frequency of sweep
#define   MAX_FREQ        3500       //max frequency of sweep
#define   STEP_FREQ         10       //step frequency
#define   SWEEP_PERIOD     200       //lasting time at one frequency
short     y0  = 0;                   //initial output
short     y_1 = -6270;               //y(-1)=-sinwT(scaled)  f=500 Hz
short     y_2 = -11585;              //y(-2_=-sin2wT(scaled) f=500 Hz
short     A = 30274;                 //A = 2*coswT scaled by 2^14
short     freq = MIN_FREQ;           //current frequency
short     sweep_count = 0;           //counter for lasting time
void      coeff_gen(short);          //function prototype generate coeff

interrupt void c_int11()             //ISR
{
 sweep_count++;                      //inc lasting time at one frequency
 if(sweep_count >= SWEEP_PERIOD)     //lasting time reaches max duration
   {
   if(freq>=MAX_FREQ)  freq=MIN_FREQ;//if current freq is max reinit
   else freq=freq + STEP_FREQ;       //incr to next higher frequency

   coeff_gen(freq);                  //function for new set of coeff
   sweep_count = 0;                  //reset counter for lasting time
   }

 y0=((A * y_1)>>14) - y_2; //y(n) = A*y(n-1) - y(n-2)
 y_2 = y_1;                          //update y(n-2)
 y_1 = y0;                           //update y(n-1)
 output_sample(y0);                  //output result
}

void coeff_gen(short freq)           //calculate new set of coeff
{
 float w;                            //angular frequency

 w = two_pi*freq;                    //w = 2*pi*f
 A = 2*cos(w*T)*two_14;              //A = 2*coswT * (2^14)
 y_1 = -sin(w*T)*two_14;             //y_1 = -sinwT *(2^14)
 y_2 = -sin(2*T*w)*two_14;           //y_2 = -sin2wT * (2^14)
 return;
}

void main()
{
 comm_intr();                        //init DSK, codec, McBSP
 while(1);                           //infinite loop
}
```

FIGURE 5.22. Program to generate a sweeping sinusoid using a difference equation (sweepDE.c).

Build and run this project as **sweepDE**. Verify that the output is a swept sinusoidal signal.

Example 5.5: IIR Inverse Filter (IIRinverse)

This example illustrates an IIR inverse filter. With noise as input, a forward IIR filter is calculated. The output of the forward filter becomes the input to an inverse IIR filter. The output of the inverse filter is the original input noise sequence. See Example 4.9, which implements an inverse FIR filter, and Example 5.1, which implements an IIR filter.

The transfer function of an IIR filter is

$$H(z) = \frac{\displaystyle\sum_{i=0}^{N-1} a_i z^{-i}}{\displaystyle\sum_{j=1}^{M-1} b_j z^{-j}}$$

The output sequence of the IIR filter is

$$y(n) = \sum_{i=0}^{N-1} a_i x(n-i) - \sum_{j=1}^{M-1} b_j y(n-j)$$

where $x(n-i)$ represents the input sequence. The input sequence $x(n)$ can then be recovered using $\hat{x}(n)$ as an estimate of $x(n)$, or

$$\hat{x}(n) = \frac{y(n) + \displaystyle\sum_{j=1}^{M-1} b_j y(n-j) - \sum_{i=1}^{N-1} a_i \hat{x}(n-i)}{a_0}$$

The program IIRinverse.c (Figure 5.23) implements the inverse IIR filter.

Build this project as **IIRinverse**. Use noise as input to the system. Run the program and verify that the resulting output is the input noise (with the slider in the default position 1).

Change the slider to position 2 and verify that the output of the forward IIR filter is an IIR bandpass filter centered at 2 kHz. The coefficient file *bp2000.cof* was used in Example 5.1 to implement an IIR filter. With the slider in position 3, verify that the output of the inverse IIR filter is the original input noise.

In this example, the forward filter's characteristics are known. This example can be extended so that the filter's characteristics are unknown. In such a case, the unknown forward filter's coefficients, a's and b's, can be estimated using Prony's method [9].

```
//IIRinverse.C  Inverse IIR Filter

#include "dsk6713_AIC23.h"            //codec-DSK support file
Uint32 fs=DSK6713_AIC23_FREQ_8KHZ;   //set sampling rate
#include "bp2000.cof"                 //BP @ 2 kHz coefficient file
short dly[stages][2] = {0};           //delay samples per stage
short out_type = 1;                   //type of output for slider
short a0, a1, a2, b1, b2;             //coefficients

interrupt void c_int11()             //ISR
{
 short i, input, input1;
 int un1, yn1, un2, input2, yn2;

 input1 = input_sample();            //input to 1st stage
 input = input1;                     //original input
 for(i = 0; i < stages; i++)         //repeat for each stage
   {
   a1 = ((a[i][1]*dly[i][0])>>15);   //a1*u(n-1)
   a2 = ((a[i][2]*dly[i][1])>>15);   //a2*u(n-2)
   b1 = ((b[i][0]*dly[i][0])>>15);   //b1*u(n-1)
   b2 = ((b[i][1]*dly[i][1])>>15);   //b2*u(n-2)
   un1 = input1 - b1 - b2;
   a0=((a[i][0]*un1)>>15);
   yn1 = a0 + a1 + a2;               //stage output
   input1 = yn1;                     //intermediate out->in next stage
   dly[i][1] = dly[i][0];            //update delays u(n-2) = u(n-1)
   dly[i][0] = un1;                  //update delays u(n-1) = u(n)
   }
 input2 = yn1;                       //out forward=in reverse filter

 for(i = stages; i > 0; i--)         //for inverse IIR filter
   {
   a1 = ((a[i][1]*dly[i][0])>>15);   //a1u(n-1)
   a2 = ((a[i][2]*dly[i][1])>>15);   //a2u(n-2)
   b1 = ((b[i][0]*dly[i][0])>>15);   //b1u(n-1)
   b2 = ((b[i][1]*dly[i][1])>>15);   //b2u(n-2)
   un2 = input2 - a1 - a2;
   yn2 = (un2 + b1 + b2);
   input2 = (yn2<<15)/a[i][0];       //intermediate out->in next stage
   }
 if(out_type == 1)                   //if slider in position 1
    output_sample(input);            //original input signal
 if(out_type == 2) output_sample((short)yn1);//forward filter
 if(out_type == 3) output_sample((short)(yn2>>9));//inverse filter
 return;                             //return from ISR
}

void main()
{
 comm_intr();                        //init DSK, codec, McBSP
 while(1);                           //infinite loop
}
```

FIGURE 5.23. Program to implement an inverse IIR filter (IIRinverse.c).

```
//Sinegencasm.c Sine generation using DE calling ASM function

#include "dsk6713_aic23.h"              //codec-DSK support file
Uint32 fs=DSK6713_AIC23_FREQ_8KHZ;      //set sampling rate
short y[3] = {0, 15137,11585};          //y(1)=sinwT (f=1.5kHz)
short A =12540;                         //A=2*coswT * 2^14
short n = 2;

interrupt void c_int11()                //interrupt service routine
{
      sinegencasmfunc(&y[0], A);        //calls ASM function
      output_sample(y[n]);
      return;
}

void main()
{
  comm_intr();                          //init DSK, codec, McBSP
  while(1);                             //infinite loop
}
```

FIGURE 5.24. C source program that calls an ASM function to generate a sine wave using a difference equation (sinegencasm.c).

```
;Sinegencasmfunc.asm ASM function to generate sine using DE
;A4 = address of y array, B4 = A

            .def    _sinegencasmfunc        ;ASM function called from C
_sinegencasmfunc:
            LDH     *+A4[0], A5             ;y[n-2]-->A5
            LDH     *+A4[1], A2             ;y[n-1]-->A2
            LDH     *+A4[2], A3             ;y[n]-->A3
            NOP     3                       ;NOP due to LDH
            MPY     B4, A2, A8              ;A*y[n-1]
            NOP     1                       ;NOP due to MPY
            SHR     A8, 14, A8              ;shift right by 14
            SUB     A8, A5, A8              ;A*y[n-1]-y[n-2]
            STH     A8, *+A4[2]             ;y[n]=A*y[n-1]-y[n-2]
            STH     A2, *+A4[0]             ;y[n-2]=y[n-1]
            STH     A8, *+A4[1]             ;y[n-1] = y[n]
            B       B3                      ;return addr to call routine
            NOP     5                       ;delays due to branching
            .end
```

FIGURE 5.25. ASM function called from C to generate a sine wave using a difference equation (sinegencasmfunc.asm).

Example 5.6: Sine Generation Using a Difference Equation with C Calling an ASM Function (Sinegencasm)

This example is based on Example 5.3 with a C source program calling an ASM function to generate a sine wave using a difference equation. Figure 5.24 shows the C source program *Sinegencasm.c* calling the ASM function *Sinegencasmfunc.asm* shown in Figure 5.25. The C source program shows the array y[3], which contains the values y(0), y(1), and y(2) and the coefficient A, calculated to generate a 1.5-kHz sine wave. The address of the array y[3], along with the coefficient A, is passed to the ASM function through A4 and B4, respectively. The values in the array y[3] and the coefficient A were scaled by 2^{14} to allow for a fixed-point implementation. As a result, within the ASM function, A8 initially containing $Ay(n-1)$ is scaled back (shifted right) by 2^{14}.

Build this project as **Sinegencasm**. Verify that a 1.5-kHz sine wave is generated. Verify that changing the array to y[3] = {0, 16384, 0} and $A = 0$ yields a 2-kHz sine wave.

5.5 ASSIGNMENTS

1. Design and implement in real time a 12th-order IIR lowpass filter using a Chebyshev type 2, with a cutoff frequency of 1700 Hz and a sampling frequency of 8 kHz. Compare/discuss the characteristics of this filter in terms of:

   (a) the order of the filter

   (b) the filter's type with an Elliptical and Butterworth design. If it has been designed with MATLAB, then illustrate with the appropriate frequency responses from MATLAB plots.

2. Write a program using a difference equation to generate a swept sinusoidal waveform with a frequency range between 400 and 3700 Hz.

3. Write a program using a difference equation to generate a 1-kHz cosine wave. Similar to the second-order difference equation used in Example 5.3 to generate a sine wave, the following equation generates a cosine wave:

$$y(n) = Ay(n-1) + By(n-2) + x(n) - (A/2) \times (n-1)$$

with zero initial conditions, and $A = 2\cos wT$ and $B = -1$. Assume an impulse at $n = 0$, such that $x(0) = 1$ and $y(-1) = y(-2) = 0$. Then

$$n = 0: \quad y(0) = Ay(-1) + By(-2) + x(0) - (A/2)x(-1) = 1$$
$$n = 1: \quad y(1) = Ay(0) + By(-1) + x(1) - (A/2)x(0) = A - A/2$$
$$n = 2: \quad y(2) = Ay(1) + By(0) + 0 = A(A - A/2) + B$$

.   .

.   .

.   .

4. Three sets of coefficients associated with a fourth-, a sixth-, and an eighth-order IIR filter were obtained using the DigiFilter package in Appendix E. The center and sampling frequencies for each filter are 1250 and 10,000 Hz, respectively. The filter design uses cascaded direct form II sections. Implement an IIR filter in real time with each set of coefficients, using a sampling rate of 8 kHz (not 10 kHz). Determine the center frequencies of these filters and compare the Butterworth filters in terms of selectivity (sharpness).

(a) Fourth-Order Elliptic

	First Stage	Second Stage
a_0	0.078371	0.143733
a_1	−0.148948	0.010366
a_2	0.078371	0.143733
b_1	−1.549070	−1.228110
b_2	0.968755	0.960698

(b) Sixth-Order Butterworth

	First Stage	Second Stage	Third Stage
a_0	0.137056	0.122159	0.122254
a_1	0.0	0.0	0.0
a_2	−0.137056	−0.122159	−0.122254
b_1	−1.490630	−1.152990	−1.256790
b_2	0.886387	0.856946	0.755492

(c) Eighth-Order Butterworth

	First Stage	Second Stage	Third Stage	Fourth Stage
a_0	0.123118	0.130612	0.127179	0.143859
a_1	0.0	0.0	0.0	0.0
a_2	−0.123118	−0.130612	−0.127179	−0.143859
b_1	−1.18334	−1.33850	−1.15014	−1.52176
b_2	0.754301	0.777976	0.884409	0.910547

REFERENCES

1. L. B. Jackson, *Digital Filters and Signal Processing*, Kluwer Academic, Norwell, MA, 1996.

2. L. B. Jackson, Roundoff noise analysis for fixed-point digital filters realized in cascade or parallel form, *IEEE Transactions on Audio and Electroacoustics*, Vol. AU-18, June 1970, pp. 107–122.

3. L. B. Jackson, An analysis of limit cycles due to multiplicative rounding in recursive digital filters, *Proceedings of the 7th Allerton Conference on Circuit and System Theory*, 1969, pp. 69–78.

4. L. B. Lawrence and K. V. Mirna, A new and interesting class of limit cycles in recursive digital filters, *Proceedings of the IEEE International Symposium on Circuit and Systems*, Apr. 1977, pp. 191–194.

5. R. Chassaing and D. W. Horning, *Digital Signal Processing with the TMS320C25*, Wiley, New York, 1990.

6. A. H. Gray and J. D. Markel, Digital lattice and ladder filter synthesis, *IEEE Transactions on Acoustics, Speech, and Signal Processing*, Vol. ASSP-21, 1973, pp. 491–500.

7. A. H. Gray and J. D. Markel, A normalized digital filter structure, *IEEE Transactions on Acoustics, Speech, and Signal Processing*, Vol. ASSP-23, 1975, pp. 268–277.

8. A. V. Oppenheim and R. Schafer, *Discrete-Time Signal Processing*, Prentice Hall, Upper Saddle River, NJ, 1989.

9. I. Progri, W. R. Michalson, and R. Chassaing, Fast and efficient filter design and implementation on the TMS320C6711 digital signal processor, *International Conference on Acoustics Speech and Signal Processing Student Forum*, 2001.

10. N. Ahmed and T. Natarajan, *Discrete-Time Signals and Systems*, Reston Publishing, Reston, VA, 1983.

11. D. W. Horning and R. Chassaing, IIR filter scaling for real-time digital signal processing, *IEEE Transactions on Education*, Feb. 1991.

6

Fast Fourier Transform

- The fast Fourier transform using radix-2 and radix-4
- Decimation or decomposition in frequency and in time
- Programming examples

The fast Fourier transform (FFT) is an efficient algorithm that is used for converting a time-domain signal into an equivalent frequency-domain signal, based on the discrete Fourier transform (DFT). Several real-time programming examples on FFT are included.

6.1 INTRODUCTION

The DFT converts time-domain sequence into an equivalent frequency-domain sequence. The inverse DFT performs the reverse operation and converts a frequency-domain sequence into an equivalent time-domain sequence. The FFT is a very efficient algorithm technique based on the DFT but with fewer computations required. The FFT is one of the most commonly used operations in DSP to provide a frequency spectrum analysis [1–6]. Two different procedures are introduced to compute an FFT: the decimation-in-frequency and the decimation-in-time. Several variants of the FFT have been used, such as the Winograd transform [7,8], the discrete cosine transform (DCT) [9], and the discrete Hartley transform [10–12]. The fast Hartley transform (FHT) is described in Appendix F. Transform methods such

Digital Signal Processing and Applications with the C6713 and C6416 DSK   By Rulph Chassaing
ISBN 0-471-69007-4   Copyright © 2005 by John Wiley & Sons, Inc.

as the DCT have become increasingly popular in recent years, especially for real-time systems. They provide a large compression ratio.

6.2 DEVELOPMENT OF THE FFT ALGORITHM WITH RADIX-2

The FFT reduces considerably the computational requirements of the DFT. The DFT of a discrete-time signal $x(nT)$ is

$$X(k) = \sum_{n=0}^{N-1} x(n)W^{nk} \qquad k = 0, 1, \ldots, N-1 \tag{6.1}$$

where the sampling period T is implied in $x(n)$ and N is the frame length. The constants W are referred to as *twiddle constants* or *factors*, which represent the phase, or

$$W = e^{-j2\pi/N} \tag{6.2}$$

and are a function of the length N. Equation (6.1) can be written for $k = 0, 1, \ldots, N-1$, as

$$X(k) = x(0) + x(1)W^k + x(2)W^{2k} + \cdots + x(N-1)W^{(N-1)k} \tag{6.3}$$

This represents a matrix of $N \times N$ terms, since $X(k)$ needs to be calculated for N values for k. Since (6.3) is an equation in terms of a complex exponential, for each specific k there are $(N - 1)$ complex additions and N complex multiplications. This results in a total of $(N^2 - N)$ complex additions and N^2 complex multiplications. Hence, the computational requirements of the DFT can be very intensive, especially for large values of N. FFT reduces computational complexity from N^2 to $N \log N$.

The FFT algorithm takes advantage of the periodicity and symmetry of the twiddle constants to reduce the computational requirements of the FFT. From the periodicity of W,

$$W^{k+N} = W^k \tag{6.4}$$

and from the symmetry of W,

$$W^{k+N/2} = -W^k \tag{6.5}$$

Figure 6.1 illustrates the properties of the twiddle constants W for $N = 8$. For example, let $k = 2$, and note that from (6.4), $W^{10} = W^2$, and from (6.5), $W^6 = -W^2$.

For a radix-2 (base 2), the FFT decomposes an N-point DFT into two $(N/2)$-point or smaller DFTs. Each $(N/2)$-point DFT is further decomposed into two $(N/4)$-point DFTs, and so on. The last decomposition consists of $(N/2)$ two-point DFTs. The

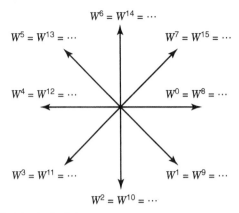

FIGURE 6.1. Periodicity and symmetry of twiddle constant W.

smallest transform is determined by the radix of the FFT. For a radix-2 FFT, N must be a power or base of 2, and the smallest transform or the last decomposition is the two-point DFT. For a radix-4, the last decomposition is a four-point DFT.

6.3 DECIMATION-IN-FREQUENCY FFT ALGORITHM WITH RADIX-2

Let a time-domain input sequence $x(n)$ be separated into two halves:

(a)
$$x(0), x(1), \ldots, x\left(\frac{N}{2}-1\right) \tag{6.6}$$

and

(b)
$$x\left(\frac{N}{2}\right), x\left(\frac{N}{2}+1\right), \ldots, x(N-1) \tag{6.7}$$

Taking the DFT of each set of the sequence in (6.6) and (6.7) gives us

$$X(k) = \sum_{n=0}^{(N/2)-1} x(n)W^{nk} + \sum_{n=N/2}^{N-1} x(n)W^{nk} \tag{6.8}$$

Let $n = n + N/2$ in the second summation of (6.8); $X(k)$ becomes

$$X(k) = \sum_{n=0}^{(N/2)-1} x(n)W^{nk} + W^{kN/2} \sum_{n=0}^{(N/2)-1} x\left(n+\frac{N}{2}\right)W^{nk} \tag{6.9}$$

where $W^{kN/2}$ is taken out of the second summation because it is not a function of n. Using

$$W^{kN/2} = e^{-jk\pi} = \left(e^{-j\pi}\right)^k = \left(\cos \pi - j\sin \pi\right)^k = (-1)^k$$

in (6.9), $X(k)$ becomes

$$X(k) = \sum_{n=0}^{(N/2)-1} \left[x(n) + (-1)^k x\left(n + \frac{N}{2}\right) \right] W^{nk} \tag{6.10}$$

Because $(-1)^k = 1$ for even k and -1 for odd k, (6.10) can be separated for even and odd k, or

1. For even k:

$$X(k) = \sum_{n=0}^{(N/2)-1} \left[x(n) + x\left(n + \frac{N}{2}\right) \right] W^{nk} \tag{6.11}$$

2. For odd k:

$$X(k) = \sum_{n=0}^{(N/2)-1} \left[x(n) - x\left(n + \frac{N}{2}\right) \right] W^{nk} \tag{6.12}$$

Substituting $k = 2k$ for even k, and $k = 2k + 1$ for odd k, (6.11) and (6.12) can be written for $k = 0, 1, \ldots, (N/2) - 1$ as

$$X(2k) = \sum_{n=0}^{(N/2)-1} \left[x(n) + x\left(n + \frac{N}{2}\right) \right] W^{2nk} \tag{6.13}$$

$$x(2k+1) = \sum_{n=0}^{(N/2)-1} \left[x(n) - x\left(n + \frac{N}{2}\right) \right] W^n W^{2nk} \tag{6.14}$$

Because the twiddle constant W is a function of the length N, it can be represented as W_N. Then W_N^2 can be written as $W_{N/2}$. Let

$$a(n) = x(n) + x(n + N/2) \tag{6.15}$$
$$b(n) = x(n) - x(n + N/2) \tag{6.16}$$

Equations (6.13) and (6.14) can be written more clearly as two $(N/2)$-point DFTs, or

$$X(2k) = \sum_{n=0}^{(N/2)-1} a(n) W_{N/2}^{nk} \tag{6.17}$$

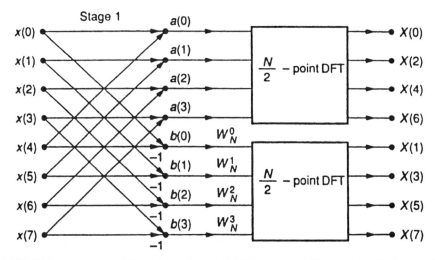

FIGURE 6.2. Decomposition of an N-point DFT into two $(N/2)$-point DFTs for $N = 8$.

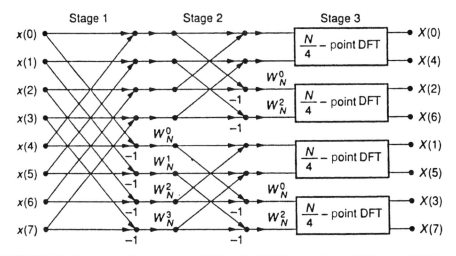

FIGURE 6.3. Decomposition of two $(N/2)$-point DFTs into four $(N/4)$-point DFTs for $N = 8$.

$$X(2k+1) = \sum_{n=0}^{(N/2)-1} b(n)W_N^n W_{N/2}^{nk} \qquad (6.18)$$

Figure 6.2 shows the decomposition of an N-point DFT into two $(N/2)$-point DFTs for $N = 8$. As a result of the decomposition process, the X's in Figure 6.2 are even in the upper half and odd in the lower half. The decomposition process can now be repeated such that each of the $(N/2)$-point DFTs is further decomposed into two $(N/4)$-point DFTs, as shown in Figure 6.3, again using $N = 8$ to illustrate.

The upper section of the output sequence in Figure 6.2 yields the sequence $X(0)$ and $X(4)$ in Figure 6.3, ordered as even. $X(2)$ and $X(6)$ from Figure 6.3 represent

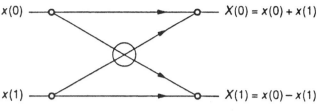

FIGURE 6.4. Two-point FFT butterfly.

the odd values. Similarly, the lower section of the output sequence in Figure 6.2 yields $X(1)$ and $X(5)$, ordered as the even values, and $X(3)$ and $X(7)$ as the odd values. This scrambling is due to the decomposition process. The final order of the output sequence $X(0), X(4), \ldots$ in Figure 6.3 is shown to be scrambled. The output needs to be resequenced or reordered. Programming examples presented later in this chapter include the appropriate function for resequencing. The output sequence $X(k)$ represents the DFT of the time sequence $x(n)$.

This is the last decomposition, since we now have a set of $(N/2)$ two-point DFTs, the lowest decomposition for a radix-2. For the two-point DFT, $X(k)$ in (6.1) can be written as

$$X(k) = \sum_{n=0}^{1} x(n)W^{nk} \qquad k = 0, 1 \tag{6.19}$$

or

$$X(0) = x(0)W^0 + x(1)W^0 = x(0) + x(1) \tag{6.20}$$

$$X(1) = x(0)W^0 - x(1)W^0 = x(0) - x(1) \tag{6.21}$$

since $W^1 = e^{-j2\pi/2} = -1$. Equations (6.20) and (6.21) can be represented by the flow graph in Figure 6.4, usually referred to as a *butterfly*. The final flow graph of an eight-point FFT algorithm is shown in Figure 6.5. This algorithm is referred to as *decimation-in-frequency* (DIF) because the output sequence $X(k)$ is decomposed (decimated) into smaller subsequences, and this process continues through M stages or iterations, where $N = 2^M$. The output $X(k)$ is complex with both real and imaginary components, and the FFT algorithm can accommodate either complex or real input values.

The FFT is not an approximation of the DFT. It yields the same result as the DFT with fewer computations required. This reduction becomes more and more important with higher-order FFT.

There are other FFT structures that have been used to illustrate the FFT. An alternative flow graph to that in Figure 6.5 can be obtained with ordered output and scrambled input.

An eight-point FFT is illustrated through the following exercise. We will see that flow graphs for higher-order FFT (larger N) can readily be obtained.

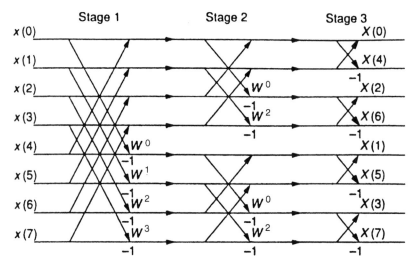

FIGURE 6.5. Eight-point FFT flow graph using DIF.

Exercise 6.1: Eight-Point FFT Using DIF

Let the input $x(n)$ represent a rectangular waveform, or $x(0) = x(1) = x(2) = x(3) = 1$ and $x(4) = x(5) = x(6) = x(7) = 0$. The eight-point FFT flow graph in Figure 6.5 can be used to find the output sequence $X(k)$, $k = 0, 1, \ldots, 7$. With $N = 8$, four twiddle constants need to be calculated, or

$$W^0 = 1$$
$$W^1 = e^{-j2\pi/8} = \cos(\pi/4) - j\sin(\pi/4) = 0.707 - j0.707$$
$$W^2 = e^{-j4\pi/8} = -j$$
$$W^3 = e^{-j6\pi/8} = -0.707 - j0.707$$

The intermediate output sequence can be found after each stage.

Stage 1

$$x(0) + x(4) = 1 \rightarrow x'(0)$$
$$x(1) + x(5) = 1 \rightarrow x'(1)$$
$$x(2) + x(6) = 1 \rightarrow x'(2)$$
$$x(3) + x(7) = 1 \rightarrow x'(3)$$
$$[x(0) - x(4)]W^0 = 1 \rightarrow x'(4)$$
$$[x(1) - x(5)]W^1 = 0.707 - j0.707 \rightarrow x'(5)$$
$$[x(2) - x(6)]W^2 = -j \rightarrow x'(6)$$
$$[x(3) - x(7)]W^3 = -0.707 - j0.707 \rightarrow x'(7)$$

where $x'(0), x'(1), \ldots, x'(7)$ represent the intermediate output sequence after the first iteration, which becomes the input to the second stage.

Stage 2

$$x'(0) + x'(2) = 2 \rightarrow x''(0)$$
$$x'(1) + x'(3) = 2 \rightarrow x''(1)$$
$$[x'(0) - x'(2)]W^0 = 0 \rightarrow x''(2)$$
$$[x'(1) - x'(3)]W^2 = 0 \rightarrow x''(3)$$
$$x'(4) + x'(6) = 1 - j \rightarrow x''(4)$$
$$x'(5) + x'(7) = (0.707 - j0.707) + (-0.707 - j0.707) = -j1.41 \rightarrow x''(5)$$
$$[x'(4) - x'(6)]W^0 = 1 + j \rightarrow x''(6)$$
$$[x'(5) - x'(7)]W^2 = -j1.41 \rightarrow x''(7)$$

The resulting intermediate, second-stage output sequence $x''(0), x''(1), \ldots, x''(7)$ becomes the input sequence to the third stage.

Stage 3

$$X(0) = x''(0) + x''(1) = 4$$
$$X(4) = x''(0) - x''(1) = 0$$
$$X(2) = x''(2) + x''(3) = 0$$
$$X(6) = x''(2) - x''(3) = 0$$
$$X(1) = x''(4) + x''(5) = (1 - j) + (-j1.41) = 1 - j2.41$$
$$X(5) = x''(4) - x''(5) = 1 + j0.41$$
$$X(3) = x''(6) + x''(7) = (1 + j) + (-j1.41) = 1 - j0.41$$
$$X(7) = x''(6) - x''(7) = 1 + j2.41$$

We now use the notation of X's to represent the final output sequence. The values $X(0), X(1), \ldots, X(7)$ form the scrambled output sequence. These results can be verified with MATLAB, as described in Appendix D. We show later how to reorder the output sequence and plot the output magnitude.

Exercise 6.2: Sixteen-Point FFT

Given $x(0) = x(1) = \ldots = x(7) = 1$, and $x(8) = x(9) = \ldots x(15) = 0$, which represents a rectangular input sequence. The output sequence can be found using the 16-point flow graph shown in Figure 6.6. The intermediate output results after each stage are found in a manner similar to that in Exercise 6.1. Eight twiddle constants W^0, $W^1, \ldots, W^7$ need to be calculated for $N = 16$.

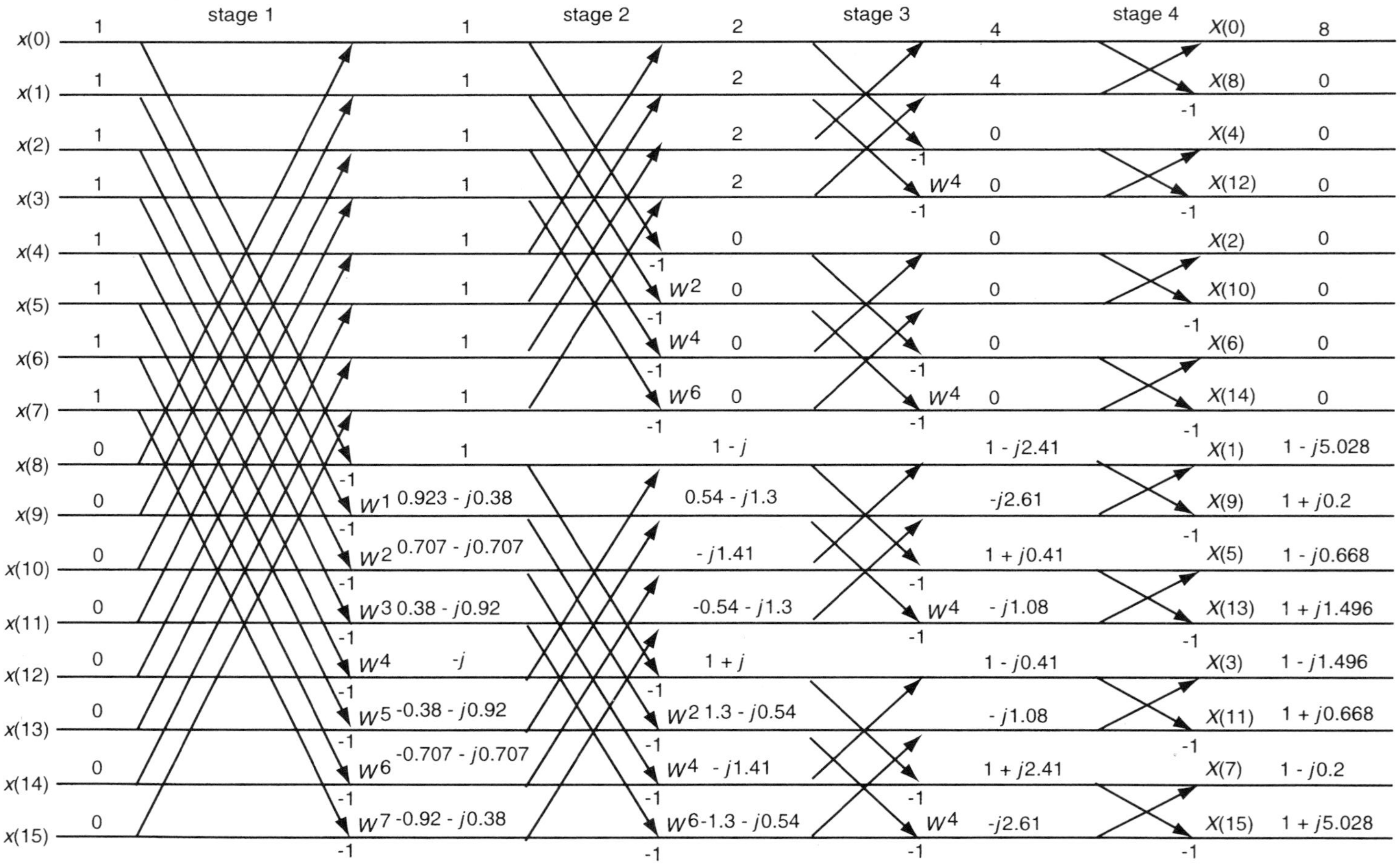

FIGURE 6.6. Sixteen-point FFT flow graph using DIF.

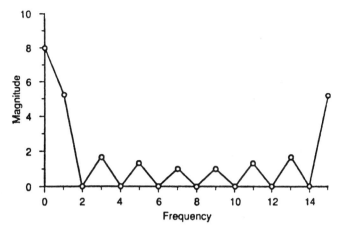

FIGURE 6.7. Output magnitude for 16-point FFT.

Verify the scrambled output sequence X's as shown in Figure 6.6. Reorder this output sequence and take its magnitude. Verify the plot in Figure 6.7, which represents a sinc function. The output $X(8)$ represents the magnitude at the Nyquist frequency. These results can be verified with MATLAB, as described in Appendix D.

6.4 DECIMATION-IN-TIME FFT ALGORITHM WITH RADIX-2

Whereas the DIF process decomposes an output sequence into smaller subsequences, *decimation-in-time* (DIT) is a process that decomposes the input sequence into smaller subsequences. Let the input sequence be decomposed into an even sequence and an odd sequence, or

$$x(0), x(2), x(4), \ldots, x(2n)$$

and

$$x(1), x(3), x(5), \ldots, x(2n+1)$$

We can apply (6.1) to these two sequences to obtain

$$X(k) = \sum_{n=0}^{(N/2)-1} x(2n)W^{2nk} + \sum_{n=0}^{(N/2)-1} x(2n+1)W^{(2n+1)k} \tag{6.22}$$

Using $W_N^2 = W_{N/2}$ in (6.22) yields

$$X(k) = \sum_{n=0}^{(N/2)-1} x(2n)W_{N/2}^{nk} + W_N^k \sum_{n=0}^{(N/2)-1} x(2n+1)W_{N/2}^{nk} \tag{6.23}$$

which represents two $(N/2)$-point DFTs. Let

$$C(k) = \sum_{n=0}^{(N/2)-1} x(2n)W_{N/2}^{nk} \tag{6.24}$$

$$D(k) = \sum_{n=0}^{(N/2)-1} X(2n+1)W_{N/2}^{nk} \tag{6.25}$$

Then $X(k)$ in (6.23) can be written as

$$X(k) = C(k) + W_N^k D(k) \tag{6.26}$$

Equation (6.26) needs to be interpreted for $k > (N/2) - 1$. Using the symmetry property (6.5) of the twiddle constant, $W^{k+N/2} = -W^k$,

$$X(k+N/2) = C(k) - W^k D(k) \qquad k = 0, 1, \ldots, (N/2) - 1 \tag{6.27}$$

For example, for $N = 8$, (6.26) and (6.27) become

$$X(k) = C(k) + W^k D(k) \qquad k = 0, 1, 2, 3 \tag{6.28}$$
$$X(k+4) = C(k) - W^k D(k) \qquad k = 0, 1, 2, 3 \tag{6.29}$$

Figure 6.8 shows the decomposition of an eight-point DFT into two four-point DFTs with the DIT procedure. This decomposition or decimation process is repeated so that each four-point DFT is further decomposed into two two-point DFTs, as shown in Figure 6.9. Since the last decomposition is $(N/2)$ two-point DFTs, this is as far as this process goes.

Figure 6.10 shows the final flow graph for an eight-point FFT using a DIT process. The input sequence is shown to be scrambled in Figure 6.10 in the same manner as the output sequence $X(k)$ was scrambled during the DIF process. With the input

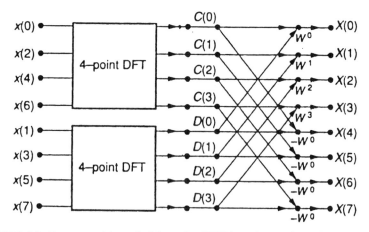

FIGURE 6.8. Decomposition of eight-point DFT into four-point DFTs using DIT.

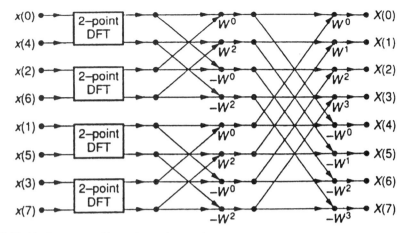

FIGURE 6.9. Decomposition of two four-point DFTs into four two-point DFTs using DIT.

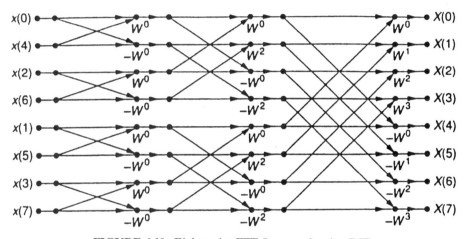

FIGURE 6.10. Eight-point FFT flow graph using DIT.

sequence $x(n)$ scrambled, the resulting output sequence $X(k)$ becomes properly ordered. Identical results are obtained with an FFT using either the DIF or the DIT process. An alternative DIT flow graph to the one shown in Figure 6.10, with ordered input and scrambled output, can also be obtained.

The following exercise shows that the same results are obtained for an eight-point FFT with the DIT process as in Exercise 6.1 with the DIF process.

Exercise 6.3: Eight-Point FFT Using DIT

Given the input sequence $x(n)$ representing a rectangular waveform as in Exercise 6.1, the output sequence $X(k)$, using the DIT flow graph in Figure 6.10, is the same as in Exercise 6.1. The twiddle constants are the same as in Exercise 6.1. Note that the twiddle constant W is multiplied with the second term only (not with the first).

Stage 1

$$x(0) + W^0 x(4) = 1 + 0 = 1 \rightarrow x'(0)$$
$$x(0) - W^0 x(4) = 1 - 0 = 1 \rightarrow x'(4)$$
$$x(2) + W^0 x(6) = 1 + 0 = 1 \rightarrow x'(2)$$
$$x(2) - W^0 x(6) = 1 - 0 = 1 \rightarrow x'(6)$$
$$x(1) + W^0 x(5) = 1 + 0 = 1 \rightarrow x'(1)$$
$$x(1) - W^0 x(5) = 1 - 0 = 1 \rightarrow x'(5)$$
$$x(3) + W^0 x(7) = 1 + 0 = 1 \rightarrow x'(3)$$
$$x(3) - W^0 x(7) = 1 - 0 = 1 \rightarrow x'(7)$$

where the sequence x's represents the intermediate output after the first iteration and becomes the input to the subsequent stage.

Stage 2

$$x'(0) + W^0 x'(2) = 1 + 1 = 2 \rightarrow x''(0)$$
$$x'(4) + W^2 x'(6) = 1 + (-j) = 1 - j \rightarrow x''(4)$$
$$x'(0) - W^0 x'(2) = 1 - 1 = 0 \rightarrow x''(2)$$
$$x'(4) - W^2 x'(6) = 1 - (-j) = 1 + j \rightarrow x''(6)$$
$$x'(1) + W^0 x'(3) = 1 + 1 = 2 \rightarrow x''(1)$$
$$x'(5) + W^2 x'(7) = 1 + (-j)(1) = 1 - j \rightarrow x''(5)$$
$$x'(1) - W^0 x'(3) = 1 - 1 = 0 \rightarrow x''(3)$$
$$x'(5) - W^2 x'(7) = 1 - (-j)(1) = 1 + j \rightarrow x''(7)$$

where the intermediate second-stage output sequence x''s becomes the input sequence to the final stage.

Stage 3

$$X(0) = x''(0) + W^0 x''(1) = 4$$
$$X(1) = x''(4) + W^1 x''(5) = 1 - j2.414$$
$$X(2) = x''(2) + W^2 x''(3) = 0$$
$$X(3) = x''(6) + W^3 x''(7) = 1 - j0.414$$
$$X(4) = x''(0) - W^0 x''(1) = 0$$
$$X(5) = x''(4) - W^1 x''(5) = 1 + j0.414$$
$$X(6) = x''(2) - W^2 x''(3) = 0$$
$$X(7) = x''(6) - W^3 x''(7) = 1 + j2.414$$

which is the same output sequence found in Exercise 6.1.

6.5 BIT REVERSAL FOR UNSCRAMBLING

A bit-reversal procedure allows a scrambled sequence to be reordered. To illustrate this bit-swapping process, let $N = 8$, represented by three bits. The first and third bits are swapped. For example, $(100)_b$ is replaced by $(001)_b$. As such, $(100)_b$ specifying the address of $X(4)$ is replaced by or swapped with $(001)_b$ specifying the address of $X(1)$. Similarly, $(110)_b$ is replaced/swapped with $(011)_b$, or the addresses of $X(6)$ and $X(3)$ are swapped. In this fashion, the output sequence in Figure 6.5 with the DIF, or the input sequence in Figure 6.10 with the DIT, can be reordered.

This bit-reversal procedure can be applied for larger values of N. For example, for $N = 64$, represented by six bits, the first and sixth bits, the second and fifth bits, and the third and fourth bits are swapped.

Several examples in this chapter illustrate the FFT algorithm, incorporating algorithms for unscrambling.

6.6 DEVELOPMENT OF THE FFT ALGORITHM WITH RADIX-4

A radix-4 (base 4) algorithm can increase the execution speed of the FFT. FFT programs on higher radices and split radices have been developed. We use a DIF decomposition process to introduce the development of the radix-4 FFT. The last or lowest decomposition of a radix-4 algorithm consists of four inputs and four outputs. The order or length of the FFT is 4^M, where M is the number of stages. For a 16-point FFT, there are only two stages or iterations, compared with four stages with the radix-2 algorithm. The DFT in (6.1) is decomposed into four summations instead of two as follows:

$$X(k) = \sum_{n=0}^{(N/4)-1} x(n)W^{nk} + \sum_{n=N/4}^{(N/2)-1} x(n)W^{nk} + \sum_{n=N/2}^{(3N/4)-1} x(n)W^{nk} + \sum_{n=3N/4}^{N-1} x(n)W^{nk} \quad (6.30)$$

Let $n = n + N/4$, $n = n + N/2$, and $n = n + 3N/4$ in the second, third, and fourth summations, respectively. Then (6.30) can be written as

$$X(k) = \sum_{n=0}^{(N/4)-1} x(n)W^{nk} + W^{kN/4} \sum_{n=0}^{(N/4)-1} x(n+N/4)W^{nk}$$

$$+ W^{kN/2} \sum_{n=0}^{(N/4)-1} x(n+N/2)W^{nk} + W^{3kN/4} \sum_{n=0}^{(N/4)-1} x(n+3N/4)W^{nk} \quad (6.31)$$

which represents four $(N/4)$-point DFTs. Using

$$W^{kN/4} = \left(e^{-j2\pi/N}\right)^{kN/4} = e^{-jk\pi/2} = (-j)^k$$

$$W^{kN/2} = e^{-jk\pi} = (-1)^k$$

$$W^{3kN/4} = (j)^k$$

(6.31) becomes

$$X(k) = \sum_{n=0}^{(N/4)-1} \left[x(n) + (-j)^k x(n+N/4) + (-1)^k x(n+N/2) + (j)^k x(n+3N/4) \right] W^{nk} \quad (6.32)$$

Let $W_N^4 = W_{N/4}$. Equation (6.32) can be written as

$$X(4k) = \sum_{n=0}^{(N/4)-1} [x(n) + x(n+N/4) + x(n+N/2) + x(n+3N/4)] W_{N/4}^{nk} \quad (6.33)$$

$$X(4k+1) = \sum_{n=0}^{(N/4)-1} [x(n) - jx(n+N/4) - x(n+N/2) + jx(n+3N/4)] W_N^n W_{N/4}^{nk} \quad (6.34)$$

$$X(4k+2) = \sum_{n=0}^{(N/4)-1} [x(n) - x(n+N/4) + x(n+N/2) - x(n+3N/4)] W_N^{2n} W_{N/4}^{nk} \quad (6.35)$$

$$X(4k+3) = \sum_{n=0}^{(N/4)-1} [x(n) + jx(n+N/4) - x(n+N/2) - jx(n+3N/4)] W_N^{3n} W_{N/4}^{nk} \quad (6.36)$$

for $k = 0, 1, \ldots, (N/4) - 1$. Equations (6.33) through (6.36) represent a decomposition process yielding four four-point DFTs. The flow graph for a 16-point radix-4 DIF FFT is shown in Figure 6.11. Note the four-point butterfly in the flow graph. The $\pm j$ and -1 are not shown in Figure 6.11. The results shown in the flow graph are for the following exercise.

Exercise 6.4: Sixteen-Point FFT with Radix-4

Given the input sequence $x(n)$ as in Exercise 6.2, representing a rectangular sequence $x(0) = x(1) = \ldots = x(7) = 1$, and $x(8) = x(9) = \ldots = x(15) = 0$, we will find the output sequence for a 16-point FFT with radix-4 using the flow graph in Figure 6.11. The twiddle constants are shown in Table 6.1.

TABLE 6.1   Twiddle Constants for 16-Point FFT with Radix-4

m	W_N^m	$W_{N/4}^m$
0	1	1
1	$0.9238 - j0.3826$	$-j$
2	$0.707 - j0.707$	-1
3	$0.3826 - j0.9238$	$+j$
4	$0 - j$	1
5	$-0.3826 - j0.9238$	$-j$
6	$-0.707 - j0.707$	-1
7	$-0.9238 - j0.3826$	$+j$

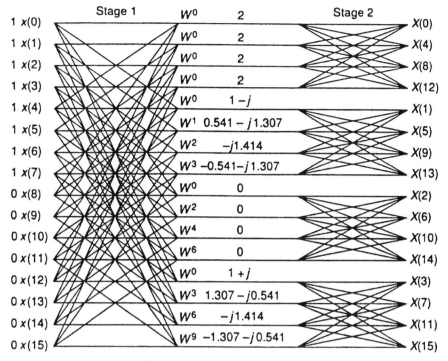

FIGURE 6.11. Sixteen-point radix-4 FFT flow graph using DIF.

The intermediate output sequence after stage 1 is shown in Figure 6.11. For example, after stage 1:

$$[x(0) + x(4) + x(8) + x(12)]W^0 = 1 + 1 + 0 + 0 = 2 \rightarrow x'(0)$$
$$[x(1) + x(5) + x(9) + x(13)]W^0 = 1 + 1 + 0 + 0 = 2 \rightarrow x'(1)$$
$$\vdots \qquad\qquad\qquad \vdots$$
$$[x(0) - jx(4) - x(8) + jx(12)]W^0 = 1 - j - 0 - 0 = 1 - j \rightarrow x'(4)$$
$$\vdots \qquad\qquad\qquad \vdots$$
$$[x(3) - x(7) + x(11) - x(15)]W^6 = 0 \rightarrow x'(11)$$
$$[x(0) + jx(4) - x(8) - jx(12)]W^0 = 1 + j - 0 - 0 = 1 + j \rightarrow x'(12)$$
$$\vdots \qquad\qquad\qquad \vdots$$
$$[x(3) + jx(7) - x(11) - jx(15)]W^9 = [1 + j - 0 - 0](-W^1)$$
$$= -1.307 - j0.541 \rightarrow x'(15)$$

For example, after stage 2:

$$X(3) = (1 + j) + (1.307 - j0.541) + (-j1.414) + (-1.307 - j0.541) = 1 - j1.496$$

and

$$X(15) = (1+j)(1) + (1.307 - j0.541)(-j) + (-j1.414)(1)$$
$$+ (-1.307 - j0.541)(-j) = 1 + j5.028$$

The output sequence $X(0), X(1), \ldots, X(15)$ is identical to the output sequence obtained with the 16-point FFT with the radix-2 in Figure 6.6. These results also can be verified with MATLAB, as described in Appendix D.

The output sequence is scrambled and needs to be resequenced or reordered. This can be done using a digit-reversal procedure, in a similar fashion as a bit reversal in a radix-2 algorithm. The radix-4 (base 4) uses the digits 0, 1, 2, 3. For example, the addresses of $X(8)$ and $X(2)$ need to be swapped because $(8)_{10}$ in base 10 or decimal is equal to $(20)_4$ in base 4. Digits 0 and 1 are reversed to yield $(02)_4$ in base 4, which is also $(02)_{10}$ in decimal.

Although mixed or higher radices can provide a further reduction in computation, programming considerations become more complex. As a result, radix-2 is still the most widely used, followed by radix-4. Two programming examples are included in Section 6.8, and two projects are described in Chapter 10.

6.7 INVERSE FAST FOURIER TRANSFORM

The inverse discrete Fourier transform (IDFT) converts a frequency-domain sequence $X(k)$ into an equivalent sequence $x(n)$ in the time domain. It is defined as

$$x(n) = \frac{1}{N} \sum_{k=0}^{N-1} X(k) W^{-nk} \qquad n = 0, 1, \ldots, N-1 \tag{6.37}$$

Comparing (6.37) with the DFT equation definition in (6.1), we see that the FFT algorithm (forward) described previously can be used to find the inverse FFT (IFFT) with the two following changes:

1. Adding a scaling factor of $1/N$
2. Replacing W^{nk} by its complex conjugate W^{-nk}

With the changes, the same FFT flow graphs can be used for the IFFT. We also develop programming examples to illustrate the inverse FFT.

A variant of the FFT, such as the FHT, can be obtained readily from the FFT. Conversely, the FFT can be obtained from the FHT [10,11]. A development of the FHT with flow graphs and exercises for 8- and 16-point FHTs can be found in Appendix F.

Exercise 6.5: Eight-Point IFFT

Let the output sequence $X(0) = 4$, $X(1) = 1 - j2.41, \ldots, X(7) = 1 + j2.41$ obtained in Exercise 6.1 become the input to an eight-point IFFT flow graph. Make the two changes (scaling and complex conjugate of W) to obtain an eight-point IFFT (reverse) flow graph from an eight-point FFT (forward) flow graph. The resulting flow graph becomes an IFFT flow graph similar to Figure 6.5. Verify that the resulting output sequence is $x(0) = 1, x(1) = 1, \ldots, x(7) = 0$, which represents the rectangular input sequence in Exercise 6.1.

6.8 PROGRAMMING EXAMPLES

Example 6.1: DFT of a Sequence of Real Numbers with Output from the CCS Window (DFT)

This example illustrates the DFT of an N-point sequence. Figure 6.12 shows a listing of the program $DFT.c$, which implements the DFT. The input sequence is x(n). The program calculates

$$X(k) = \text{DFT}\{x(n)\} = \sum_{n=0}^{N-1} x(n)W^{nk} \qquad k = 0, 1, \ldots, N-1$$

where $W = e^{-j2\pi/N}$ are the twiddle constants. This can be decomposed into a sum of real components and a sum of imaginary components, or

$$\text{Re}\{X(k)\} = \sum_{n=0}^{N-1} x(n)\cos(2\pi nk/N)$$

$$\text{IM}\{X(k)\} = \sum_{n=0}^{N-1} x(n)\sin(2\pi nk/N)$$

Using a sequence of real numbers with an integer number of cycles m, $X(k) = 0$ for all k, except at $k = m$ and at $k = N - m$.

Build this project as **DFT**. The input x(n) is a cosine with $N = 8$ data points. To test the results, load the program. Then:

1. Select View → Watch Window and insert the two expressions *j* and *out* (right click on the Watch window). Click on +*out* to expand and view out[0] and out[1], which represent the real and imaginary components, respectively.
2. Place a breakpoint at the bracket "}" that follows the DFT function call.
3. Select Debug → Animate (Animation speed can be controlled through Options). Verify that the real component value out[0] is large (3996) at j = 1 and at j = 7, while small otherwise. Since x(n) is a one-cycle sequence, m = 1. Since the number of points is N = 8, a "spike" occurs at j = m = 1 and at j = N - m = 7. The flowing two MATLAB commands can be used to verify these results (see also Appendix D):

```
//DFT.c DFT of N-point from lookup table. Output from watch window

#include <stdio.h>
#include <math.h>
void dft(short *x, short k, int *out);  //function prototype
#define N 8                             //number of data values
float pi = 3.1416;
int out[2] = {0,0};                     //init Re and Im results
short x[N] = {1000,707,0,-707,-1000,-707,0,707}; //1-cycle cosine

//short x[N]={0,602,974,974,602,0,-602,-974,-974,-602,
//            0,602,974,974,602,0,-602,-974,-974,-602};//2-cycles sine

void dft(short *x, short k, int *out)   //DFT function
{
 int sumRe = 0, sumIm = 0;              //init real/imag components
 float cs = 0, sn = 0;                  //init cosine/sine components
 int i = 0;
 for (i = 0; i < N; i++)                //for N-point DFT
    {
      cs = cos(2*pi*(k)*i/N);           //real component
      sn = sin(2*pi*(k)*i/N);           //imaginary component
      sumRe = sumRe + x[i]*cs;          //sum of real components
      sumIm = sumIm - x[i]*sn;          //sum of imaginary components
    }
 out[0] = sumRe;                        //sum of real components
 out[1] = sumIm;                        //sum of imaginary components
}

void main()
{
 int j;
 for (j = 0; j < N; j++)
    {
      dft(x,j,out);                     //call DFT function
    }
}
```

FIGURE 6.12. DFT implementation program with input from a lookup table (DFT.c).

$$x = [1000 \quad 707 \quad 0 \quad -707 \quad -1000 \quad -707 \quad 0 \quad 707];$$
$$y = fft(x)$$

Note that the data values in the table are rounded (yielding a spike with a maximum value of 3996 in lieu of 4000). Since it is a cosine, the imaginary component out[1] is zero (small). In a real-time implementation, with $F_s = 8\,kHz$, the frequency generated would be at $f = F_s$ (number of cycles)/ $N = 1\,kHz$.

4. Use the two-cycle sine data table (in the program) with 20 points as input x(n). Within the program, change N to 20, comment the table that corresponds

to the cosine (first input), and instead use the sine table values. Rebuild and Animate again. Verify a large negative value at j = m = 2 (-10,232) and a large positive value at j = N - m = 18 (10,232). For a real-time implementation, the magnitude of X(k), k = 0, 1, ... can be found. With F_s = 8 kHz, the frequency generated would correspond to f = 800 Hz.

Example 6.2: FFT of a Real-Time Input Signal Using an FFT Function in C (FFT256C)

Figure 6.13 shows a listing of the program *FFT256c.c* which implements a 256-point FFT in real time, using an external input signal. It calls a generic FFT function in C, FFT.c (on the accompanying CD). This FFT function, used with the C31 DSK and the C30 EVM, is listed and described in Refs. 13 and 14.

The twiddle constants are generated within the program. The imaginary components of the input data are set to zero to illustrate this implementation. The magnitude of the resulting FFT (scaled) is taken for output to the codec. Three buffers are used:

1. *samples*: contains the data to be transformed
2. *iobuffer*: used to output processed data as well as acquiring new input sampled data
3. *x1*: contains the magnitude (scaled) of the transformed (processed) data

In every sample period, an output value from a buffer (*iobuffer*) is sent to the codec's DAC and an input value is acquired and stored into the same buffer. An index (*buffercount*) to this buffer is used to set a flag when this buffer is full. When this buffer is full, it is copied to another buffer (*samples*), which will be used when calling the FFT function. The magnitude (scaled) of the processed FFT data, contained in a buffer *x1*, can now be copied to the I/O buffer, *iobuffer*, for output. In a filtering algorithm, processing can be done as each new sample is acquired. On the other hand, an FFT algorithm requires that an entire frame of data be available for processing.

Build and run this project as **FFT256c**. Input a 2-kHz sine wave with an amplitude of approximately 1 V p-p. Figure 6.14 shows a time-domain representation of the magnitude of the transformed data obtained with an HP dynamic signal analyzer (you can use an oscilloscope). The two negative spikes are $256(T_s)$ = 32 ms apart, as shown in Figure 6.14. This interval also represents the sampling frequency F_s. The location of the first positive spike then corresponds to a frequency of 2 kHz (the mid-distance between the two spikes corresponds to 4 kHz). The location of the second positive spike corresponds to the folding frequency of $F_s - f$ = 6 kHz. Increase the frequency of the input signal and observe the convergence of the two spikes toward the 4-kHz Nyquist frequency.

```
//FFT256c.c FFT implementation calling a C-coded FFT function

#include "dsk6713_aic23.h"
Uint32 fs=DSK6713_AIC23_FREQ_8KHZ;
#include <math.h>
#define PTS 256                         //# of points for FFT
#define PI 3.14159265358979
typedef struct {float real,imag;} COMPLEX;
void FFT(COMPLEX *Y, int n);            //FFT prototype
float iobuffer[PTS];                    //as input and output buffer
float x1[PTS];                          //intermediate buffer
short i;                                //general purpose index variable
short buffercount = 0;                  //number of new samples in iobuffer
short flag = 0;                         //set to 1 by ISR when iobuffer full
COMPLEX w[PTS];                         //twiddle constants stored in w
COMPLEX samples[PTS];                   //primary working buffer

main()
{
  for (i = 0 ; i<PTS ; i++)             // set up twiddle constants in w
    {
     w[i].real = cos(2*PI*i/512.0); //Re component of twiddle constants
     w[i].imag =-sin(2*PI*i/512.0); //Im component of twiddle constants
    }
  comm_intr();                          //init DSK, codec, McBSP
  while(1)                              //infinite loop
    {
     while (flag == 0) ;                //wait until iobuffer is full
     flag = 0;                          //reset flag
     for (i = 0 ; i < PTS ; i++)        //swap buffers
       {
        samples[i].real=iobuffer[i]; //buffer with new data
        iobuffer[i] = x1[i];          //processed frame to iobuffer
       }
     for (i = 0 ; i < PTS ; i++)
       samples[i].imag = 0.0;          //imag components = 0
     FFT(samples,PTS);                 //call function FFT.c
     for (i = 0 ; i < PTS ; i++)       //compute magnitude
       {
        x1[i] = sqrt(samples[i].real*samples[i].real
              + samples[i].imag*samples[i].imag)/16;
       }
     x1[0] = 32000.0;                   //negative spike for reference
    }                                   //end of infinite loop
}                                       //end of main

interrupt void c_int11()                //ISR
{
output_sample((short)(iobuffer[buffercount]));//output from iobuffer
iobuffer[buffercount++]=(float)((short)input_sample());//input>iobuffer
if (buffercount >= PTS)                          //if iobuffer full
 {
     buffercount = 0;                             //reinit buffercount
     flag = 1;                                    //set flag
 }
}
```

FIGURE 6.13. FFT program of real-time input calling a C-coded FFT function (FFT256c.c).

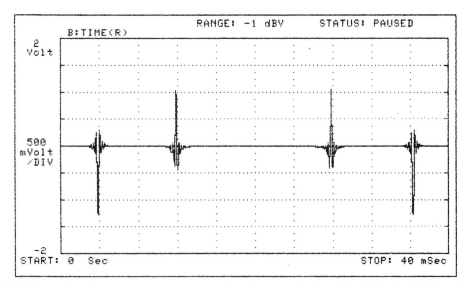

FIGURE 6.14. Time-domain plot representing the magnitude of the FFT of a 2 kHz real-time input sinusoid.

A project application in Chapter 10 makes use of this example to display a spectrum to LCDs, connected to the DSK through the EMIF 80-pin connector.

Example 6.3: FFT of a Sinusoidal Signal from a Table Using TI's C Callable Optimized FFT Function (*FFTsinetable*)

Figure 6.15 shows a listing of the program *FFTsinetable.c*, which illustrates a C program calling TI's optimized floating-point FFT function *cfftr2_dit.sa*, available at TI's Web site (also on CD). The twiddle constants are calculated within the program. The imaginary components of the twiddle constants are negated, as required (assumed) by the FFT function. The FFT function also assumes $N/2$ complex twiddle constants. It is important to align the data in memory (on an 8-byte boundary). Both the input data and the twiddle constants are structured as "complex."

The input signal consists of sine data values set in a table as real input data. The imaginary components of the input sine data are set to zero. The input data are arranged in memory as successive real and imaginary number pairs, as required (assumed) by the FFT function. The resulting output is still complex.

The FFT function *cfftr2_dit.sa* uses a DIT, radix 2, and takes the FFT of a "complex" input signal. Two support functions, *digitrev_index.c* and *bitrev.sa*, are used in conjunction with the complex FFT function for bit reversal. These two support files are also available through TI's Web site (also on CD). The FFT function *cfftr2_dit.sa* assumes that the input data x are in normal

```
//FFTsinetable.c FFT{sine}from table. Calls TI float-point FFT function

#include "dsk6713_aic23.h"              //codec-DSK support file
Uint32 fs=DSK6713_AIC23_FREQ_8KHZ;     //set sampling rate
#include <math.h>
#define N 32                           //number of FFT points
#define FREQ 8                         //select # of points/cycle
#define RADIX 2                        //radix or base 2
#define DELTA (2*PI)/N                 //argument for sine/cosine
#define TAB_PTS 32                     //# of points in sine_table
#define PI 3.14159265358979
short i = 0;
short iTwid[N/2];                      //index for twiddle constants
short iData[N];                        //index for bitrev X
float Xmag[N];                         //magnitude spectrum of x
typedef struct Complex_tag {float re,im;}Complex;
Complex W[N/RADIX];                    //array for twiddle constants
Complex x[N];                          //N complex data values
#pragma DATA_ALIGN(W,sizeof(Complex))  //align W
#pragma DATA_ALIGN(x,sizeof(Complex))  //align x

short sine_table[TAB_PTS] = {0,195,383,556,707,831,924,981,1000,
981,924,831,707,556,383,195,-0,-195,-383,-556,-707,-831,-924,-981,
-1000,-981,-924,-831,-707,-556,-383,-195};

void main()
{
 for( i = 0 ; i < N/RADIX ; i++ )
  {
   W[i].re = cos(DELTA*i);             //real component of W
   W[i].im = sin(DELTA*i);             //neg imag component
  }                                    //see cfftr2_dit
 for( i = 0 ; i < N ; i++ )
  {
   x[i].re=3*sine_table[FREQ*i % TAB_PTS]; //wrap when i=TAB_PTS
   x[i].im = 0 ;                       //zero imaginary part
  }
 digitrev_index(iTwid, N/RADIX, RADIX); //produces index for bitrev() W
 bitrev(W, iTwid, N/RADIX);            //bit reverse W
 cfftr2_dit(x, W, N ) ;                //TI floating-pt complex FFT

 digitrev_index(iData, N, RADIX);      //produces index for bitrev() X
 bitrev(x, iData, N);                  //freq scrambled->bit-reverse X
 for(i = 0 ; i < N ; i++)
    Xmag[i] = sqrt(x[i].re*x[i].re+x[i].im*x[i].im ); //magnitude of X

 comm_poll( ) ;                        //init DSK,codec,McBSP
 while (1)                             //infinite loop
  {
   output_sample(32000) ;             //negative spike as reference
   for (i = 1; i < N; i++)
     output_sample((short)Xmag[i]);    //output magnitude samples
  }
}
```

FIGURE 6.15. FFT program of input data from a table using TI's optimized complex FFT function (FFTsinetable.c).

order, while the FFT coefficients or twiddle constants are in reverse order. As a result, the support function $digitrev\_index.c$, to produce the index for bit reversal, and $bitrev.sa$, to perform the bit reversal on the twiddle constants, are called before the FFT function is invoked. These two support files for bit reversal are again called to bit-reverse the resulting scrambled output.

N is the number of complex input (note that the input data consist of 2N elements) or output data, so that an N-point FFT is performed. FREQ determines the frequency of the input sine data by selecting the number of points per cycle within the data table. With FREQ set at 8, every eighth point from the table is selected, starting with the first data point. The modulo operator is used as a flag to reinitialize the index. The following four points (scaled) within one period are selected: 0, 1000, 0, and –1000. Example 2.4 (sine2sliders) illustrates this indexing scheme to select different number of data points within a table.

The magnitude of the resulting FFT is taken. The line of code

```
output_sample (32000);
```

outputs a negative spike (not positive, due to the 2's-complement format of the AIC23 codec). It is used as a reference scheme. The input data are scaled so that the output magnitude is positive (again due to the codec data format). The sampling rate is achieved through polling.

Build and run this project as **FFTsinetable**. The two support files for bit reversal and the complex FFT function also are included in the Source project. Figure 6.16 shows a time-domain plot of the resulting output (obtained with an HP dynamic

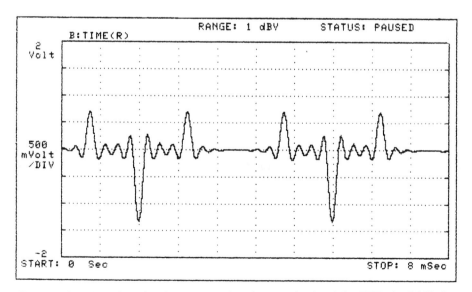

FIGURE 6.16. Time-domain plot representing the magnitude of the FFT of a 2-kHz input data from a table obtained using TI's FFT function.

signal analyzer). Since an output occurs every T_s, the time interval for 32 points corresponds to $32T_s$, or $32(0.125\,\text{ms}) = 4\,\text{ms}$. A negative spike is then repeated every 4 ms. This provides a reference, since the time interval between the two negative spikes corresponds to the sampling frequency of 8 kHz. The center of this time interval then corresponds to the Nyquist frequency of 4 kHz (2 ms from the negative spike). The first positive spike occurs at 1 ms from the first negative spike. This corresponds to a frequency of $f = F_s/4 = 2\,\text{kHz}$. The second positive spike occurs at 3 ms from the first negative spike and corresponds to the folding frequency of $F_s - f = 6\,\text{kHz}$.

Change FREQ to 4 in order to select eight sine data values within the table. Verify that the output is a 1-kHz signal (obtain a plot similar to that in Figure 6.14 from an oscilloscope). A FREQ value of 12 produces an output of 3 kHz. A FREQ value of 15 shows the two positive spikes at the center (between the two negative spikes). Note that aliasing occurs for frequencies larger than 4 kHz. To illustrate that, change FREQ to a value of 20. Verify that the output is an aliased signal at 3 kHz, in lieu of 5 kHz. A FREQ value of 24 shows an aliased signal of 2 kHz in lieu of 6 kHz.

The number of cycles is documented within the function *cfftr2_dit.sa* (by TI) as

```
Cycles = ((2N) + 23)log₂(N) + 6
```

For a 1024-point FFT, the number of cycles would be $(2071)(10) + 6 = 20{,}716$. This corresponds to a time of $t = 20{,}716$ cycles$/(225\,\text{MHz}) = 92\,\mu\text{s}$.

Example 6.4: FFT of Real-Time Input Using TI's C-Callable Optimized Radix-2 FFT Function (FFTr2)

This example expands Example 6.3 for real-time external input in lieu of a sine table as input. Figure 6.17 shows a listing of the C source program *FFTr2.c* that implements this project. The same FFT support files are used as in Example 6.3: TI's radix-2 optimized FFT function (*cfftr2_dit*), the function for generating the index for bit reversal (*digitrev_index*), and the function for the bit reversal procedure (*bitrev*). Since the FFT function assumes that the twiddle constants are in reverse order while the input data are in normal order, the index generation and bit reversal associated with the twiddle constants are performed (as in Example 6.3) before the complex FFT function is invoked.

Build this project as **FFTr2**. Input a 2-kHz sinusoidal signal with an approximate amplitude of 2 V p-p and verify the results in Figure 6.18. These results are similar to those in Example 6.2.

A project application in Chapter 10 makes use of this example to display a spectrum to a bank of LEDs connected to the DSK through the EMIF 80-pin connector.

```
//FFTr2.c FFT using TI's optimized FFT function and real-time input

#include "dsk6713_aic23.h"              //codec-DSK support file
Uint32 fs=DSK6713_AIC23_FREQ_8KHZ;     //set sampling rate
#include <math.h>
#define N 256                          //number of FFT points
#define RADIX 2                        //radix or base
#define DELTA (2*PI)/N                 //argument for sine/cosine
#define PI 3.14159265358979
short i = 0;
short iTwid[N/2];                      //index for twiddle constants W
short iData[N];                        //index for bitrev X
float Xmag[N];                         //magnitude spectrum of x
typedef struct Complex_tag {float re,im;}Complex;
Complex W[N/RADIX];                    //array for twiddle constants
Complex x[N];                          //N complex data values
#pragma DATA_ALIGN(W,sizeof(Complex))  //align W on boundary
#pragma DATA_ALIGN(x,sizeof(Complex))  //align input x on boundary

void main()
{
 for( i = 0 ; i < N/RADIX ; i++ )
  {
   W[i].re = cos(DELTA*i);             //real component of W
   W[i].im = sin(DELTA*i);             //neg imag component
  }                                    //see cfftr2_dit
 digitrev_index(iTwid, N/RADIX, RADIX); //obtain index for bitrev()W
 bitrev(W, iTwid, N/RADIX);            //bit reverse W
 comm_poll();                          //init DSK, codec, McBSP
 for(i=0; i<N; i++)
     Xmag[i] = 0;                      //init output magnitude
 while (1)                             //infinite loop
 {
  output_sample(32000);               //negative spike for reference
  for( i = 0 ; i < N ; i++ )
   {
   x[i].re = (float)((short)input_sample()); //external input
   x[i].im = 0.0 ;                     //zero imaginary part
   if(i>0) output_sample((short)Xmag[i]);//output magnitude
   }
  cfftr2_dit(x, W, N ) ;              //TI floating-pt complex FFT
  digitrev_index(iData, N, RADIX);    //produces index for bitrev()X
  bitrev(x, iData, N);                //freq scrambled->bit-reverse x
  for (i =0; i<N; i++)
    Xmag[i]=sqrt(x[i].re*x[i].re+x[i].im*x[i].im)/32; //magnitude of x
 }
}
```

FIGURE 6.17. FFT program of real-time input using TI's optimized radix-2 FFT function (FFTr2.c).

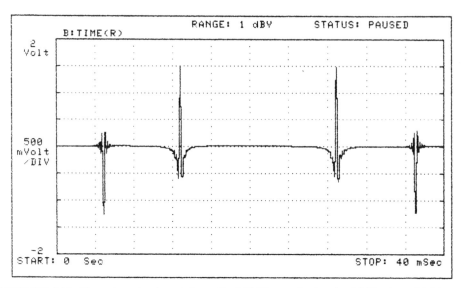

FIGURE 6.18. Time-domain plot of the radix-2 FFT magnitude of a 2-kHz sinusoidal input signal using *FFTr2.c*.

Example 6.5: Radix-4 FFT of Input from a Lookup Table Using TI's C-Callable Optimized FFT Function (*FFTr4_sim*)

Figure 6.19 shows the C source program *FFTr4_sim.c* that calls TI's optimized FFT function *cfftr4_dif.sa* (included on the CD) that performs a radix-4 complex FFT. The program takes a 16-point FFT of an input sinusoidal signal from a lookup table. The FFT support functions for this example are included on the CD. They are:

1. *cfftr4_dif.sa*: performs a DIF FFT in assembly code (not in linear ASM).
2. *R4DigitRevIndexTableGen*: generates the index for digit reversal.
3. *digit_reverse*: performs the digit reversal.

Note: The support files *digitrev_index* and *bitrev*, included in TI's Web site, for digit reversal do not work with the radix-4 FFT function. Use the functions included on the CD.

Build this project as **FFTr4_sim** and *load* the program:

1. View the watch window for the array x, arranged as successive real and imaginary number pairs. Expand x and verify that the real component data values are as in the lookup table and the imaginary components are all zeros. Place a breakpoint at the line following the first call to the FFT function. Run the program (execution stops at the set breakpoint). Verify that the FFT output is scrambled since the digit-reversal function *digit_reverse* is not yet executed. For example, $x(4) = 0 - 8000i$ should be swapped with $x(1)$, $x(6) = 0 + 1.23i$ should be swapped with $x(9)$, $x(7) = 0 + 0.68i$ should be swapped with $x(13)$, and so on (see Figure 6.11). Note that $x(5)$ is in the proper address sequence.
2. Remove the set breakpoint and place one after the first call to the digit reverse function (so that it is executed). Run the program again (to the set break-

```
//FFTr4_sim.c Radix-4 FFT and IFFT using a sine table
//Uses TI's optimized linear ASM radix-4 complex FFT function

#include <math.h>
#define N 16                          //# of complex FFT points
float x[2*N];                         //input real/imag aligned
float w[3*N/2];                       //complex twiddle factors
unsigned short J[4*N];                //index for digit reversal
unsigned short I[4*N];                //index for digit reversal
#pragma DATA_ALIGN(x,8);              //align x for unscrambling
#pragma DATA_ALIGN(w,8);              //align w for unscrambling
double delta = 2*3.14159265359/N;
short sine_table[16]={0,383,707,924,1000,924,707,383,
                -0,-383,-707,-924,-1000,-924,-707,-383};//sine

void main(void)
{
 int i, count;
 R4DigitRevIndexTableGen(N,&count,I,J);//indexes for digit reversal
 for(i = 0; i < 3*N/4; i++)            //generate twiddle constants
    {
      w[2*i+1] = sin(i * delta);      //real component of w
      w[2*i] = cos(i * delta);        //imag component of w negated
    }
 for (i=0 ; i<N ; i++)
    {
      x[2*i] = sine_table[i];         //input from sine table
      x[2*i+1] = 0;                   //set imag component to zero
    }
 cfftr4_dif(x, w, N);                 //call ASM FFT function
 digit_reverse((double*)x,I,J,count); //digit reverse FFT output
 for(i = 0; i < N; i++)
      x[2*i + 1] = -x[2*i + 1];       //for IFFT conjugate input
 cfftr4_dif(x, w, N);                 //to perform IFFT call FFT
 digit_reverse((double *)x,I,J,count); //digit reverse to unscramble
 for (i=0; i <(2*N); i++)
      x[i] /= N;                      //scale to get original input
}
```

FIGURE 6.19. FFT program that calls TI's optimized radix-4 FFT function with input from a lookup table (*FFTr4_sim.c*).

point), and verify that the output sequence is now unscrambled or in normal order. These results can be also verified using MATLAB (see Appendix D).

3. Run the program again (with the breakpoint in the same location). Verify that the output is now the same as the original input. By taking the complex conjugate of the input, again calling the same FFT function, unscrambling the output sequence, and scaling by *N* yields the original input.

4. In lieu of using a sine as input, let the input be a rectangular signal consisting of eight values of 1000 followed by eight values of 0. Repeat the previous procedures and verify the results with those in Figure 6.6 or with MATLAB. Note that the magnitude of the FFT output after the digit reversal is a sinc function, as in Figure 6.7.

Example 6.6: Radix-4 FFT of Real-Time Input Using TI's C-Callable Optimized FFT Function (FFTr4)

This example expands Example 6.5 for a real-time input signal. Figure 6.20 shows the C source program *FFTr4.c* that calls a radix-4 FFT function to take the FFT of a real-time input signal. The same FFT support functions are used in this example

```
//FFTr4.c FFT using TI's optimized FFT function with real-time input

#include "dsk6713_aic23.h"                //codec-DSK support file
Uint32 fs=DSK6713_AIC23_FREQ_8KHZ;        //set sampling rate
#include <math.h>
#define N 256                             // Specifies number of
complex FFT points
unsigned short JIndex[4*N];               //index for digit reversal
unsigned short IIndex[4*N];               //index for digit reversal
int i, count;
float Xmag[N];                            //magnitude spectrum of x
typedef struct Complex_tag {float re,im;}Complex;
Complex W[3*N/2];                         //array for twiddle constants
Complex x[N];                             //N complex data values
double delta = 2*3.14159265359/N;
#pragma DATA_ALIGN(x,sizeof(Complex));    //align x on boundary
#pragma DATA_ALIGN(W,sizeof(Complex));    //align W on boundary

void main()
{
 R4DigitRevIndexTableGen(N,&count,IIndex,JIndex);//for digit rev index
 for(i = 0; i < 3*N/4; i++)
    {
      W[i].re = cos(delta*i);             //real component of W
      W[i].im = sin(delta*i);             //Im component of W
    }
 comm_poll();                             //init DSK,codec,McBSP
 for(i=0; i<N; i++)
     Xmag[i] = 0;                         //init output magnitude
 while (1)                                //infinite loop
  {
   output_sample(32000);                  //negative spike as reference
   for( i = 0 ; i < N ; i++ )
    {
      x[i].re = (float)((short)input_sample()); //external input
      x[i].im = 0.0 ;                      //zero imaginary part
      if(i>0)   output_sample((short)Xmag[i]);  //output magnitude
    }
   cfftr4_dif(x, W, N);                    //radix-4 FFT function
   digit_reverse((double *)x,IIndex,JIndex,count);//unscramble
   for (i =0; i<N; i++)
     Xmag[i] = sqrt(x[i].re*x[i].re+x[i].im*x[i].im)/32;   //magnitude
  }
}
```

FIGURE 6.20. FFT program that calls TI's optimized radix-4 FFT function using real-time input (*FFTr4.c*).

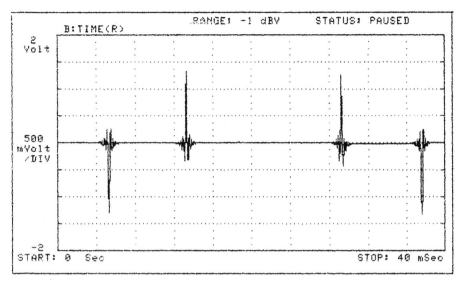

FIGURE 6.21. Time-domain plot of the radix-4 FFT magnitude of a 2-kHz input sinusoidal signal using *FFTr4.c*.

as in Example 6.5. This includes the FFT function as well as the function for generating the digit reversal index and the digit reversal function.

Build this project as **FFTr4**. Input a 2-kHz sinusoidal signal with an approximate amplitude of 2 V p-p. Verify the output in Figure 6.21. These results are similar to those obtained with the radix-2 FFT function in Example 6.4 and the radix-2 C-coded FFT function in Example 6.2.

A project application in Chapter 10 makes use of the real-time radix-4 FFT function with frequency-domain filtering.

6.8.1 Fast Convolution

The following examples show how the FFT enables signals to be processed in the frequency domain. Fast convolution [19,20] takes less computational effort and is potentially more accurate than time-domain implementation of FIR filters having very large numbers of coefficients.

Example 6.7: Fast Convolution with Overlap-Add for FIR Implementation Using TI's Floating-Point FFT Functions (fastconvo)

Figure 6.22 shows a listing of the program fastconvo.c to implement an FIR filter and illustrate the fast convolution's overlap-add scheme [19,20]. TI's floating-point FFT support functions, bitrev, digitrev_index, and cfftr2_dit were

```
//FastConvo.c FIR filter implemented using overlap-add fast convolution

#include "DSK6713_AIC23.h"          //codec-DSK support file
Uint32 fs=DSK6713_AIC23_FREQ_8KHZ;//set sampling rate
#include <math.h>
#include "coeffs.h"                 //time domain FIR coefficients
#define PI 3.14159265358979
#define PTS 256                     //number of points for FFT
#define SQRT_PTS 16                 //used in twiddle factor calc.
#define RADIX 2                     //passed to TI FFT routines
#define DELTA (2*PI)/PTS
typedef struct Complex_tag {float real, imag;} COMPLEX ;
#pragma DATA_ALIGN(W, sizeof(COMPLEX))
#pragma DATA_ALIGN(samples, sizeof(COMPLEX))
#pragma DATA_ALIGN(h, sizeof(COMPLEX))
COMPLEX W[PTS/RADIX] ;              //twiddle factor array
COMPLEX samples[PTS];               //processing buffer
COMPLEX h[PTS];                     //FIR filter coefficients
short buffercount = 0;             //buffer count for iobuffer samples
float iobuffer[PTS/2];             //primary input/output buffer
float overlap[PTS/2];             //intermediate result buffer
short i;                           //index variable
short flag = 0;                    //set to indicate iobuffer full
float a, b;                        //variables used in complex multiply
short NUMCOEFFS = sizeof(coeffs)/sizeof(float);
short iTwid[SQRT_PTS] ;            //PTS/2 + 1 > sqrt(PTS)

interrupt void c_int11(void)       //ISR
{
 output_sample((short)(iobuffer[buffercount]));
 iobuffer[buffercount++] = (float)((short)input_sample());
 if (buffercount >= PTS/2)         //for overlap-add method iobuffer
  {                                //is half size of FFT used
   buffercount = 0;
   flag = 1;
  }
}


main()
{                                  //set up array of twiddle factors
 digitrev_index(iTwid, PTS/RADIX, RADIX);
 for(i = 0 ; i < PTS/RADIX ; i++)
  {
   W[i].real = cos(DELTA*i);
   W[i].imag = sin(DELTA*i);
  }
 bitrev(W, iTwid, PTS/RADIX);      //bit reverse order W
 for (i = 0 ; i<PTS ; i++)         //initialise PTS element
  {                                //of COMPLEX to hold real-valued
    h[i].real = 0.0;               //time domain FIR filter coefficients
    h[i].imag = 0.0;
  }
 for (i = 0 ; i < NUMCOEFFS ; i++)
  {                                //read FIR filter coeffs
    h[i].real = coeffs[i];         //NUMCOEFFS should be less than PTS/2
  }
 cfftr2_dit(h,W,PTS);              //transform filter coeffs
```

FIGURE 6.22. Fast convolution program using overlap-add with TI's floating-point FFT functions (fastconvo.c).

```
comm_intr();                          //initialise DSK, codec, McBSP
while(1)                              //frame processing infinite loop
 {
  while (flag == 0);                  //wait for iobuffer full
     flag = 0;
  for (i = 0 ; i<PTS/2 ; i++)         //iobuffer into first half of
    {                                 //samples buffer
     samples[i].real = iobuffer[i];
     iobuffer[i] = overlap[i];        //previously processed output
    }                                 //to iobuffer
  for (i = 0 ; i<PTS/2 ; i++)
    {                                 //second half of samples to overlap
     overlap[i] = samples[i+PTS/2].real;
     samples[i+PTS/2].real = 0.0;//zero-pad input from iobuffer
    }
  for (i=0 ; i<PTS ; i++)
     samples[i].imag = 0.0;           //init imag parts in samples buffer
  cfftr2_dit(samples,W,PTS);          //complex FFT function from TI
  for (i=0 ; i<PTS ; i++)             //frequency-domain representation
    {                                 //complex multiply samples by h
     a = samples[i].real;
     b = samples[i].imag;
     samples[i].real = h[i].real*a - h[i].imag*b;
     samples[i].imag = h[i].real*b + h[i].imag*a;
    }
  icfftr2_dif(samples,W,PTS);         //inverse FFT function from TI
  for (i=0 ; i<PTS ; i++)
     samples[i].real /= PTS;
  for (i=0 ; i<PTS/2 ; i++)           //add first half of samples
     overlap[i] += samples[i].real; //to overlap
  }                                   //end of while(1)
}                                     //end of main()
```

FIGURE 6.22. (*Continued*)

introduced in Examples 6.3 and 6.4. In addition, TI's inverse complex FFT function icfftr2_dif (radix-2, DIF) is used here. This function expects its input to be scrambled or to be in bit-reversed order. As a result, the bit-reversed output of the complex FFT function cfftr2_dit need not be reordered, and the support files for bit reversal, digitrev_index.c and bitrev.sa, are not needed after the FFT section of the program. Both data (samples) and filter coefficients (h) are in bit-reversed order and may be multiplied together in that order.

Build this project as **Fastconvo** (use compiler optimization level -o1, or no optimization). The time-domain filter coefficients are read from the file coeffs.h. Verify that the output yields a 2-kHz bandpass filter. The filter coefficients are the same as BP55.cof, with a center frequency at $F_s/4$, introduced in Example 4.4.

The coefficient file coeffs.h also contains a set of coefficients identical to LP55.cof, which represents a lowpass FIR filter with a cutoff frequency at $F_s/4$, also introduced in Example 4.4. Edit the file coeffs.h to implement/verify this lowpass filter.

Several buffers are used, and `iobuffer` is the primary input/output buffer. At each sampling interval, the ISR is executed. The next output value is read from `iobuffer`, output to the codec, and then replaced by a new input sample. After PTS/2 sampling instants, `iobuffer` contains a new frame of PTS/2 input samples. This situation is signaled by setting `flag` to 1.

The main program waits for this flag signal using

```
while (flag == 0);
```

and subsequently carries out the following operations:

1. Resets `flag` to 0
2. Copies the contents of the buffer `iobuffer` (frame of new input samples) to the first PTS/2 locations of the buffer `samples`
3. Copies the contents of the buffer `overlap` (previously computed frame of output samples) to the buffer `iobuffer`
4. Processes the new frame of input samples to compute the next frame of output samples

The frame processing operation (within an infinite loop) has PTS/2 sampling periods in which to execute and comprises the following steps:

1. The contents of the last PTS/2 locations of the `samples` buffer (real parts) are copied to the `overlap` buffer. These time-domain data may be thought of as the overlapping latter half (PTS/2 samples) of the *previous* frame processing operation.
2. The last PTS/2 locations of the buffer `samples` are zero-padded. The buffer `samples` now contains PTS/2 new samples followed by PTS/2 zeros.
3. The buffer `samples` is transformed in-place into the frequency domain using a PTS-point FFT.
4. The complex frequency-domain sample values are multiplied by the complex frequency-domain filter coefficients stored in `h`.
5. The results are transformed back into the time domain by applying a PTS-point IFFT to the contents of the samples buffer. The resulting PTS time-domain samples will be real-valued.
6. The contents of the first PTS/2 locations of the buffer `samples` (i.e., the former half of the *current* frame processing result) are added to the contents of the `overlap` buffer.

Since the input and output signals are real-valued, so are the buffers `iobuffer` and `overlap`. However, since the frequency-domain representation of these signals is complex, the buffer `samples` and the array of filter coefficients `h` are complex, requiring two floating-point values (real and imaginary parts) per sample.

A faster and more efficient implementation of buffering is possible using pointers rather than copying data from one buffer to another, but the latter approach is adopted for purposes of clarity.

Example 6.8: Fast Convolution with Overlap-Add Simulation for FIR Implementation Using a C-Coded FFT Function (`fastconvo_sim`)

This example further illustrates the overlap-add fast convolution scheme. The program `fastconvo_sim.c` (on the CD) is a non-real-time version of the program `fastconvo.c`, which processes a prestored sequence of input samples. In lieu of using TI's FFT and support functions for bit reversal, the C-coded FFT function introduced in Example 6.2 is invoked. The program also performs an inverse FFT by first taking the conjugate of the samples and then invoking the FFT function.

Build this project as **Fastconvo_sim**. Using breakpoints (several breakpoint locations are specified within the program), the user can step through the various stages in the overlap-add process, viewing the contents of each of the buffers at each step. Figure 6.23 shows a typical view of the contents of the buffers (obtained with CCS): *h*, *iobuffer*, *overlap*, and *samples* at an intermediate stage in the process.

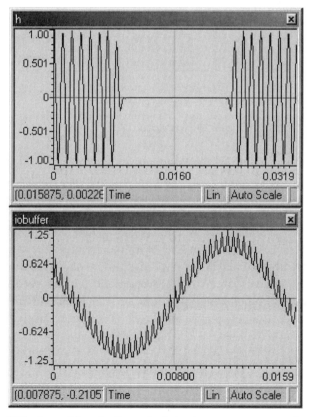

FIGURE 6.23. CCS plots of four buffers—h, iobuffer, overlap, and samples—at an intermediate processing stage using the simulation version program fastconvo_sim.c.

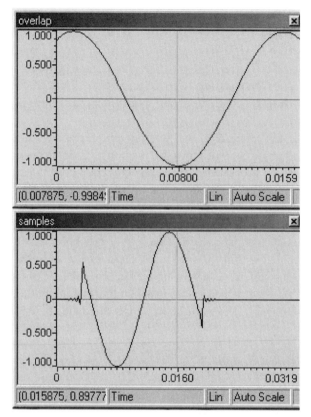

FIGURE 6.23. (*Continued*)

Example 6.9: Graphic Equalizer (graphicEQ)

Figure 6.24 shows a listing of the program graphicEQ.c, which implements a three-band graphic equalizer. TI's floating-point complex radix-2 FFT and IFFT support functions are used again in this project (see also Examples 6.3, 6.4, and 6.7).

The coefficient file graphicEQcoeff.h contains three sets of coefficients; lowpass at 1.3 kHz, bandpass between 1.3 and 2.6 kHz, and highpass at 2.6 kHz, designed with MATLAB's function fir1. Both the input samples and the three sets of coefficients are transformed into the frequency domain. The filtering is performed in the frequency domain based on the overlap-add scheme used in Example 6.7 [19,20]. Note that the complex multiplication (H)(X), where H represents the transfer function and X the input sample, yields

$$(H_R + jH_I)(X_R + jX_I) = (H_R X_R - H_I X_I) + j(H_R X_I + H_I X_R)$$

as used in the program, where $j = \sqrt{-1}$

ISR continuously (every sample period T_s) outputs a value from the buffer iobuffer, then inputs a new value until iobuffer is full. At this time a new frame

```
//GraphicEQ.c Graphic Equalizer using TI floating-point FFT functions

#include "DSK6713_AIC23.h"          //codec-DSK support file
Uint32 fs=DSK6713_AIC23_FREQ_8KHZ;//set sampling rate
#include <math.h>
#include "GraphicEQcoeff.h"         //time-domain FIR coefficients
#define PI 3.14159265358979
#define PTS 256                     //number of points for FFT
#define RADIX 2
#define DELTA (2*PI)/PTS
typedef struct Complex_tag {float real,imag;} COMPLEX;
#pragma DATA_ALIGN(W,sizeof(COMPLEX))
#pragma DATA_ALIGN(samples,sizeof(COMPLEX))
#pragma DATA_ALIGN(h,sizeof(COMPLEX))
COMPLEX W[PTS/RADIX] ;              //twiddle array
COMPLEX samples[PTS];
COMPLEX h[PTS];
COMPLEX bass[PTS], mid[PTS], treble[PTS];
short buffercount = 0;             //buffer count for iobuffer samples
float iobuffer[PTS/2];             //primary input/output buffer
float overlap[PTS/2];             //intermediate result buffer
short i;                            //index variable
short flag = 0;                     //set to indicate iobuffer full
float a, b;                         //variables for complex multiply
short NUMCOEFFS = sizeof(lpcoeff)/sizeof(float);
short iTwid[PTS/2] ;
float bass_gain = 1.0;              //initial gain values
float mid_gain = 0.0;              //change with GraphicEQ.gel
float treble_gain = 1.0;

interrupt void c_int11(void)        //ISR
{
 output_sample((short)(iobuffer[buffercount]));
 iobuffer[buffercount++] = (float)((short)input_sample());
 if (buffercount >= PTS/2)         //for overlap-add method iobuffer
  {                                 //is half size of FFT used
   buffercount = 0;
   flag = 1;
  }
}

main()
{
 digitrev_index(iTwid, PTS/RADIX, RADIX);
 for( i = 0; i < PTS/RADIX; i++ )
  {
   W[i].real = cos(DELTA*i);
   W[i].imag = sin(DELTA*i);
  }
 bitrev(W, iTwid, PTS/RADIX);        //bit reverse W
 for (i=0 ; i<PTS ; i++)
  {
   bass[i].real = 0.0;
   bass[i].imag = 0.0;
```

FIGURE 6.24. Equalizer program using TI's floating-point FFT support functions (graphicEQ.c).

```
    mid[i].real = 0.0;
    mid[i].imag = 0.0;
    treble[i].real = 0.0;
    treble[i].imag = 0.0;
  }
for (i=0; i<NUMCOEFFS; i++)       //same # of coeff for each filter
  {
    bass[i].real = lpcoeff[i];    //lowpass coeff
    mid[i].real =  bpcoeff[i];    //bandpass coeff
    treble[i].real = hpcoeff[i];  //highpass coef
  }
cfftr2_dit(bass,W,PTS);           //transform each band
cfftr2_dit(mid,W,PTS);            //into frequency domain
cfftr2_dit(treble,W,PTS);
comm_intr();                      //initialise DSK, codec, McBSP
while(1)                          //frame processing infinite loop
  {
    while (flag == 0);            //wait for iobuffer full
          flag = 0;
    for (i=0 ; i<PTS/2 ; i++)     //iobuffer into samples buffer
      {
        samples[i].real = iobuffer[i];
        iobuffer[i] = overlap[i]; //previously processed output
      }                           //to iobuffer
    for (i=0 ; i<PTS/2 ; i++)
      {                                 //upper-half samples to overlap
        overlap[i] = samples[i+PTS/2].real;
        samples[i+PTS/2].real = 0.0; //zero-pad input from iobuffer
      }
    for (i=0 ; i<PTS ; i++)
        samples[i].imag = 0.0;    //init samples buffer
    cfftr2_dit(samples,W,PTS);
    for (i=0 ; i<PTS ; i++)       //construct freq domain filter
      {                           //sum of bass,mid,treble coeffs
        h[i].real = bass[i].real*bass_gain + mid[i].real*mid_gain
                + treble[i].real*treble_gain;
        h[i].imag = bass[i].imag*bass_gain + mid[i].imag*mid_gain
                + treble[i].imag*treble_gain;
      }
    for (i=0; i<PTS; i++)                 //frequency-domain representation
      {                                   //complex multiply samples by h
        a = samples[i].real;
        b = samples[i].imag;
        samples[i].real = h[i].real*a - h[i].imag*b;
        samples[i].imag = h[i].real*b + h[i].imag*a;
      }
    icfftr2_dif(samples,W,PTS);
    for (i=0 ; i<PTS ; i++)
        samples[i].real /= PTS;
    for (i=0 ; i<PTS/2 ; i++)     //add 1st half to overlap
        overlap[i] += samples[i].real;
  }                               //end of infinite loop
}                                 //end of main()
```

FIGURE 6.24. (*Continued*)

of input data is available. The `iobuffer` index is initialized and the flag is set. The main program waits for this flag to be set, then resets it.

Build this project as **graphicEQ** (use the optimization level -o1). Test this project using an input voice file such as `TheForce.wav` (see Example 4.9) or noise. Verify that the low- and high-frequency components are accentuated, while the midrange frequency components are attenuated. This is because the filter coefficients are scaled in the program by `bass_gain` and `treble_gain`, initially set to 1, and by `mid_gain`, initially set to 0. The slider file `graphicEQ.gel` (on the CD) allows you to control the three frequency bands independently. Figure 6.25 shows the output spectrum obtained with a signal analyzer using noise as input and three different gain settings.

6.9 ASSIGNMENTS

1. Implement a 128-point radix-2 FFT of a real-time input sinusoid with a frequency of 3 kHz and an approximate amplitude of 2 V p-p. Use a sampling frequency of 16 kHz. Obtain a plot of the output (similar to Figure 6.18) and explain the results in terms of the distance between the two negative spikes and the location of the positive spikes. What is the output frequency when the input signal frequency is 6 kHz and 10 kHz? Explain.

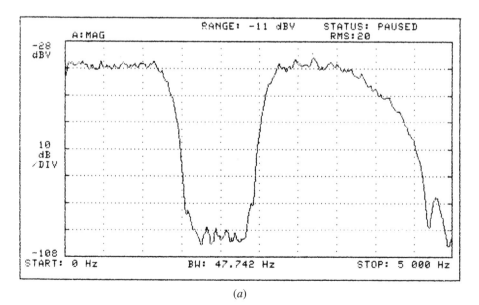

(*a*)

FIGURE 6.25. Output spectrum of a graphic equalizer obtained with a signal analyzer: (*a*) bass_gain = treble_gain = 1, mid_gain = 0; (*b*) bass_gain = treble_gain = 0, mid_gain = 1; (*c*) bass_gain = mid_gain = 1, treble_gain = 0.

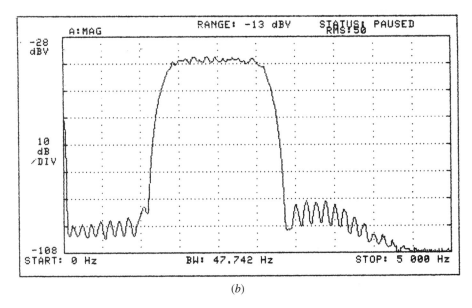

(b)

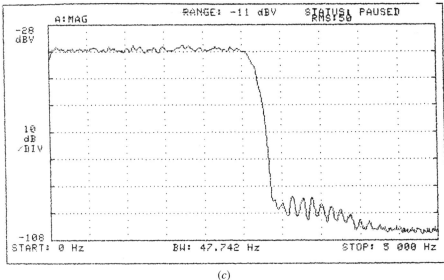

(c)

FIGURE 6.25. (*Continued*)

2. Implement a radix-2 FFT of an input using a lookup table with 32 sine data values over one cycle. Select an 8 kHz sample rate. Obtain a plot of the output magnitude. What is the output frequency?

3. In Example 6.8, the program *Fastconvo_sim.c* implements a simulated version of the fast convolution based on Example 6.7. It incorporates the C-coded FFT function in Example 6.2 in lieu of TI's radix-2 FFT function and the two support functions for bit reversal used in Example 6.7. Modify this C

source code so that it calls TI's FFT and support functions, as in Example 6.7. Verify the same results, obtained with the C-coded FFT function.

4. Example 6.7 implements a fast convolution using TI's optimized radix-2 FFT function and two support functions for bit reversal. Modify the C source program so that it calls the C-coded FFT function in Example 6.2. See Example 6.8 and the previous assignment. Note: the inverse FFT can be achieved by first taking the conjugate of the samples and then taking the FFT.

REFERENCES

1. J. W. Cooley and J. W. Tukey, An algorithm for the machine calculation of complex Fourier series, *Mathematics of Computation*, Vol. 19, 1965, pp. 297–301.

2. J. W. Cooley, How the FFT gained acceptance, *IEEE Signal Processing*, Jan. 1992, pp. 10–13.

3. J. W. Cooley, The structure of FFT and convolution algorithms, from a tutorial, *IEEE 1990 International Conference on Acoustics, Speech, and Signal Processing*, Apr. 1990.

4. C. S. Burrus and T. W. Parks, *DFT/FFT and Convolution Algorithms: Theory and Implementation*, Wiley, New York, 1988.

5. G. D. Bergland, A guided tour of the fast Fourier transform, *IEEE Spectrum*, Vol. 6, 1969, pp. 41–51.

6. E. O. Brigham, *The Fast Fourier Transform*, Prentice Hall, Upper Saddle River, NJ, 1974.

7. S. Winograd, On computing the discrete Fourier transform, *Mathematics of Computation*, Vol. 32, 1978, pp. 175–199.

8. H. F. Silverman, An introduction to programming the Winograd Fourier transform algorithm (WFTA), *IEEE Transactions on Acoustics, Speech, and Signal Processing*, Vol. ASSP-25, Apr. 1977, pp. 152–165.

9. P. E. Papamichalis, ed., *Digital Signal Processing Applications with the TMS320 Family: Theory, Algorithms, and Implementations*, Vol. 3, Texas Instruments, Dallas, TX, 1990.

10. R. N. Bracewell, Assessing the Hartley transform, *IEEE Transactions on Acoustics, Speech, and Signal Processing*, Vol. ASSP-38, 1990, pp. 2174–2176.

11. R. N. Bracewell, *The Hartley Transform*, Oxford University Press, New York, 1986.

12. H. V. Sorensen, D. L. Jones, M. T. Heidman, and C. S. Burrus, Real-valued fast Fourier transform algorithms, *IEEE Transactions on Acoustics, Speech, and Signal Processing*, Vol. ASSP-35, 1987, pp. 849–863.

13. R. Chassaing, *Digital Signal Processing Laboratory Experiments Using C and the TMS320C31 DSK*, Wiley, New York, 1999.

14. R. Chassaing, *Digital Signal Processing with C and the TMS320C30*, Wiley, New York, 1992.

15. P. M. Embree and B. Kimble, *C Language Algorithms for Digital Signal Processing*, Prentice Hall, Upper Saddle River, NJ, 1990.

16. S. Kay and R. Sudhaker, A zero crossing spectrum analyzer, *IEEE Transactions on Acoustics, Speech, and Signal Processing*, Vol. ASSP-34, Feb. 1986, pp. 96–104.

17. P. Kraniauskas, A plain man's guide to the FFT, *IEEE Signal Processing*, Apr. 1994.

18. J. R. Deller, Jr., Tom, Dick, and Mary discover the DFT, *IEEE Signal Processing*, Apr. 1994.

19. A. V. Oppenheim and R. Schafer, *Discrete-Time Signal Processing*, Prentice Hall, Upper Saddle River, NJ, 1989.

20. J. G. Proakis and D. G. Manolakis, *Digital Signal Processing: Principles, Algorithms and Applications*, Prentice Hall, Upper Saddle River, NJ, 2002.

7

Adaptive Filters

- Adaptive structures
- The linear adaptive combiner
- The least mean squares (LMS) algorithm
- Programming examples for noise cancellation and system identification using C code

Adaptive filters are best used in cases where signal conditions or system parameters are slowly changing and the filter is to be adjusted to compensate for this change. A very simple but powerful filter is called the *linear adaptive combiner*, which is nothing more than an adjustable FIR filter. The LMS criterion is a search algorithm that can be used to provide the strategy for adjusting the filter coefficients. Programming examples are included to give a basic intuitive understanding of adaptive filters.

7.1 INTRODUCTION

In conventional FIR and IIR digital filters, it is assumed that the process parameters to determine the filter characteristics are known. They may vary with time, but the nature of the variation is assumed to be known. In many practical problems, there may be a large uncertainty in some parameters because of inadequate prior test data about the process. Some parameters might be expected to change with time, but the exact nature of the change is not predictable. In such cases it is highly

Digital Signal Processing and Applications with the C6713 and C6416 DSK By Rulph Chassaing
ISBN 0-471-69007-4 Copyright © 2005 by John Wiley & Sons, Inc.

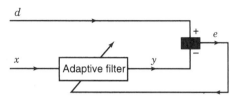

FIGURE 7.1. Basic adaptive filter structure.

desirable to design the filter to be self-learning so that it can adapt itself to the situation at hand.

The coefficients of an adaptive filter are adjusted to compensate for changes in input signal, output signal, or system parameters. Instead of being rigid, an adaptive system can learn the signal characteristics and track slow changes. An adaptive filter can be very useful when there is uncertainty about the characteristics of a signal or when these characteristics change.

Conceptually, the adaptive scheme is fairly simple. Most of the adaptive schemes can be described by the structure shown in Figure 7.1. This is a basic adaptive filter structure in which the adaptive filter's output y is compared with a desired signal d to yield an error signal e, which is fed back to the adaptive filter. The error signal is input to the adaptive algorithm, which adjusts the variable filter to satisfy some predetermined criteria or rules. The desired signal is usually the most difficult one to obtain. One of the first questions that probably comes to mind is: Why are we trying to generate the desired signal at y if we already know it? Surprisingly, in many applications the desired signal does exist somewhere in the system or is known a priori. The challenge in applying adaptive techniques is to figure out where to get the desired signal, what to make the output y, and what to make the error e.

The coefficients of the adaptive filter are adjusted, or optimized, using an LMS algorithm based on the error signal. Here we discuss only the LMS searching algorithm with a linear combiner (FIR filter), although there are several strategies for performing adaptive filtering. The output of the adaptive filter in Figure 7.1 is

$$y(n) = \sum_{k=0}^{N-1} w_k(n)x(n-k) \tag{7.1}$$

where $w_k(n)$ represent N weights or coefficients for a specific time n. The convolution equation (7.1) was implemented in Chapter 4 in conjunction with FIR filtering. It is common practice to use the terminology of weights w for the coefficients associated with topics in adaptive filtering and neural networks.

A performance measure is needed to determine how good the filter is. This measure is based on the error signal,

$$e(n) = d(n) - y(n) \tag{7.2}$$

which is the difference between the desired signal $d(n)$ and the adaptive filter's output $y(n)$. The weights or coefficients $w_k(n)$ are adjusted such that a mean squared error function is minimized. This mean squared error function is $E[e^2(n)]$, where E represents the expected value. Since there are k weights or coefficients, a gradient of the mean squared error function is required. An estimate can be found instead using the gradient of $e^2(n)$, yielding

$$w_k(n+1) = w_k(n) + 2\beta e(n)x(n-k) \qquad k = 0, 1, \ldots, N-1 \qquad (7.3)$$

which represents the LMS algorithm [1–3]. Equation (7.3) provides a simple but powerful and efficient means of updating the weights, or coefficients, without the need for averaging or differentiating, and will be used for implementing adaptive filters. The input to the adaptive filter is $x(n)$, and the rate of convergence and accuracy of the adaptation process (adaptive step size) is β.

For each specific time n, each coefficient, or weight, $w_k(n)$ is updated or replaced by a new coefficient, based on (7.3), unless the error signal $e(n)$ is zero. After the filter's output $y(n)$, the error signal $e(n)$ and each of the coefficients $w_k(n)$ are updated for a specific time n, a new sample is acquired (from an ADC) and the adaptation process is repeated for a different time. Note that from (7.3), the weights are not updated when $e(n)$ becomes zero.

The linear adaptive combiner is one of the most useful adaptive filter structures and is an adjustable FIR filter. Whereas the coefficients of the frequency-selective FIR filter discussed in Chapter 4 are fixed, the coefficients, or weights, of the adaptive FIR filter can be adjusted based on a changing environment such as an input signal. Adaptive IIR filters (not discussed here) can also be used. A major problem with an adaptive IIR filter is that its poles may be updated during the adaptation process to values outside the unit circle, making the filter unstable.

The programming examples developed later will make use of equations (7.1)–(7.3). In (7.3) we simply use the variable β in lieu of 2β.

7.2 ADAPTIVE STRUCTURES

A number of adaptive structures have been used for different applications in adaptive filtering.

1. *For noise cancellation.* Figure 7.2 shows the adaptive structure in Figure 7.1 modified for a noise cancellation application. The desired signal d is corrupted by uncorrelated additive noise n. The input to the adaptive filter is a noise n' that is correlated with the noise n. The noise n' could come from the same source as n but modified by the environment. The adaptive filter's output y is adapted to the noise n. When this happens, the error signal approaches the desired signal d. The overall output is this error signal and not the adaptive filter's output y. If d is uncorrelated with n, the strategy is to minimize $E(e^2)$,

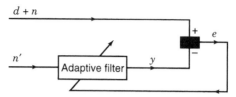

FIGURE 7.2. Adaptive filter structure for noise cancellation.

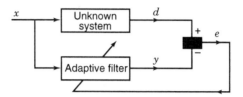

FIGURE 7.3. Adaptive filter structure for system identification.

where $E()$ is the expected value. The expected value is generally unknown; therefore, it is usually approximated with a running average or with the instantaneous function itself. Its signal component, $E(d^2)$, will be unaffected and only its noise component $E[(n - y)^2]$ will be minimized. A more complete discussion is found in Widrow and Stearns [1]. This structure will be further illustrated with programming examples using C code.

2. *For system identification.* Figure 7.3 shows an adaptive filter structure that can be used for system identification or modeling. The same input is to an unknown system in parallel with an adaptive filter. The error signal e is the difference between the response of the unknown system d and the response of the adaptive filter y. This error signal is fed back to the adaptive filter and is used to update the adaptive filter's coefficients until the overall output $y = d$. When this happens, the adaptation process is finished, and e approaches zero. If the unknown system is linear and not time-varying, then after the adaptation is complete, the filter's characteristics no longer change. In this scheme, the adaptive filter models the unknown system. This structure is illustrated later with three programming examples.

3. *Adaptive predictor.* Figure 7.4 shows an adaptive predictor structure that can provide an estimate of an input. This structure is illustrated later with a programming example.

4. Additional structures have been implemented, such as:

(a) *Notch with two weights,* which can be used to notch or cancel/reduce a sinusoidal noise signal. This structure has only two weights or coefficients. It is shown in Figure 7.5 and is illustrated in Refs. 1, 3, and 4.

(b) Adaptive channel equalization, used in a modem to reduce channel distortion resulting from the high speed of data transmission over telephone channels.

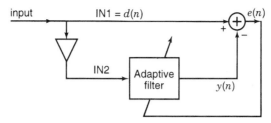

FIGURE 7.4. Adaptive predictor structure.

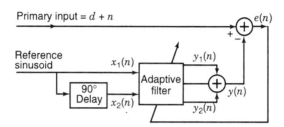

FIGURE 7.5. Adaptive notch structure with two weights.

The LMS is well suited for a number of applications, including adaptive echo and noise cancellation, equalization, and prediction.

Other variants of the LMS algorithm have been employed, such as the sign-error LMS, the sign-data LMS, and the sign-sign LMS.

1. For the sign-error LMS algorithm, (7.3) becomes

$$w_k(n+1) = w_k(n) + \beta \, \text{sgn}[e(n)]x(n-k) \tag{7.4}$$

where sgn is the signum function,

$$\text{sgn}(u) = \begin{cases} 1 & \text{if } u \geq 0 \\ -1 & \text{if } u < 0 \end{cases} \tag{7.5}$$

2. For the sign-data LMS algorithm, (7.3) becomes

$$w_k(n+1) = w_k(n) + \beta e(n) \, \text{sgn}[x(n-k)] \tag{7.6}$$

3. For the sign-sign LMS algorithm, (7.3) becomes

$$w_k(n+1) = w_k(n) + \beta \, \text{sgn}[e(n)] \, \text{sgn}[x(n-k)] \tag{7.7}$$

which reduces to

$$w_k(n+1) = \begin{cases} w_k(n)+\beta & \text{if } \text{sgn}[e(n)] = \text{sgn}[x(n-k)] \\ w_k(n)-\beta & \text{otherwise} \end{cases} \qquad (7.8)$$

which is more concise from a mathematical viewpoint because no multiplication operation is required for this algorithm.

The implementation of these variants does not exploit the pipeline features of the TMS320C6x processor. The execution speed on the TMS320C6x for these variants can be slower than for the basic LMS algorithm due to additional decision-type instructions required for testing conditions involving the sign of the error signal or the data sample.

The LMS algorithm has been quite useful in adaptive equalizers, telephone cancelers, and so forth. Other methods, such as the recursive least squares (RLS) algorithm [4], can offer faster convergence than the basic LMS but at the expense of more computations. The RLS is based on starting with the optimal solution and then using each input sample to update the impulse response in order to maintain that optimality. The right step size and direction are defined over each time sample.

Adaptive algorithms for restoring signal properties become useful when an appropriate reference signal is not available. The filter is adapted in such a way as to restore some property of the signal lost before reaching the adaptive filter. Instead of the desired waveform as a template, as in the LMS or RLS algorithms, this property is used for the adaptation of the filter. When the desired signal is available, a conventional approach such as the LMS can be used; otherwise, a priori knowledge about the signal is used.

7.3 ADAPTIVE LINEAR COMBINER

We will consider one of the most useful adaptive filter structures—the linear adaptive combiner. Two cases occur when using the linear combiner: (1) multiple inputs and (2) a single input.

Multiple Inputs

The case of multiple inputs is described in Figure 7.6. The configuration consists of K independent input signals, each of which is weighted by $w(k)$ and combined to form the output,

$$y(n) = \sum_{k=0}^{K} w(k,n)x(k,n) \qquad (7.9)$$

The input can be represented as a $(K+1)$-dimensional vector,

$$\mathbf{X}(n) = [x(0,n) \quad x(1,n) \quad \cdots \quad x(K,n)]^{\mathrm{T}} \qquad (7.10)$$

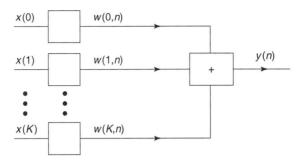

FIGURE 7.6. Linear combiner with multiple inputs.

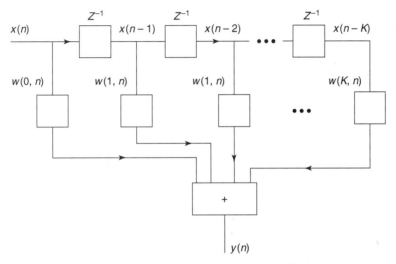

FIGURE 7.7. Adaptive linear combiner with single input.

where n is the time index and the transpose T is used so that the vector can be written on one line.

Single Input

In the case of a single input, the structure reduces to a $(K + 1)$-tap FIR filter with adjustable coefficients as shown in Figure 7.7. Each delayed input is weighted and summed to produce the output,

$$y(n) = \sum_{k=0}^{K} w(k,n)x(n-k) \tag{7.11}$$

The single input and the weights can also be written as vectors,

$$\mathbf{X}(n) = [x(n) \quad x(n-1) \quad \cdots \quad x(n-K)]^{\mathrm{T}} \tag{7.12}$$

$$\mathbf{W}(n) = [w(0,n) \quad w(1,n) \quad w(2,n) \quad \cdots \quad w(K,n)]^{\mathrm{T}} \tag{7.13}$$

where n is the time index, which will frequently be dropped from the notation for both w and x.

Using the vector notation, (7.11) is cast as

$$y(n) = \mathbf{X}^{\mathrm{T}}(n)\mathbf{W}(n) = \mathbf{W}^{\mathrm{T}}(n)\mathbf{X}(n) \tag{7.14}$$

Equations (7.9), (7.11), and (7.14), as well as Figures 7.6 and 7.7, all contain the same information. To become more familiar with the notation, let us examine a filter with two weights and a single input.

Exercise 7.1: Two Weights

Verify that equations (7.11) and (7.14) and Figure 7.8 give the same y for a two-weight filter.

Solution

For $K = 1$, equation (7.11) reduces to

$$y(n) = \sum_{k=0}^{1} w(k,n)x(n-k) = w(0,n)x(n) + w(1,n)x(n-1)$$

or with the time index n implied on the weights,

$$y(n) = w(0)x(n) + w(1)x(n-1)$$

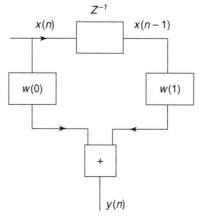

FIGURE 7.8. Two-weight linear combiner.

The equation above can also be obtained using (7.14),

$$y(n) = [x(n) \quad x(n-1)]\begin{bmatrix} w(0) \\ w(1) \end{bmatrix} = [w(0) \quad w(1)]\begin{bmatrix} x(n) \\ x(n-1) \end{bmatrix}$$

which reduces to

$$y(n) = x(n)w(0) + x(n-1)w(1)$$

which can also be obtained by summing the signals at the node of the two-weight diagram shown in Figure 7.8.

As can be seen in Figure 7.8, the linear combiner with a single input is just an FIR filter with adjustable coefficients. Although this is a very simple configuration, it can handle many of the adaptive applications.

7.4 PERFORMANCE FUNCTION

In the preceding section we provided a structure for the filter whose characteristics may be changed by adjusting the weights. However, we still need a way to judge how well the filter is operating—a performance measure is needed. The performance function will be based on the error, which is obtained from the block diagram in Figure 7.1, with the time index incorporated:

$$e(n) = d(n) - y(n) \tag{7.15}$$

The square of this function is

$$e^2(n) = d^2(n) - 2d(n)y(n) + y^2(n) \tag{7.16}$$

which is the instantaneous squared-error function. In terms of the weights, it becomes

$$e^2(n) = d^2(n) - 2d(n)\mathbf{X}^{\mathrm{T}}(n)\mathbf{W} + \mathbf{W}^{\mathrm{T}}\mathbf{X}(n)\mathbf{X}^{\mathrm{T}}(n)\mathbf{W} \tag{7.17}$$

where the time index on the $\mathbf{W}$ has been dropped. Equation (7.17) represents a quadratic surface in $\mathbf{W}$, which means that the highest power of the weights is the squared power. The strategy will be to adjust the weights so that the squared-error function will be a minimum.

To understand the performance surface equation (7.17), consider the case of one weight. The error surface then becomes

$$e^2(n) = d^2(n) - 2d(n)x(n)w(0) + x^2(n)w^2(0) \tag{7.18}$$

which is a second-order function in $w(0)$. To find the minimum, set the derivative of

(7.18) with respect to $w(0)$ equal to zero, or

$$\frac{de^2(n)}{dw(0)} = -2d(n)x(n) + 2x^2(n)w(0) = 0 \qquad (7.19)$$

resulting in

$$w(0) = \frac{d(n)}{x(n)} \qquad (7.20)$$

which is the value of $w(0)$ that yields the desired minimum.

Since the signals d and x are functions of time, the minimum and the performance surface also fluctuate with the signals. This is not desirable; we would feel more comfortable with a rigid performance function. To eliminate this problem, we can take the expected value of the squared-error function, which for one weight becomes

$$E[e^2(n)] = E[d^2(n)] - 2E[d(n)x(n)]w(0) + E[x^2(n)]w^2(0) \qquad (7.21)$$

This performance function is called the *mean-squared error.*

Note that the expected value of any sum is the sum of the expected values. The expected value of a product is the product of the expected values only if the variables are statistically independent. The signals $d(n)$ and $x(n)$ are generally not statistically independent. If the signals d and x are statistically time invariant, the expected values of the signal products of d and x are constants, and (7.21) is rewritten as

$$E[e^2(n)] = A - 2Bw(0) + Cw^2(0) \qquad (7.22)$$

where A, B, and C are constants.

Using (7.21) as the performance function for one weight results in a fixed minimum point on a rigid performance function,

$$w(0) = B/C \qquad (7.23)$$

A plot of the one-dimensional error function with respect to $w(0)$ is shown in Figure 7.9. This is a simple second-order curve in two dimensions ($E[e^2]$, $w(0)$) with a single minimum at $w(0) = B/C$. If we examine two weights, a three-dimensional second-order surface that resembles a bowl will result. With more weights, a higher-dimensional second-order surface will result that cannot be visualized by humans. In practice, the weights (the weight in this case) will start at some initial value w_i and are adjusted in increments toward the minimum value of the performance function. The procedure for adjusting the weights is a subject of the next section.

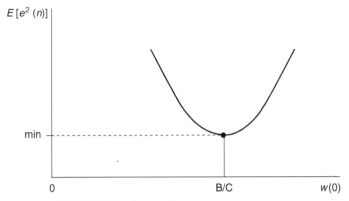

$E[e^2(n)]$

min

0   B/C $w(0)$

FIGURE 7.9. One-weight performance curve.

Taking the mean of the general squared-error function, (7.17), results in a general mean-squared-error performance function:

$$E[e^2(n)] = E[d^2(n)] - 2E[d(n)\mathbf{X}^T(n)]\mathbf{W} + \mathbf{W}^T E[\mathbf{X}(n)\mathbf{X}^T(n)]\mathbf{W} \qquad (7.24)$$

Again notice that the mean value of any sum is the sum of the mean values. The product values of d and $\mathbf{X}$ and $\mathbf{X}$ with $\mathbf{X}^T$ cannot be further reduced since the mean value of a product is the product of mean values only when the two variables are statistically independent; d and $\mathbf{X}$ are generally not independent. This is still the same second-order performance surface as before, but now it is not fluctuating with d and $\mathbf{X}$ but is rigid. However, if d and $\mathbf{X}$ are statistically time varying, the error surface will wiggle as the statistics of d and $\mathbf{X}$ change.

7.5 SEARCHING FOR THE MINIMUM

In this section we deal with how the weights should be adjusted to find the minimum in a reasonably efficient fashion. Of course, the weights could be adjusted randomly, but life is too short. Since we will be dealing with real-time events and changes that must be tracked, we need a relatively fast way of reaching the minimum.

Consider the one-weight system again to get an idea of how this search can be conducted. Initially, the weight will equal some arbitrary value $w(0, n)$, and it will be adjusted in a stepwise fashion until the minimum is reached (Figure 7.10), The size and direction of the step are the two things that must be chosen when making a step. Each step will consist of adding an increment to $w(0, n)$. Notice that if the current value of $w(0, n)$ is to the right of the minimum, the step must be negative (but the derivative of the curve is positive); similarly, if the current value is to the left of the minimum, the increment must be positive (but the derivative is negative). This observation leads to the conclusion that the negation of the derivative indicates the proper direction of the increment. Since the derivative vanishes at the minimum, it can also be used to adjust the step size. With these observations we conclude that the step size and direction can be made proportional to the negative

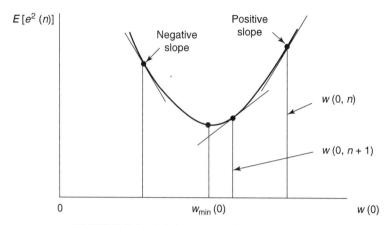

FIGURE 7.10. Minimum search on one weight.

of the derivative and the iteration for the weights can be expressed as

$$w(0, n+1) = w(0, n) - \beta \frac{dE[e^2]}{dw(0)} \qquad (7.25)$$

where β is an arbitrary positive constant. As shown in Figure 7.10, repeated application of (7.25) will cause $w(0)$ to move by steps from its initial value until it reaches the minimum.

The derivative of the function used in the one-dimensional search can be extended to an N-dimensional surface by replacing it with the gradient of the function. The gradient is a vector of first derivatives with respect to each of the weights:

$$\text{grad}\{E[e^2]\} = \text{grad}\{P\} = \left[\frac{\partial P}{\partial w(0)} \frac{\partial P}{\partial w(1)} \cdots \frac{\partial P}{\partial w(K)} \right]^{\mathrm{T}} \qquad (7.26)$$

The gradient points in the direction in which the function, in this case P, increases most rapidly. Therefore, the step size and direction can be made proportional to the gradient of the performance function.

Similarly. the minimum of the N-dimensional performance curve occurs when the gradient vanishes,

$$\text{grad}\{P\} = 0 \qquad (7.27)$$

or when the partial derivative with respect to each weight vanishes,

$$\frac{\partial P}{\partial w(0)} = 0, \quad \frac{\partial P}{\partial w(1)} = 0, \quad \cdots \quad \frac{\partial P}{\partial w(K)} = 0 \qquad (7.28)$$

Replacing the single weight with a vector of weights and the derivative with the gradient in (7.25) gives the multiple weight iteration rule,

$$\mathbf{W}(n+1) = \mathbf{W}(n) - \beta \text{grad}\{P\} \tag{7.29}$$

The only issue left to resolve is how to find grad $\{P\}$. To get a simple yet practical way to find grad $\{P\}$, we will use an estimate for it rather than the exact gradient. Instead of using the gradient of the expected squared error, we will approximate it with the grad $\{e^2\}$:

$$\text{grad}\{P\} \simeq \text{grad}\{e^2\} \tag{7.30}$$

To get a workable expression, let us perform the gradient operation on the squared-error function,

$$\text{grad}\{e^2\} = 2e\,\text{grad}\{e\} \tag{7.31}$$

where

$$e(n) = [d(n) - \mathbf{X}^{\mathrm{T}}(n)\mathbf{W}(n)] \tag{7.32}$$

Substitution yields

$$\text{grad}\{e^2\} = 2e\,\text{grad}[d(n) - \mathbf{X}^{\mathrm{T}}(n)\mathbf{W}(n)] \tag{7.33}$$

Expanding the gradient term gives

$$\text{grad}\{e^2\} = 2e \begin{bmatrix} \dfrac{\partial e}{\partial w(0)} \\ \dfrac{\partial e}{\partial w(1)} \\ \vdots \\ \dfrac{\partial e}{\partial w(K)} \end{bmatrix} = -2e \begin{bmatrix} x(0) \\ x(1) \\ \vdots \\ x(K) \end{bmatrix} \tag{7.34}$$

and

$$\text{grad}\{e^2(n)\} = -2e(n)\mathbf{X}(n) \tag{7.35}$$

Substituting this result for grad $\{P\}$ in equation (7.29) results in

$$\mathbf{W}(n+1) = \mathbf{W}(n) + 2\beta e(n)\mathbf{X}(n) \tag{7.36}$$

The time index n has been included in the last two equations, implying that e will be updated every sample time, Notice that if e goes to zero, then $\mathbf{W}(n + 1) = \mathbf{W}(n)$ and the weights remain constant.

Equation (7.36) forms the single most important result of this chapter, and it is the basis for the LMS algorithm. This equation allows the weights to be updated without squaring, averaging, or differentiating, yet it is powerful and efficient. This equation, as in (7.3), will be used in the following examples.

7.6 PROGRAMMING EXAMPLES FOR NOISE CANCELLATION AND SYSTEM IDENTIFICATION

The following programming examples illustrate adaptive filtering using the LMS algorithm. It is instructive to read the first example even though it does not use the DSK, since it illustrates the steps in the adaptive process.

Example 7.1: Adaptive Filter Using C Code Compiled with Borland C/C++ (Adaptc)

This example applies the LMS algorithm using a C-coded program compiled with Borland C/C++. It illustrates the following steps for the adaptation process using the adaptive structure in Figure 7.1:

1. Obtain a new sample for each, the desired signal d and the reference input to the adaptive filter x, which represents a noise signal.
2. Calculate the adaptive FIR filter's output y, applying (7.1) as in Chapter 4 with an FIR filter. In the structure of Figure 7.1, the overall output is the same as the adaptive filter's output y.
3. Calculate the error signal applying (7.2).
4. Update/replace each coefficient or weight applying (7.3).
5. Update the input data samples for the next time n with the data move scheme used in Chapter 4. Such a scheme moves the data instead of a pointer.
6. Repeat the entire adaptive process for the next output sample point.

Figure 7.11 shows a listing of the program adaptc.c, which implements the LMS algorithm for the adaptive filter structure in Figure 7.1. A desired signal is chosen as $2\cos(2n\pi f/F_s)$, and a reference noise input to the adaptive filter is chosen as $\sin(2n\pi f/F_s)$, where f is 1 kHz and $F_s = 8$ kHz. The adaptation rate, filter order, and number of samples are 0.01, 22, and 40, respectively.

The overall output is the adaptive filter's output y, which adapts or converges to the desired cosine signal d.

The source file was compiled with Borland's C/C++ compiler. Execute this program. Figure 7.12 shows a plot of the adaptive filter's output (y_out) converging to the desired cosine signal. Change the adaptation or convergence rate β to 0.02 and verify a faster rate of adaptation.

Interactive Adaptation

A version of the program adaptc.c in Figure 7.11, with graphics and interactive capabilities to plot the adaptation process for different values of β, is on the accom-

```
//Adaptc.c - Adaptation using LMS WITHOUT TI compiler

#include <stdio.h>
#include <math.h>
#define beta 0.01                      //convergence rate
#define N   21                         //order of filter
#define NS  40                         //number of samples
#define Fs  8000                       //sampling frequency
#define pi  3.1415926
#define DESIRED 2*cos(2*pi*T*1000/Fs)  //desired signal
#define NOISE sin(2*pi*T*1000/Fs)      //noise signal

main()
{
 long I, T;
 double D, Y, E;
 double W[N+1] = {0.0};
 double X[N+1] = {0.0};
 FILE *desired, *Y_out, *error;
 desired = fopen ("DESIRED", "w++"); //file for desired samples
 Y_out = fopen ("Y_OUT", "w++");     //file for output samples
 error = fopen ("ERROR", "w++");     //file for error samples
 for (T = 0; T < NS; T++)            //start adaptive algorithm
   {
    X[0] = NOISE;                     //new noise sample
    D = DESIRED;                      //desired signal
    Y = 0;                            //filter'output set to zero
    for (I = 0; I <= N; I++)
     Y += (W[I] * X[I]);              //calculate filter output
    E = D - Y;                        //calculate error signal
    for (I = N; I >= 0; I--)
      {
       W[I] = W[I] + (beta*E*X[I]);   //update filter coefficients
       if (I != 0)
       X[I] = X[I-1];                 //update data sample
      }
    fprintf (desired, "\n%10g    %10f", (float) T/Fs, D);
    fprintf (Y_out, "\n%10g    %10f", (float) T/Fs, Y);
    fprintf (error, "\n%10g    %10f", (float) T/Fs, E);
   }
 fclose (desired);
 fclose (Y_out);
 fclose (error);
}
```

FIGURE 7.11. Adaptive filter program compiled with Borland C/C++ (adaptc.c).

panying CD as adaptive.c, compiled with Borland C/C++. The executable file is also on the CD. It uses a desired cosine signal with an amplitude of 1 and a filter order of 31. Execute this program, enter a β value of 0.01, and verify the results in Figure 7.13. Note that the output converges to the desired cosine signal. Press F2 to execute this program again with a different beta value.

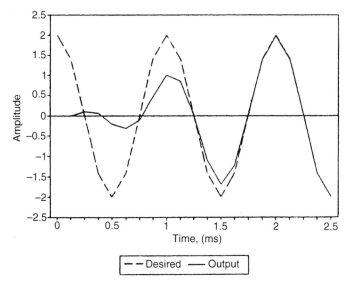

FIGURE 7.12. Plot of an adaptive filter's output converging to the desired cosine signal using *adaptc.c.*

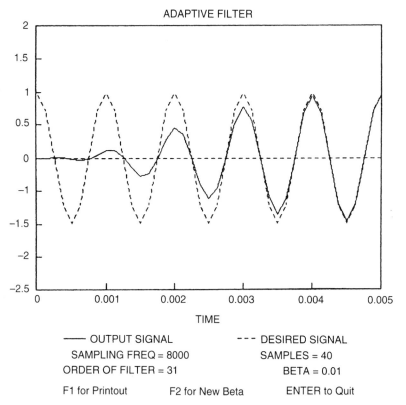

FIGURE 7.13. Plot of an adaptive filter's output converging to the desired cosine signal using interactive capability with the progam `adaptive.c.`

Example 7.2: Adaptive Filter for Sinusoidal Noise Cancellation (adaptnoise)

This example illustrates the application of the LMS criterion to cancel an undesirable sinusoidal noise. Figure 7.14 shows a listing of the program *adaptnoise.c*, which implements an adaptive FIR filter using the structure in Figure 7.2.

A desired sine wave of 1500 Hz with an additive (undesired) sine wave noise of 312 Hz forms one of two inputs to the adaptive filter structure. A reference (template) cosine signal, with a frequency of 312 Hz, is the input to a 30-coefficient adaptive FIR filter. The 312-Hz reference cosine signal is correlated with the 312-Hz additive sine noise but not with the 1500-Hz desired sine signal.

For each time n, the output of the adaptive FIR filter is calculated and the 30 weights or coefficients are updated along with the delay samples. The "error" signal E is the overall desired output of the adaptive structure. This error signal is the difference between the desired signal and additive noise (*dplusn*) and the adaptive filter's output, $y(n)$.

All signals used are from a lookup table generated with MATLAB. No external inputs are used in this example. Figure 7.15 shows a MATLAB program *adaptnoise.m* (a more complete version is on the CD) that calculates the data values for the desired sine signal of 1500 Hz, the additive noise as a sine of 312 Hz, and the reference signal as a cosine of 312 Hz. The appropriate files generated (on the CD) are:

1. *dplusn*: sine(1500 Hz) + sine(312 Hz)
2. *refnoise*: cosine(312 Hz)

Figure 7.16 shows the file *sin1500.h* with sine data values that represent the 1500-Hz sine-wave signal desired. The frequency generated associated with sin1500.h is

$$f = F_s(\text{\# of cycles})/(\text{\# of points}) = 8000(24)/128 = 1500 \text{ Hz}$$

The constant *beta* determines the rate of convergence.

Build and run this project as **adaptnoise**. Verify the following output result: The undesired 312-Hz sinusoidal signal is being gradually reduced (canceled), while the desired 1500-Hz signal remains. Note that in this application the output desired is the error signal E, which adapts (converges) to the desired signal. A faster rate of cancellation can be observed with a larger value of *beta*. However, if *beta* is too large, the adaptation process will not be observed since the output would be shown as the 1500-Hz signal. With the slider in position 2, the output is (*dplusn*), the desired 1500-Hz sinusoidal signal with the additive 312-Hz noise signal.

```
//Adaptnoise.c  Adaptive FIR filter for noise cancellation

#include "DSK6713_AIC23.h"              //codec-DSK support file
Uint32 fs= DSK6713_AIC23_FREQ_8KHZ;    //set sampling rate
#include "refnoise.h"                   //cosine 312 Hz
#include "dplusn.h"                     //sin(1500) + sin(312)
#define beta 1E-10                      //rate of convergence
#define N 30                            //# of weights (coefficients)
#define NS 128                          //# of output sample points
float w[N];                            //buffer weights of adapt filter
float delay[N];                        //input buffer to adapt filter
short output;                          //overall output
short out_type = 1;                    //output type for slider

interrupt void c_int11()               //ISR
{
 short i;
 static short buffercount=0;            //init count of # out samples
 float yn, E;                          //output filter/"error" signal

 delay[0] = refnoise[buffercount];     //cos(312Hz) input to adapt FIR
 yn = 0;                               //init output of adapt filter
 for (i = 0; i < N; i++)               //to calculate out of adapt FIR
     yn += (w[i] * delay[i]);          //output of adaptive filter
 E = dplusn[buffercount] - yn;         //"error" signal=(d+n)-yn
 for (i = N-1; i >= 0; i--)            //to update weights and delays
  {
   w[i] = w[i] + beta*E*delay[i];      //update weights
   delay[i] = delay[i-1];              //update delay samples
  }
 buffercount++;                        //increment buffer count
 if (buffercount >= NS)                //if buffercount=# out samples
   buffercount = 0;                    //reinit count
 if (out_type == 1)                    //if slider in position 1
   output = ((short)E*10);            //"error" signal overall output
 else if (out_type == 2)               //if slider in position 2
   output=dplusn[buffercount]*10;      //desired(1500)+noise(312)
 output_sample(output);                //overall output result
 return;                               //return from ISR
}

void main()
{
 short T=0;

 for (T = 0; T < 30; T++)
   {
     w[T] = 0;                         //init buffer for weights
     delay[T] = 0;                     //init buffer for delay samples
   }
 comm_intr();                          //init DSK, codec, McBSP
 while(1);                             //infinite loop
}
```

FIGURE 7.14. Adaptive FIR filter program for sinusoidal noise cancellation (adaptnoise.c).

```
%Adaptnoise.m Generates: dplusn.h, refnoise.h, and sin1500.h

for i=1:128
  desired(i) = round(100*sin(2*pi*(i-1)*1500/8000)); %sin(1500)
  addnoise(i) = round(100*sin(2*pi*(i-1)*312/8000)); %sin(312)
  refnoise(i) = round(100*cos(2*pi*(i-1)*312/8000)); %cos(312)
end

dplusn = addnoise + desired;                    %sin(312)+sin(1500)

fid=fopen('sin1500.h','w');                     %desired sin(1500)
fprintf(fid,'short sin1500[128]={');
fprintf(fid,'%d, ' ,desired(1:127));
fprintf(fid,'%d' ,desired(128));
fprintf(fid,'};\n');
fclose(fid);

%       fid=fopen('dplusn.h','w');              %desired + noise
%       fid=fopen('refnoise.h','w');            %reference noise
```

FIGURE 7.15. MATLAB program to generate data values for sine(1500), sine(1500)+ sine(312), and cosine(312) (adaptnoise.m).

```
short sin1500[128]={0, 92, 71, -38, -100, -38, 71, 92, 0, -92, -71, 38,
100, 38, -71, -92, 0, 92, 71, -38, -100, -38, 71, 92, 0, -92, -71, 38,
100, 38, -71, -92, 0, 92, 71, -38, -100, -38, 71, 92, 0, -92, -71, 38,
100, 38, -71, -92, 0, 92, 71, -38, -100, -38, 71, 92, 0, -92, -71, 38,
100, 38, -71, -92, 0, 92, 71, -38, -100, -38, 71, 92, 0, -92, -71, 38,
100, 38, -71, -92, 0, 92, 71, -38, -100, -38, 71, 92, 0, -92, -71, 38,
100, 38, -71, -92, 0, 92, 71, -38, -100, -38, 71, 92, 0, -92, -71, 38,
100, 38, -71, -92, 0, 92, 71, -38, -100, -38, 71, 92, 0, -92, -71, 38,
100, 38, -71, -92};
```

FIGURE 7.16. MATLAB's header file generated for sine(1500Hz) with 128 points (sin1500.h).

Example 7.3: Adaptive FIR Filter for Noise Cancellation Using External Inputs (adaptnoise_2IN)

This example extends the previous one to cancel an undesirable sinusoidal noise using external inputs. Figure 7.17 shows the source program *adaptnoise_2IN.c* that allows two external inputs: a desired signal and a sinusoidal interference. The program uses the union structure introduced in Chapter 2 with the project example *loop_stereo*. A 32-bit signal is captured using this structure that allows an external 16-bit input signal through each channel. The 16-bit desired signal is input through the left channel and the undesirable 16-bit signal through the right channel. An adapter with two connectors at one end for each input signal and one connector at the other end, which connects to the DSK, was introduced in Chapter 2 with the *loop_stereo* project and is required to implement this example. The basic adaptive structure in Figure 7.2 is applied here along with the LMS algorithm.

//**Adaptnoise_2IN.c** Adaptive FIR for sinusoidal noise interference

```
#include "DSK6713_AIC23.h"              //codec-DSK support file
Uint32 fs=DSK6713_AIC23_FREQ_48KHZ;    //set sampling rate
#define beta 1E-13                     //rate of convergence
#define N 30                           //# of weights (coefficients)
#define LEFT 0                         //left channel
#define RIGHT 1                        //right channel
float w[N];                            //weights for adapt filter
float delay[N];                        //input buffer to adapt filter
short output;                          //overall output
short out_type = 1;                    //output type for slider
volatile union{unsigned int uint; short channel[2];}AIC23_data;

interrupt void c_int11()               //ISR
  {
  short i;
  float yn=0, E=0, dplusn=0, desired=0, noise=0;

  AIC23_data.uint = input_sample();    //input 32-bit from both channels
  desired =(AIC23_data.channel[LEFT]);//input left channel
  noise = (AIC23_data.channel[RIGHT]);//input right channel

  dplusn = desired + noise;            //desired+noise
  delay[0] = noise;                    //noise as input to adapt FIR

  for (i = 0; i < N; i++)              //to calculate out of adapt FIR
      yn += (w[i] * delay[i]);         //output of adaptive filter
  E = (desired + noise) - yn;          //"error" signal=(d+n)-yn
  for (i = N-1; i >= 0; i--)           //to update weights and delays
    {
      w[i] = w[i] + beta*E*delay[i];   //update weights
      delay[i] = delay[i-1];           //update delay samples
    }
  if(out_type == 1)                    //if slider in position 1
      output=((short)E);               //error signal as overall output
  else if(out_type==2)                 //if slider in position 2
      output=((short)dplusn);          //output (desired+noise)
  output_sample(output);               //overall output result
  return;
}

 void main()
{
  short T=0;
  for (T = 0; T < 30; T++)
    {
      w[T] = 0;                        //init buffer for weights
      delay[T] = 0;                    //init buffer for delay samples
    }
  comm_intr();                         //init DSK, codec, McBSP
  while(1);                            //infinite loop
}
```

FIGURE 7.17. Adaptive filter program for noise cancellation using external inputs (*adaptnoise_2IN.c*).

Build this project as **adaptnoise_2IN**.

1. Desired: 1.5 kHz; undesired: 2 kHz. Input a desired sinusoidal signal (with a frequency such as 1.5 kHz) into the left channel and an undesired sinusoidal noise signal of 2 kHz into the right channel. Run the program. Verify that the 2-kHz noise signal is being canceled gradually. You can adjust the rate of convergence by changing *beta* by a factor of 10 in the program. Access/load the slider program *adaptnoise_2IN.gel* and change the slider position from 1 to 2. Verify the output as the two original sinusoidal signals at 1.5 and at 2 kHz.

2. Desired: wideband random noise; undesired: 2 kHz. Input random noise (from a noise generator, Goldwave, etc.) as the desired wideband signal into the left input channel and the undesired 2-kHz sinusoidal noise signal into the right input channel. Restart/run the program. Verify that the 2-kHz sinusoidal noise signal is being canceled gradually, with the wideband random noise signal remaining. With the slider in position 2, observe that both the undesired and desired input signals are as shown in Figure 7.18a. Figure 7.18b shows only the desired wideband random noise signal after the adaptation process.

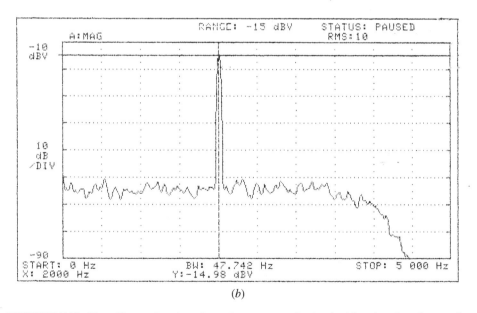

(*b*)

FIGURE 7.18. Plots illustrating the adaptation process obtained with a signal analyzer using adaptnoise_2IN.c; (*a*): 2-kHz undesired sinusoidal interference and desired wideband noise signal before adaptation; (*b*) cancellation of 2-kHz interference after adaptation.

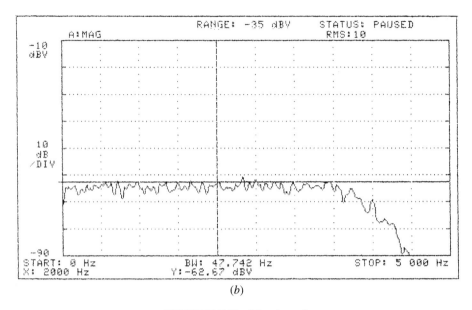

(*b*)

FIGURE 7.18. (*Continued*)

Example 7.4: Adaptive FIR Filter for System ID of a Fixed FIR as an Unknown System (adaptIDFIR)

Figure 7.19 shows a listing of the program *adaptIDFIR.c*, which models or identifies an unknown system. See also Examples 7.2 and 7.3, which implement an adaptive FIR for noise cancellation.

To test the adaptive scheme, the unknown system to be identified is chosen as an FIR bandpass filter with 55 coefficients centered at $F_s/4 = 2\,\text{kHz}$. The coefficients of this fixed FIR filter are in the file *bp55.cof*, introduced in Chapter 4. A 60-coefficient adaptive FIR filter models the fixed unknown FIR bandpass filter.

A pseudorandom noise sequence is generated within the program (see Examples 2.16 and 4.4) and becomes the input to both the fixed (unknown) and the adaptive FIR filters. This input signal represents a training signal. The adaptation process continues until the error signal is minimized. This feedback error signal is the difference between the output of the fixed unknown FIR filter and the output of the adaptive FIR filter.

An extra memory location is used in each of the two delay sample buffers (fixed and adaptive FIR). This is used to update the delay samples (see method B in Example 4.8).

Build and run this project as **adaptIDFIR**. Verify that the output (*adapt-fir_out*) of the adaptive FIR filter converges to a bandpass filter centered at $2\,\text{kHz}$ (with the slider in position 1 by default). With the slider in position 2, verify the

```
//AdaptIDFIR.c Adaptive FIR for system ID of an FIR

#include "DSK6713_AIC23.h"                  //codec-DSK support file
Uint32 fs=DSK6713_AIC23_FREQ_8KHZ;         //set sampling rate
#include "bp55.cof"                         //fixed FIR filter coefficients
#include "noise_gen.h"                      //support  noise generation file
#define beta 1E-14                          //rate of convergence
#define WLENGTH 60                          //# of coefffor adaptive FIR
float w[WLENGTH+1];                         //buffer coeff for adaptive FIR
int dly_adapt[WLENGTH+1];                   //buffer samples of adaptive FIR
int dly_fix[N+1];                           //buffer samples of fixed FIR
short out_type = 1;                         //output for adaptive/fixed FIR
int fb;                                     //feedback variable
shift_reg sreg;                             //shift register

int prand(void)                            //pseudo-random sequence {-1,1}
{
  int prnseq;
  if(sreg.bt.b0)  prnseq = -8000;          //scaled negative noise level
  else            prnseq =  8000;          //scaled positive noise level
  fb =(sreg.bt.b0)^(sreg.bt.b1);           //XOR bits 0,1
  fb^=(sreg.bt.b11)^(sreg.bt.b13);         //with bits 11,13 -> fb
  sreg.regval<<=1;
  sreg.bt.b0=fb;                           //close feedback path
  return prnseq;                           //return noise sequence
}

interrupt void c_int11()                   //ISR
{
 int i;
 int fir_out = 0;                          //init output of fixed FIR
 int adaptfir_out = 0;                     //init output of adapt FIR
 float E;                                  //error=diff of fixed/adapt out
 dly_fix[0] = prand();                     //input noise to fixed FIR
 dly_adapt[0]=dly_fix[0];                  //as well as to adaptive FIR
 for (i = N-1; i>= 0; i--)
  {
   fir_out +=(h[i]*dly_fix[i]);            //fixed FIR filter output
   dly_fix[i+1] = dly_fix[i];              //update samples of fixed FIR
  }
 for (i = 0; i < WLENGTH; i++)
   adaptfir_out +=(w[i]*dly_adapt[i]);     //adaptive FIR filter output
E = fir_out - adaptfir_out;                //error signal
for (i = WLENGTH-1; i >= 0; i--)
  {
   w[i] = w[i]+(beta*E*dly_adapt[i]);      //update weights of adaptive FIR
   dly_adapt[i+1] = dly_adapt[i];          //update samples of adaptive FIR
  }
 if (out_type == 1)                        //slider position for adapt FIR
   output_sample((short)adaptfir_out);     //output of adaptive FIR filter
 else if (out_type == 2)                   //slider position for fixed FIR
   output_sample((short)fir_out);          //output of fixed FIR filter
 return;
}
```

FIGURE 7.19. Program to implement an adaptive FIR filter that models (identifies) a fixed
FIR filter (adaptIDFIR.c).

```
void main()
{
 int T=0, i=0;
 for (i = 0; i < WLENGTH; i++)
   {
     w[i] = 0.0;                        //init coeff for adaptive FIR
     dly_adapt[i] = 0;                  //init buffer for adaptive FIR
   }
 for (T = 0; T < N; T++)
     dly_fix[T] = 0;                    //init buffer for fixed FIR
 sreg.regval=0xFFFF;                    //initial seed value
 fb = 1;                               //initial feedback value
 comm_intr();                          //init DSK, codec, McBSP
 while (1);                            //infinite loop
}
```

FIGURE 7.19. (*Continued*)

output (*fir_out*) of the fixed FIR bandpass filter centered at 2 kHz and represented by the coefficient file *bp55.cof*. It can be observed that this output is practically identical to the adaptive filter's output.

Edit the main program to include the coefficient file *BS55.cof* (introduced in Example 4.4), which represents an FIR bandstop filter with 55 coefficients centered at 2 kHz. The FIR bandstop filter represents the unknown system to be identified.

Rebuild/run and verify that the output of the adaptive FIR filter (with the slider in position 1) is practically identical to the FIR bandstop filter (with the slider in position 2). Increase (decrease) *beta* by a factor of 10 to observe a faster (slower) rate of convergence. Change the number of weights (coefficients) from 60 to 40 and verify a slight degradation of the identification process.

Example 7.5: Adaptive FIR for System ID of a Fixed FIR as an Unknown System with Weights of an Adaptive Filter Initialized as an FIR Bandpass (adaptIDFIRw)

The program *adaptIDFIR.c* in Example 7.4 is modified slightly to create the program *adaptIDFIRW.c* (on the CD). This new program initializes the weights of the adaptive FIR filter with the coefficients of an FIR bandpass filter centered at 3 kHz and represented by the coefficient file *bp3000.cof* (on the CD). The weights *w[i]* within the function *main* are initialized with the coefficients in the file *bp3000.cof* in lieu of zero.

Build this project as **adaptIDFIRw**. Initially, the spectrum of the output of the adaptive FIR filter shows the FIR bandpass filter centered at 3 kHz. Then, gradually, the output spectrum adapts (converges) to the fixed (unknown) FIR bandpass filter centered at 2 kHz (represented by *bp55.cof*), while the reference filter

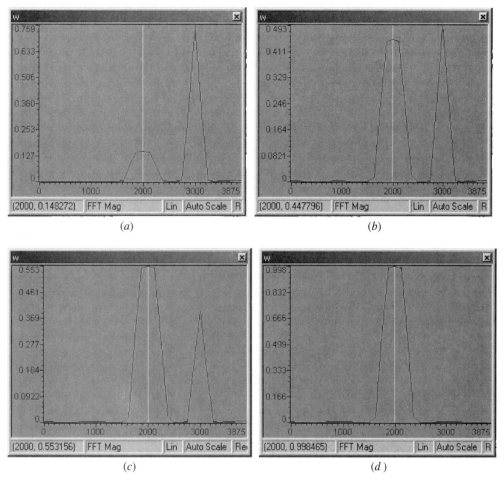

(a)   (b)

(c)   (d)

FIGURE 7.20. CCS plots illustrating the adaptation process of an adaptive filter: (a) weights set initially as a 3-kHz bandpass filter; (b) weights starting to converge to a 2-kHz filter; (c) weights almost converged to 2 kHz with the 3-kHz filter reduced; (d) adaptation completed with convergence to the 2-kHz bandpass filter.

gradually phases out. As the adaptation process takes place, one can observe at some time the two bandpass filters. You may wish to increase slightly the rate of adaptation (`beta`).

The adaptation process is illustrated with the CCS plots in Figure 7.20. Figure 7.21 illustrates the real-time adaptation process using an HP dynamic signal analyzer.

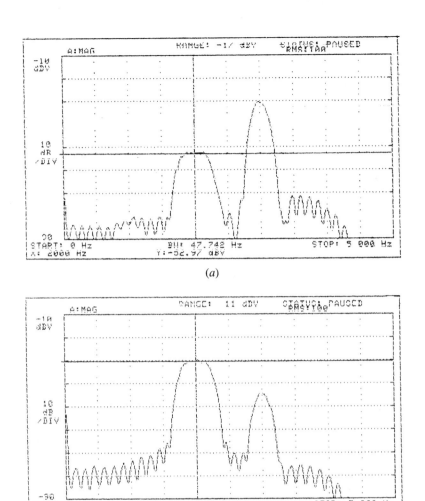

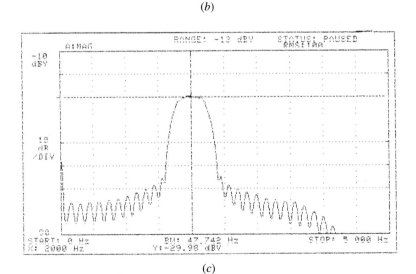

FIGURE 7.21. Real-time adaptation process with an adaptive filter converging to 2 kHz, obtained with an HP signal analyzer: (*a*) showing both the 3- and 2-kHz filters; (*b*) converging further to the 2-kHz filter; (*c*) adapted to the 2-kHz fixed filter.

274

Example 7.6: Adaptive FIR for System ID of Fixed IIR as an Unknown System (adaptIDIIR)

Figure 7.22 shows a listing of the program *adaptIDIIR.c*, which uses an adaptive FIR filter to model or identify a system (fixed unknown IIR). See Example 5.1, which implements an IIR filter, and Examples 7.4 and 7.5, which implement an adaptive FIR filter to model a fixed FIR filter.

To test the adaptive scheme, the unknown system to be identified is chosen as a 36th-order IIR bandpass filter with *18* second-order stages centered at 2 kHz. The coefficients of this fixed IIR filter are in the file *bp2000.cof*, introduced in Example 5.1. A 200-coefficient adaptive FIR filter is used to model the fixed unknown IIR bandpass filter. A larger number of coefficients or weights than for the adaptive FIR filter are necessary for a good model of the IIR filter.

A pseudorandom noise sequence is generated (see Example 2.16) and becomes the input to both the fixed IIR filter and the adaptive FIR filter. The adaptation process continues until the error signal is minimized. This feedback error signal is the difference between the output of the fixed unknown IIR filter and the output of the adaptive FIR filter.

Build and run this project as **adaptIDIIR**. Verify that the output of the adaptive filter (*adaptfir_out*) converges to (models) the IIR bandpass filter centered at 2 kHz, as shown in Figure 7.23 (with the slider initially in position 1). Verify that the output (*iir_out*) is the fixed IIR bandpass filter with the slider in position 2.

Include the coefficient file *lp2000.cof* in lieu of *bp2000.cof*. The coefficient file *lp2000.cof* represents an eighth-order (four second-order stages) IIR lowpass filter with a cutoff frequency of 2 kHz, introduced in Example 5.1. Verify that the adaptive FIR filter now adapts to the IIR lowpass filter with a cutoff frequency of 2 kHz.

Example 7.7: Adaptive Predictor for Cancellation of Narrowband Interference Added to a Desired Wideband Signal (adaptpredict)

The program *adaptpredict.c*, shown in Figure 7.24, implements an adaptive FIR predictor for the cancellation of a narrowband interference in the presence of a wideband signal. The desired wideband signal with an additive narrowband interference is delayed and becomes the input to a 60-coefficient adaptive FIR filter.

The desired wideband signal is generated with a MATLAB program *wbsignal.m*, shown in Figure 7.25. This MATLAB program generates a 256-point lookup table in the file *wbsignal.h* (on the CD). A random sequence {−1, 1} is generated, scaled, and written into the file *wbsignal.h*. Since the random sequence is for a length of 128 with a bit rate of 4 kHz, it is up-sampled to a 256-point sequence with a bit rate of 8 kHz. The wideband random sequence generated (with the file *wbsignal.h*) represents the signal desired.

The narrowband interference is an external signal. The bandwidth of the interference is narrow compared with the bandwidth of the random sequence generated

```
//AdaptIDIIR.c Adaptive FIR for system ID of fixed IIR using C67x tools
#include "DSK6713_AIC23.h"              //codec-DSK support file
Uint32 fs=DSK6713_AIC23_FREQ_8KHZ;     //set sampling rate
#include "bp2000.cof"                   //BP @ 2kHz fixed IIR coeff
#include "noise_gen.h"                  //support file noise sequence
#define beta 1E-11                      //rate of convergence
#define WLENGTH 200                     //# of coeff for adaptive FIR
float w[WLENGTH+1];                     //buffer coeff for adaptive FIR
int dly_adapt[WLENGTH+1];              //buffer samples of adaptive FIR
int dly_fix[stages][2] = {0};          //delay samples of fixed IIR
int a[stages][3], b[stages][2];        //coefficients of fixed IIR
short out_type = 1;                     //slider adaptive FIR/fixed IIR
int fb;                                 //feedback variable for noise
shift_reg sreg;                         //shift register for noise

int prand(void)                         //pseudo-random sequence {-1,1}
{
  int prnseq;
  if(sreg.bt.b0) prnseq = -4000;        //scaled negative noise level
  else           prnseq =  4000;        //scaled positive noise level
  fb =(sreg.bt.b0)^(sreg.bt.b1);        //XOR bits 0,1
  fb^=(sreg.bt.b11)^(sreg.bt.b13);      //with bits 11,13 ->fb
  sreg.regval<<=1;
  sreg.bt.b0=fb;                         //close feedback path
  return prnseq;                         //return noise sequence
}
interrupt void c_int11()                //ISR
{
 int i, un, input, yn;
 int iir_out=0;                          //init output of fixed IIR
 int adaptfir_out=0;                     //init output of adaptive FIR
 float E;                                //error signal
 dly_fix[0][0] = prand();                //input noise to fixed IIR
 dly_adapt[0] = dly_fix[0][0];           //same input to adaptive FIR
 input = prand();                        //noise as input to fixed IIR
 for (i = 0; i < stages; i++)            //repeat for each stage
  {
  un=input-((b[i][0]*dly_fix[i][0])>>15)-((b[i][1]*dly_fix[i][1])>>15);
  yn = ((a[i][0]*un)>>15)+((a[i][1]*dly_fix[i][0])>>15)
     + ((a[i][2]*dly_fix[i][1])>>15);
  dly_fix[i][1] = dly_fix[i][0];         //update delays of fixed IIR
  dly_fix[i][0] = un;                    //update delays of fixed IIR
  input = yn;                            //in next stage=out previous
  }
 iir_out = yn;                           //output of fixed IIR
 for (i = 0; i < WLENGTH; i++)
   adaptfir_out +=(w[i]*dly_adapt[i]);   //output of adaptive FIR
 E = iir_out - adaptfir_out;             //error as difference of outputs
 for (i = WLENGTH; i > 0; i--)
  {
  w[i] = w[i]+(beta*E*dly_adapt[i]);     //update weights of adaptive FIR
  dly_adapt[i] = dly_adapt[i-1];         //update samples of adaptive FIR
  }
 if (out_type == 1)                      //slider position->adaptive FIR
output_sample((short)adaptfir_out);      //output of adaptive FIR
 else if (out_type == 2)                 //slider position->fixed IIR
   output_sample((short)iir_out);        //output of fixed IIR
```

FIGURE 7.22. Program to implement an adaptive FIR filter that models (identifies) a fixed
IIR filter (adaptIDIIR.c).

```
 return;                                //return to main
}
void main()
{
 int i=0;
 for (i = 0; i < WLENGTH; i++)
   {
    w[i] = 0.0;                         //init coeff of adaptive FIR
    dly_adapt[i] = 0.0;                 //init samples of adaptive FIR
   }
 sreg.regval=0xFFFF;                    //initial seed value
 fb = 1;                                //initial feedback value
 comm_intr();                           //init DSK, codec, McBSP
 while (1);                             //infinite loop
}
```

FIGURE 7.22. (*Continued*)

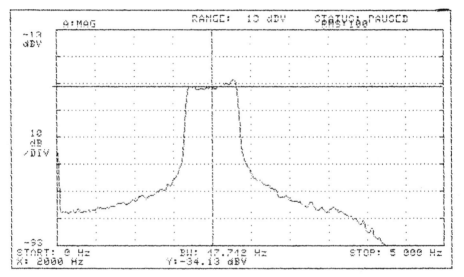

FIGURE 7.23. Adaptive FIR filter converged to a 2-kHz IIR bandpass filter obtained with an HP signal analyzer.

(the wideband signal desired). As a result, the samples of the interference are highly correlated. On the other hand, the samples of the wideband signal are relatively uncorrelated.

The characteristics of the narrowband interference permit the estimation of the narrowband interference from past samples of *splusn* in the program. The signal *splusn*, which represents the desired wideband signal with an additive narrowband interference, is delayed before becoming the input to the adaptive FIR filter. The delay is sufficiently long so that the delayed wideband signal is uncorrelated with the undelayed sample.

```
//Adaptpredict.C Adaptive predictor to cancel interference

#include "DSK6713_AIC23.h"        //codec-DSK support file
Uint32 fs=DSK6713_AIC23_FREQ_8KHZ;//set sampling rate
#include "wbsignal.h"             //wide-band signal table look-up
#define beta 1E-14                //rate of convergence
#define N 60                      //# of coefficients of adapt FIR
const short bufferlength = NS;    //buffer length for wideband signal
short splusn[N+1];                //buffer wideband signal+interference
float w[N+1];                     //buffer for weights of adapt FIR
float delay[N+1];                 //buffer for input to adapt FIR

interrupt void c_int11()          //ISR
{
 static short buffercount=0;      //init buffer
 int i;
 float yn, E;                     //yn=out adapt FIR, error signal
 short wb_signal;                 //wideband desired signal
 short noise;                     //external interference

 wb_signal=wbsignal[buffercount]; //wideband signal from look-up table
 noise = input_sample();          //external input as interference
 splusn[0] = wb_signal + noise;   //wideband signal+interference
 delay[0] = splusn[3];            //delayed input to adaptive FIR
 yn = 0;                          //init output of adaptive FIR
 for (i = 0; i < N; i++)
   yn += (w[i] * delay[i]);       //output of adaptive FIR filter
 E = splusn[0] - yn;              //(wideband+noise)-out adapt FIR
 for (i = N-1; i >= 0; i--)
  {
   w[i] = w[i]+(beta*E*delay[i]); //update weights of adapt FIR
   delay[i+1] = delay[i];         //update buffer delay samples
   splusn[i+1] = splusn[i];       //update buffer corrupted wideband
  }
 buffercount++;                   //incr buffer count of wideband
 if (buffercount >= bufferlength) //if buffer count=length of buffer
   buffercount = 0;               //reinit count
 output_sample((short)E);         //overall output
 return;
}

void main()
{
 int T = 0;
 for (T = 0; T < N; T++)          //init variables
  {
   w[T] = 0.0;                    //buffer for weights of adaptive FIR
   delay[T] = 0.0;                //buffer for delay samples
   splusn[T] = 0;                 //buffer for wideband+interference
  }
 comm_intr();                     //init DSK, codec, McBSP
 while(1);                        //infinite loop
}
```

FIGURE 7.24. Adaptive predictor program for cancellation of narrowband interference in the presence of a wideband signal (adaptpredict.c).

`%wbsignal.m` Generates wide band random sequence.Represents one info bit

```
len_code = 128;                          %length of random sequence
code = 2*round(rand(1,len_code))-1;      %generates random sequence {1,-1}
sample_rate = 2;                         %up-sampling from 4 to 8 kHz
NS = len_code * sample_rate;             %length of up-sampled sequence
sig = zeros(1,NS);                       %initialize random sequence
for i = 1:len_code                       %obtain up-sampled random sequence
   sig((i-1)*sample_rate + 1:i*sample_rate) = code(i);
end;
wbsignal = sig*5000;                     %scale for p-p amplitude of 500 mV
fid=fopen('wbsignal.h','w');             %open file for wideband signal
fprintf(fid,'#define NS 256 //number of output sample points\n\n');
fprintf(fid,'short wbsignal[256]={');
fprintf(fid,'%d, ' ,wbsignal(1:NS-1));
fprintf(fid,'%d' ,wbsignal(NS));
fprintf(fid,'};\n\n');
fclose(fid);
return;
```

FIGURE 7.25. MATLAB program that generates a desired wideband random sequence (`wbsignal.m`).

The output of the adaptive FIR filter is an estimate of the correlated narrowband interference. As a result, the error signal E is an estimate of the wideband signal desired.

Build and run this project as **adaptpredict**. Apply a sinusoidal input signal between 1 and 3 kHz, representing the narrowband interference. Run the program and verify that the output spectrum of the error signal E adapts (converges) to the desired wideband signal, showing the input interference being gradually reduced.

Change the frequency of the input sinusoidal external interference and observe the adaptation process repeated to cancel the undesirable external interference. A faster rate of convergence can be observed by increasing beta by 10.

The wideband signal desired can be observed by outputting *wb_signal* (in lieu of E). Furthermore, the wideband signal with additive interference can be observed using output_sample(*splusn[0]*). Better results are obtained when the amplitude of the external sinusoidal interference is about three times the amplitude of the wideband signal desired.

In the next example, an external wideband signal is used in lieu of a software-generated random sequence.

Example 7.8: Adaptive Predictor for Cancellation of Narrowband Interference Added to a Desired Wideband Signal Using External Inputs (`adaptpredict_2IN`)

This example extends the previous one, which implements an adaptive FIR predictor for the cancellation of narrowband interference in the presence of a wideband signal. The program `adaptpredict_2IN.c`, shown in Figure 7.26, implements the adaptive predictor using two external signals as inputs: a desired wideband signal and an undesired narrowband sinusoidal signal. The desired wideband signal with an additive narrowband interference is delayed and becomes the input to a 60-coefficient adaptive FIR filter. See also Example 7.7, where the desired wideband signal is software-generated.

The desired wideband signal is obtained from an HP signal analyzer. The bandwidth of the undesired interference is narrow compared with the bandwidth of the desired random noise signal. As a result, the samples of the interference are highly correlated. By contrast, the samples of the wideband signal are relatively uncorrelated.

The characteristics of the narrowband interference permit the estimation of the narrowband interference from past samples of `splusn` in the program. The signal `splusn`, which represents the desired wideband signal with an additive narrowband interference, is delayed before becoming the input to the adaptive FIR filter. The delay is sufficiently long so that the delayed wideband signal is uncorrelated with the undelayed sample. The output of the adaptive FIR filter is an estimate of the correlated narrowband interference. As a result, the error signal E is an estimate of the wideband signal desired.

Build and run this project as **adaptpredict_2IN**. The adapter introduced in Example 2.3 (*loop_stereo*) in Chapter 2 and used again in Example 7.3 is required to implement this example. Apply random noise into the left input channel, representing the desired wideband signal. Apply a sinusoidal signal with a frequency of 2 kHz into the right input channel, representing the undesired narrowband interference. See also Example 7.3. Run the program and verify that the output spectrum of the error signal E adapts (converges) to the desired wideband signal, showing the 2-kHz input narrowband sinusoidal interference being gradually canceled.

Change slightly the frequency of the input sinusoidal external interference and observe the adaptation process repeated to cancel the undesirable external interference. A faster rate of convergence can be observed by increasing beta by 10.

The wideband signal with the additive interference can be observed using `output_sample(splusn[0])`. Figure 7.27a shows the output spectrum of *splusn[0],* displaying both the desired wideband noise signal and the undesired narrowband interference. Figure 7.27b shows the output spectrum of the error signal E after the adaptation process converged to the desired wideband signal. Note that the desired wideband input signal can be observed by outputting `wb_signal`.

```
//Adaptpredict_2IN.c Adaptive predictor->cancel narrowband interference
#include "DSK6713_AIC23.h"        //codec-DSK support file
Uint32 fs=DSK6713_AIC23_FREQ_8KHZ;//set sampling rate
#define beta 1E-13               //rate of convergence
#define N 60                     //# of coefficients of adapt FIR
#define NS 256                   //size of wideband's buffer
#define LEFT 0                   //left channel
#define RIGHT 1                  //right channel
const short bufferlength = NS;   //buffer length for wideband signal
float splusn[N+1];               //buffer wideband signal+interference
float w[N+1];                    //buffer for weights of adapt FIR
float delay[N+1];                //buffer for input to adapt FIR
volatile union {unsigned int uint; short channel[2];}AIC23_data;

interrupt void c_int11()         //ISR
{
 static short buffercount=0;     //init buffer
 short i;
 float yn, E;                    //yn=out adapt FIR, error signal
 float wb_signal;               //wideband desired signal
 float noise;                   //external interference

 AIC23_data.uint = input_sample();//input left and right as 32-bit
 wb_signal = AIC23_data.channel[LEFT];  //desired on left channel
 noise = AIC23_data.channel[RIGHT];      //noise on right channel
 splusn[0] = (wb_signal + noise); //wideband signal + interference
 delay[0] = splusn[3];           //delayed input to adaptive FIR
 yn = 0;                         //init output of adaptive FIR
 for (i = 0; i < N; i++)
   yn += (w[i] * delay[i]);      //output of adaptive FIR filter
 E = splusn[0] - yn;             //(wideband + noise)-out adapt FIR
 for (i = N-1; i >= 0; i--)
   {
   w[i] = w[i]+(beta*E*delay[i]); //update weights of adapt FIR
   delay[i+1] = delay[i];        //update buffer delay samples
   splusn[i+1] = splusn[i];      //update buffer corrupted wideband
   }
 buffercount++;                  //incr buffer count of wideband
 if(buffercount>=bufferlength) buffercount=0;     //reinit count
 output_sample((short)E);        //overall output from left channel
 return;
}

void main()
{
 int T = 0;
 for (T = 0; T < N; T++)         //init variables
   {
   w[T] = 0.0;                   //init weights of adaptive FIR
   delay[T] = 0.0;              //init buffer for delay samples
   splusn[T] = 0;               //init wideband+interference
   }
 comm_intr();                    //init DSK, codec, McBSP
 while(1);                       //infinite loop
}
```

FIGURE 7.26. Adaptive predictor program for cancellation of narrowband interference in the presence of a wideband signal using external inputs (adaptpredict_2IN.c).

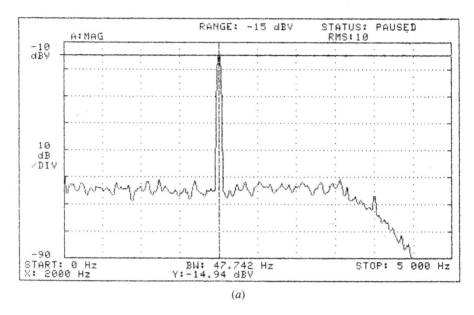

(a)

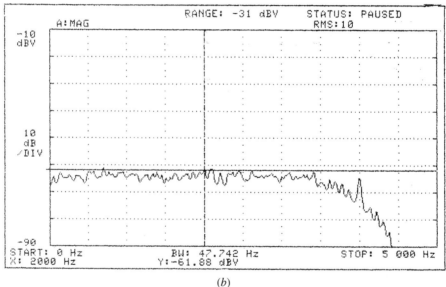

(b)

FIGURE 7.27. Plots illustrating the adaptation process obtained with a signal analyzer using adaptpredict_2IN.c; (*a*): 2-kHz undesired sinusoidal interference and a desired wideband noise signal before adaptation; (*b*) cancellation of 2-kHz interference after adaptation.

REFERENCES

1. B. Widrow and S. D. Stearns, *Adaptive Signal Processing*, Prentice Hall, Upper Saddle River, NJ, 1985.

2. B. Widrow and M. E. Hoff, Jr., Adaptive switching circuits, *IRE WESCON*, 1960, pp. 96–104.

3. B. Widrow, J. R. Glover, J. M. McCool, J. Kaunitz, C. S. Williams, R. H. Hearn, J. R. Zeidler, E. Dong, Jr., and R. C. Goodlin, Adaptive noise cancelling: principles and applications, *Proceedings of the IEEE*, Vol. 63, 1975, pp. 1692–1716.

4. R. Chassaing, *Digital Signal Processing with C and the TMS320C30*, Wiley, New York, 1992.

5. D. G. Manolakis, V. K. Ingle, and S. M. Kogon, *Statistical and Adaptive Signal Processing*, McGraw-Hill, New York, 2000.

6. S. Haykin, *Adaptive Filter Theory*, Prentice Hall, Upper Saddle River, NJ, 1986.

7. J. R. Treichler, C. R. Johnson, Jr., and M. G. Larimore, *Theory and Design of Adaptive Filters*, Wiley, New York, 1987.

8. S. M. Kuo and D. R. Morgan, *Active Noise Control Systems*, Wiley, New York, 1996.

9. K. Astrom and B. Wittenmark, *Adaptive Control*, Addison-Wesley, Reading, MA, 1995.

10. J. Tang, R. Chassaing, and W. J. Gomes III, Real-time adaptive PID controller using the TMS320C31 DSK, *Proceedings of the 2000 Texas Instruments DSPS Fest Conference*, 2000.

11. R. Chassaing, *Digital Signal Processing Laboratory Experiments Using C and the TMS320C31 DSK*, Wiley, New York, 1999.

12. R. Chassaing et al., Student projects on applications in digital signal processing with C and the TMS320C30, *Proceedings of the 2nd Annual TMS320 Educators Conference*, Texas Instruments, Dallas, TX, 1992.

13. C. S. Linquist, *Adaptive and Digital Signal Processing*, Steward and Sons, 1989.

14. S. D. Stearns and D. R. Hush, *Digital Signal Analysis*, Prentice Hall, Upper Saddle River, NJ, 1990.

15. J. R. Zeidler, Performance analysis of LMS adaptive prediction filters, *Proceedings of the IEEE*, Vol. 78, 1990, pp. 1781–1806.

16. S. T. Alexander, *Adaptive Signal Processing: Theory and Applications*, Springer-Verlag, New York, 1986.

17. C. F. Cowan and P. F. Grant, eds., *Adaptive Filters*, Prentice Hall, Upper Saddle River, NJ, 1985.

18. M. L. Honig and D. G. Messerschmitt, *Adaptive Filters: Structures, Algorithms and Applications*, Kluwer Academic, Norwell, MA, 1984.

19. V. Solo and X. Kong, *Adaptive Signal Processing Algorithms: Stability and Performance*, Prentice Hall, Upper Saddle River, NJ, 1995.

20. S. Kuo, G. Ranganathan, P. Gupta, and C. Chen, Design and implementation of adaptive filters, *IEEE 1988 International Conference on Circuits and Systems*, June 1988.

21. M. G. Bellanger, *Adaptive Digital Filters and Signal Analysis*, Marcel Dekker, New York, 1987.

22. R. Chassaing and B. Bitler, Adaptive filtering with C and the TMS320C30 digital signal processor, *Proceedings of the 1992 ASEE Annual Conference*, June 1992.

23. R. Chassaing, D. W. Horning, and P. Martin, Adaptive filtering with the TMS320C25, *Proceedings of the 1989 ASEE Annual Conference*, June 1989.

8

Code Optimization

- · Optimization techniques for code efficiency
- · Intrinsic C functions
- · Parallel instructions
- · Word-wide data access
- · Software pipelining

In this chapter we illustrate several schemes that can be used to optimize and drastically reduce the execution time of your code. These techniques include the use of instructions in parallel, word-wide data, intrinsic functions, and software pipelining.

8.1 INTRODUCTION

Begin at a workstation level; for example, use C code on a PC. While code written in assembly (ASM) is processor-specific, C code can readily be ported from one platform to another. However, optimized ASM code runs faster than C and requires less memory space.

Before optimizing, make sure that the code is functional and yields correct results. After optimizing, the code can be so reorganized and resequenced that the optimization process makes it difficult to follow. One needs to realize that if a C-coded algorithm is functional and its execution speed is satisfactory, there is no need to optimize further.

Digital Signal Processing and Applications with the C6713 and C6416 DSK   By Rulph Chassaing
ISBN 0-471-69007-4   Copyright © 2005 by John Wiley & Sons, Inc.

After testing the functionality of your C code, transport it to the C6x platform. A floating-point implementation can be modeled first, then converted to a fixed-point implementation if desired. If the performance of the code is not adequate, use different compiler options to enable software pipelining (discussed later), reduce redundant loops, and so on. If the performance desired is still not achieved, you can use loop unrolling to avoid overhead in branching. This generally improves the execution speed but increases code size. You also can use word-wide optimization by loading/accessing 32-bit word (int) data rather than 16-bit half-word (short) data. You can then process lower and upper 16-bit data independently.

If performance is still not satisfactory, you can rewrite the time-critical section of the code in linear assembly, which can be optimized by the assembler optimizer. The profiler can be used to determine the specific function(s) that need to be optimized further.

The final optimization procedure that we discuss is a software pipelining scheme to produce hand-coded ASM instructions [1,2]. It is important to follow the procedure associated with software pipelining to obtain an efficient and optimized code.

8.2 OPTIMIZATION STEPS

If the performance and results of your code are satisfactory after any particular step, you are done.

1. Program in C. Build your project without optimization.
2. Use intrinsic functions when appropriate as well as the various optimization levels.
3. Use the profiler to determine/identify the function(s) that may need to be further optimized. Then convert these function(s) to linear ASM.
4. Optimize code in ASM.

8.2.1 Compiler Options

When the optimizer is invoked, the following steps are performed. A C-coded program is first passed through a parser that performs preprocessing functions and generates an intermediate file (.if) that becomes the input to an optimizer. The optimizer generates an .opt file that becomes the input to a code generator for further optimizations and generates an ASM file.

The options:

1. -o0 optimizes the use of registers.
2. -o1 performs a local optimization in addition to the optimizations performed by the previous option: -o0.
3. -o2 performs a global optimization in addition to the optimizations performed by the previous options: -o0 and -o1.

4. -o3 performs a file optimization in addition to the optimizations performed by the three previous options: -o0, -o1, and -o2.

The options -o2 and -o3 attempt to do software optimization.

8.2.2 Intrinsic C Functions

There are a number of available C intrinsic functions that can be used to increase the efficiency of code (see also Example 3.1):

1. *int_mpy()* has the equivalent ASM instruction MPY, which multiplies the 16 LSBs of a number by the 16 LSBs of another number.

2. *int_mpyh()* has the equivalent ASM instruction MPYH, which multiplies the 16 MSBs of a number by the 16 MSBs of another number.

3. *int_mpylh()* has the equivalent ASM instruction MPYLH, which multiplies the 16 LSBs of a number by the 16 MSBs of another number.

4. *int_mpyhl()* has the equivalent instruction MPYHL, which multiplies the 16 MSBs of a number by the 16 LSBs of another number.

5. *void_nassert(int)* generates no code. It tells the compiler that the expression declared with the assert function is true. This conveys information to the compiler about alignment of pointers and arrays and of valid optimization schemes, such as word-wide optimization.

6. *uint_lo(double)* and *uint_hi(double)* obtain the low and high 32 bits of a double word, respectively (available on C67x or C64x).

8.3 PROCEDURE FOR CODE OPTIMIZATION

1. Use instructions in parallel so that multiple functional units can be operated within the same cycle.

2. Eliminate NOPs or delay slots, placing code where the NOPs are located.

3. Unroll the loop to avoid overhead with branching.

4. Use word-wide data to access a 32-bit word (int) in lieu of a 16-bit half-word (short).

5. Use software pipelining, illustrated in Section 8.5.

8.4 PROGRAMMING EXAMPLES USING CODE OPTIMIZATION TECHNIQUES

Several examples are developed to illustrate various techniques to increase the efficiency of code. Optimization using software pipelining is discussed in Section 8.5.

The dot product is used to illustrate the various optimization schemes. The dot product of two arrays can be useful for many DSP algorithms, such as filtering and correlation. The examples that follow assume that each array consists of 200 numbers. Several programming examples using mixed C and ASM code, which provide necessary background, were given in Chapter 3.

Example 8.1: Sum of Products with Word-Wide Data Access for Fixed-Point Implementation Using C Code (twosum)

Figure 8.1 shows the C code *twosum.c*, which obtains the sum of products of two arrays accessing 32-bit word data. Each array consists of 200 numbers. Separate sums of products of even and odd terms are calculated within the loop. Outside the loop, the final summation of the even and odd terms is obtained.

For a floating-point implementation, the function and the variables *sum*, *suml*, and *sumh* in Figure 8.1 are cast as *float* in lieu of *int*:

```
float   dotp (float a[ ], float b [ ])
{
        float suml, sumh, sum;
        int i;
        .
        .
        .
}
```

```
//twosum.c Sum of Products with separate accumulation of even/odd terms
//with word-wide data for fixed-point implementation

int   dotp (short a[ ], short b [ ])
{
   int suml, sumh, sum, i;
   suml = 0;
   sumh = 0;
   sum = 0;
   for (i = 0; i < 200; i +=2)
     {
       suml  += a[i] * b[i];           //sum of products of even terms
       sumh += a[i + 1] * b[i + 1];    //sum of products of odd terms
     }
     sum = suml + sumh;               //final sum of odd and even terms
     return (sum);
}
```

FIGURE 8.1. C code for sum of products using word-wide data access for separate accumulation of even and odd sum of product terms (twosum.c).

```
//dotpintrinsic.c Sum of products with C intrinsic functions using C

for (i = 0; i < 100; i++)
      {
            suml = suml + _mpy(a[i], b[i]);
            sumh = sumh + _mpyh(a[i], b[i]);
      }
return (suml + sumh);
```

FIGURE 8.2. Separate sum of products using C intrinsic functions (dotpintrinsic.c).

Example 8.2: Separate Sum of Products with C Intrinsic Functions Using C Code (dotpintrinsic)

Figure 8.2 shows the C code *dotpintrinsic.c* to illustrate the separate sum of products using two C intrinsic functions, _mpy and _mpyh, which have the equivalent ASM instructions MPY and MPYH, respectively. Whereas the even and odd sums of products are calculated within the loop, the final summation is taken outside the loop and returned to the calling function.

Example 8.3: Sum of Products with Word-Wide Access for Fixed-Point Implementation Using Linear ASM Code (twosumlasmfix.sa)

Figure 8.3 shows the linear ASM code *twosumlasmfix.sa*, which obtains two separate sums of products for a fixed-point implementation. It is not necessary to specify the functional units. Furthermore, symbolic names can be used for registers. The LDW instruction is used to load a 32-bit word-wide data value (which must be word-aligned in memory when using LDW). Lower and upper 16-bit products are calculated separately. The two ADD instructions accumulate separately the even and odd sum of products.

```
;twosumlasmfix.sa Sum of Products. Separate accum of even/odd terms
;With word-wide data for fixed-point implementation using linear ASM

loop:     LDW      *aptr++, ai        ;32-bit word ai
          LDW      *bptr++, bi        ;32-bit word bi
          MPY      ai, bi, prodl      ;lower 16-bit product
          MPYH     ai, bi, prodh      ;higher 16-bit product
          ADD      prodl, suml, suml  ;accum even terms
          ADD      prodh, sumh, sumh  ;accum odd terms
          SUB      count, 1, count    ;decrement count
    [count] B      loop               ;branch to loop
```

FIGURE 8.3. Separate sum of products using linear ASM code for fixed-point implementation (twosumlasmfix.sa).

```
;twosumlasmfloat.sa Sum of products.Separate accum of even/odd terms
;Using double-word load LDDW for floating-point implementation

loop:    LDDW    *aptr++, ai1:ai0    ;64-bit word ai0 and ai1
         LDDW    *bptr++, bi1:bi0    ;64-bit word  bi0 and bi1
         MPYSP   ai0, bi0, prodl     ;lower 32-bit product
         MPYSP   ai1, bi1, prodh     ;higher 32-bit product
         ADDSP   prodl, suml, suml   ;accum 32-bit even terms
         ADDSP   prodh, sumh, sumh   ;accum 32-bit odd terms
         SUB     count, 1, count     ;decrement count
[count]  B       loop                ;branch to loopa
```

FIGURE 8.4. Separate sum of products with LDDW using ASM code for floating-point implementation (twosumlasmfloat.sa).

Example 8.4: Sum of Products with Double-Word Load for Floating-Point Implementation Using Linear ASM Code (twosumlasmfloat)

Figure 8.4 shows the linear ASM code *twosumlasmfloat.sa* used to obtain two separate sums of products for a floating-point implementation. The double-word load instruction LDDW loads a 64-bit data value and stores it in a pair of registers. Each single-precision multiply instruction MPYSP performs a 32×32 multiplication. The sums of products of the lower and upper 32 bits are performed to yield a sum of both even and odd terms as 32 bits.

Example 8.5: Dot Product with No Parallel Instructions for Fixed-Point Implementation Using ASM Code (dotpnp)

Figure 8.5 shows the ASM code *dotpnp.asm* for the dot product with no instructions in parallel for a fixed-point implementation. A fixed-point implementation can

```
;dotpnp.asm ASM Code, no parallel instructions, fixed-point

         MVK    .S1   200, A1      ;count into A1
         ZERO   .L1   A7           ;init A7 for accum
LOOP     LDH    .D1   *A4++,A2     ;A2=16-bit data pointed by A4
         LDH    .D1   *A8++,A3     ;A3=16-bit data pointed by A8
         NOP          4            ;4 delay slots for LDH
         MPY    .M1   A2,A3,A6     ;product in A6
         NOP                       ;1 delay slot for MPY
         ADD    .L1   A6,A7,A7     ;accum in A7
         SUB    .S1   A1,1,A1      ;decrement count
[A1]     B      .S2   LOOP         ;branch to LOOP
         NOP          5            ;5 delay slots for B
```

FIGURE 8.5. ASM code with no parallel instructions for fixed-point implementation (dotpnp.asm).

be performed with all C6x devices, whereas a floating-point implementation requires a C67x platform such as the C6713 DSK.

The loop iterates 200 times. With a fixed-point implementation, each pointer register A4 and A8 increments to point at the next half-word (16 bits) in each buffer, whereas with a floating-point implementation, a pointer register increments the pointer to the next 32-bit word. The load, multiply, and branch instructions must use the .D, .M, and .S units, respectively; the add and subtract instructions can use any unit (except .M). The instructions within the loop consume 16 cycles per iteration. This yields $16 \times 200 = 3200$ cycles. Table 8.4 shows a summary of several optimization schemes for both fixed- and floating-point implementations.

Example 8.6: Dot Product with Parallel Instructions for Fixed-Point Implementation Using ASM Code (dotpp)

Figure 8.6 shows the ASM code *dotpp.asm* for the dot product with a fixed-point implementation with instructions in parallel. With code in lieu of NOPs, the number of NOPs is reduced.

The MPY instruction uses a cross-path (with .M1x) since the two operands are from different register files or different paths. The instructions SUB and B are moved up to fill some of the delay slots required by LDH. The branch instruction occurs after the ADD instruction. Using parallel instructions, the instructions within the loop now consume eight cycles per iteration, to yield $8 \times 200 = 1600$ cycles.

Example 8.7: Two Sums of Products with Word-Wide (32-Bit) Data for Fixed-Point Implementation Using ASM Code (twosumfix)

Figure 8.7 shows the ASM code *twosumfix.asm*, which calculates two separate sums of products using word-wide access of data for a fixed-point implementation. The loop count is initialized to 100 (not 200) since two sums of products are obtained

```
;dotpp.asm ASM Code with parallel instructions, fixed-point

        MVK    .S1    200, A1       ;count into A1
   ||   ZERO   .L1    A7            ;init A7 for accum
LOOP    LDH    .D1    *A4++,A2      ;A2=16-bit data pointed by A4
   ||   LDH    .D2    *B4++,B2      ;B2=16-bit data pointed by B4
        SUB    .S1    A1,1,A1       ;decrement count
 [A1]   B      .S1    LOOP          ;branch to LOOP (after ADD)
        NOP           2             ;delay slots for LDH and B
        MPY    .M1x   A2,B2,A6      ;product in A6
        NOP                         ;1 delay slot for MPY
        ADD    .L1    A6,A7,A7      ;accum in A7,then branch
;branch occurs here
```

FIGURE 8.6. ASM code with parallel instructions for fixed-point implementation.

```
;twosumfix.asm ASM code for two sums of products with word-wide data
;for fixed-point implementation

            MVK    .S1   100, A1       ;count/2 into A1
     ||     ZERO   .L1   A7            ;init A7 for accum of even terms
     ||     ZERO   .L2   B7            ;init B7 for accum of odd terms
LOOP        LDW    .D1   *A4++,A2      ;A2=32-bit data pointed by A4
     ||     LDW    .D2   *B4++,B2      ;A3=32-bit data pointed by B4
            SUB    .S1   A1,1,A1       ;decrement count
     [A1]   B      .S1   LOOP          ;branch to LOOP (after ADD)
            NOP           2            ;delay slots for both LDW and B
            MPY    .M1x  A2,B2,A6      ;lower 16-bit product in A6
     ||     MPYH   .M2x  A2,B2,B6      ;upper 16-bit product in B6
            NOP           1            ;1 delay slot for MPY/MPYH
            ADD    .L1   A6,A7,A7      ;accum even terms in A7
     ||     ADD    .L2   B6,B7,B7      ;accum odd terms in B7
;branch occurs here
```

FIGURE 8.7. ASM code for two sums of products with 32-bit data for fixed-point implementation (twosumfix.asm).

per iteration. The instruction LDW loads a word or 32-bit data. The multiply instruction MPY finds the product of the lower 16×16 data, and MPYH finds the product of the upper 16×16 data. The two ADD instructions accumulate separately the even and odd sums of products. Note that an additional ADD instruction is needed outside the loop to accumulate A7 and B7. The instructions within the loop consume eight cycles, now using 100 iterations (not 200), to yield $8 \times 100 = 800$ cycles.

Example 8.8: Dot Product with No Parallel Instructions for Floating-Point Implementation Using ASM Code (dotpnpfloat)

Figure 8.8 shows the ASM code dotpnpfloat.asm for the dot product with a floating-point implementation using no instructions in parallel. The loop iterates 200

```
;dotpnpfloat.asm ASM Code with no parallel instructions for floating-pt

            MVK      .S1   200, A1       ;count into A1
            ZERO     .L1   A7            ;init A7 for accum
LOOP        LDW      .D1   *A4++,A2      ;A2=32-bit data pointed by A4
            LDW      .D1   *A8++,A3      ;A3=32-bit data pointed by A8
            NOP             4            ;4 delay slots for LDW
            MPYSP    .M1   A2,A3,A6      ;product in A6
            NOP             3            ;3 delay slots for MPYSP
            ADDSP    .L1   A6,A7,A7      ;accum in A7
            SUB      .S1   A1,1,A1       ;decrement count
     [A1]   B        .S2   LOOP          ;branch to LOOP
            NOP             5            ;5 delay slots for B
```

FIGURE 8.8. ASM code with no parallel instructions for floating-point implementation (dotpnpfloat.asm).

```
;dotppfloat.asm  ASM Code with parallel instructions for floating-point

            MVK    .S1    200, A1    ;count into A1
      ||    ZERO   .L1    A7         ;init A7 for accum
LOOP        LDW    .D1    *A4++,A2   ;A2=32-bit data pointed by A4
      ||    LDW    .D2    *B4++,B2   ;B2=32-bit data pointed by B4
            SUB    .S1    A1,1,A1    ;decrement count
            NOP           2          ;delay slots for both LDW and B
    [A1]    B      .S2    LOOP       ;branch to LOOP (after ADDSP)
            MPYSP  .M1x   A2,B2,A6   ;product in A6
            NOP           3          ;3 delay slots for MPYSP
            ADDSP  .L1    A6,A7,A7   ;accum in A7,then branch
;branch occurs here
```

FIGURE 8.9. ASM code with parallel instructions for floating-point implementation (dotppfloat.asm).

times. The single-precision floating-point instruction MPYSP performs a 32×32 multiply. Each MPYSP and ADDSP requires three delay slots. The instructions within the loop consume a total of 18 cycles per iteration (without including three NOPs associated with ADDSP). This yields a total of $18 \times 200 = 3600$ cycles. (See Table 8.4 for a summary of several optimization schemes for both fixed- and floating-point implementations.)

Example 8.9: Dot Product with Parallel Instructions for Floating-Point Implementation Using ASM Code (dotppfloat)

Figure 8.9 shows the ASM code *dotppfloat.asm* for the dot product with a floating-point implementation using instructions in parallel. The loop iterates 200 times. By moving the SUB and B instructions up to take the place of some NOPs, the number of instructions within the loop is reduced to 10. Note that three additional NOPs would be needed outside the loop to retrieve the result from ADDSP. The instructions within the loop consume a total of 10 cycles per iteration. This yields a total of $10 \times 200 = 2000$ cycles.

Example 8.10: Two Sums of Products with Double-Word-Wide (64-Bit) Data for Floating-Point Implementation Using ASM Code (twosumfloat)

Figure 8.10 shows the ASM code *twosumfloat.asm*, which calculates two separate sums of products using double-word-wide access of 64-bit data for a floating-point implementation. The loop count is initialized to 100 since two sums of products are obtained per iteration. The instruction LDDW loads a 64-bit double-word data value into a register pair. The multiply instruction MPYSP performs a 32×32 multiply. The two ADDSP instructions accumulate separately the even and odd sums of products. The additional ADDSP instruction is needed outside the loop to accumulate A7 and

```
;twosumfloat.asm ASM Code with two sums of products for floating-pt

              MVK   .S1   100, A1      ;count/2 into A1
         ||   ZERO  .L1   A7           ;init A7 for accum of even terms
         ||   ZERO  .L2   B7           ;init B7 for accum of odd terms
LOOP          LDDW  .D1   *A4++,A3:A2  ;64-bit-> register pair A2,A3
         ||   LDDW  .D2   *B4++,B3:B2  ;64-bit-> register pair B2,B3
              SUB   .S1   A1,1,A1      ;decrement count
              NOP         2            ;delay slots for LDW
      [A1]    B     .S2   LOOP         ;branch to LOOP
              MPYSP .M1x  A2,B2,A6     ;lower 32-bit product in A6
         ||   MPYSP .M2x  A3,B3,B6     ;upper 32-bit product in B6
              NOP         3            ;3 delay slot for MPYSP
              ADDSP .L1   A6,A7,A7     ;accum even terms in A7
         ||   ADDSP .L2   B6,B7,B7     ;accum odd terms in B7
;branch occurs here
              NOP         3            ;delay slots for last ADDSP
              ADDSP .L1x  A7,B7,A4     ;final sum of even and odd terms
              NOP         3            ;delay slots for ADDSP
```

FIGURE 8.10. ASM code with two sums of products for floating-point implementation (twosumfloat.asm).

B7. The instructions within the loop consume a total of 10 cycles, using 100 iterations (not 200), to yield a total of $10 \times 100 = 1000$ cycles.

8.5 SOFTWARE PIPELINING FOR CODE OPTIMIZATION

Software pipelining is a scheme to write efficient code in ASM so that all the functional units are utilized within one cycle. Optimization levels -o2 and -o3 enable code generation to generate (or attempt to generate) software-pipelined code.

There are three stages associated with software pipelining:

1. *Prolog (warm-up)*. This stage contains instructions needed to build up the loop kernel (cycle).
2. *Loop kernel (cycle)*. Within this loop, all instructions are executed in parallel. The entire loop kernel can be executed in *one* cycle, since all the instructions within the loop kernel stage are in parallel.
3. *Epilog (cool-off)*. This stage contains the instructions necessary to complete all iterations.

8.5.1 Procedure for Hand-Coded Software Pipelining

1. Draw a dependency graph.
2. Set up a scheduling table.
3. Obtain code from the scheduling table.

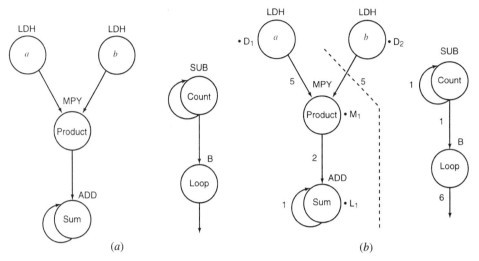

FIGURE 8.11. Dependency graph for dotp product: (*a*) initial stage; (*b*) final stage.

8.5.2 Dependency Graph

Figure 8.11 shows a dependency graph. A procedure for drawing a dependency graph follows.

1. Draw the nodes and paths.
2. Write the number of cycles to complete an instruction.
3. Assign functional units associated with each node.
4. Separate the data path so that the maximum number of units are utilized.

A node has one or more data paths going into and/or out of the node. The numbers next to each node represent the number of cycles required to complete the associated instruction. A parent node contains an instruction that writes to a variable, whereas a child node contains an instruction that reads a variable written by the parent.

The LDH instructions are considered to be the parents of the MPY instruction since the results of the two load instructions are used to perform the MPY instruction. Similarly, the MPY is the parent of the ADD instruction. The ADD instruction is fed back as input for the next iteration; similarly with the SUB instruction.

Figure 8.12 shows another dependency graph associated with two sums of products for a fixed-point implementation. The length of the prolog section is the longest path from the dependency graph in Figure 8.12. Since the longest path is 8, the length of the prolog is 7 before entering the loop kernel (cycle) at cycle 8.

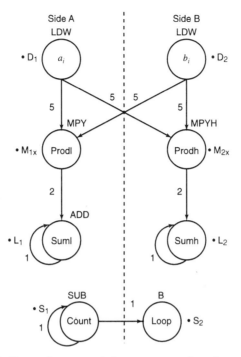

FIGURE 8.12. Dependency graph for two sums of products per iteration.

A similar dependency graph for a floating-point implementation can be obtained using LDDW, MPYSP, and ADDSP in lieu of LDW, MPY/MPYH, and ADD, respectively, in Figure 8.12. Note that the single-precision instructions ADDSP and MPYSP both take four cycles to complete (three delay slots each).

8.5.3 Scheduling Table

Table 8.1 shows a scheduling table drawn from the dependency graph.

1. LDW starts in cycle 1.
2. MPY and MPYH must start five cycles after the LDWs due to the four delay slots. Therefore, MPY and MPYH start in cycle 6.
3. ADD must start two cycles after MPY/MPYH due to the one delay slot of MPY/MPYH. Therefore, ADD starts in cycle 8.
4. B has five delay slots and starts in cycle 3, since branching occurs in cycle 9, after the ADD instruction.
5. SUB instruction must start one cycle before the branch instruction, since the loop count is decremented before branching occurs. Therefore, SUB starts in cycle 2.

TABLE 8.1   Schedule Table of Dot Product Before Software Pipelining for Fixed-Point Implementation

Cycles Units	1, 9, ...	2, 10, ...	3, 11, ...	4, 12, ...	5, 13, ...	6, 14, ...	7, 15, ...	8, 16, ...
.D1	LDW							
.D2	LDW							
.M1						MPY		
.M2						MPYH		
.L1								ADD
.L2								ADD
.S1		SUB						
.S2			B					

TABLE 8.2   Schedule Table of Dot Product After Software Pipelining for Fixed-Point Implementation

Cycles Units	Prolog							Loop Kernel
	1	2	3	4	5	6	7	8
.D1	LDW	LDW	LDW	LDW	LDW	LDW	LDW	LDW
.D2	LDW	LDW	LDW	LDW	LDW	LDW	LDW	LDW
.M1						MPY	MPY	MPY
.M2						MPYH	MPYH	MPYH
.L1								ADD
.L2								ADD
.S1		SUB	SUB	SUB	SUB	SUB	SUB	SUB
.S2			B	B	B	B	B	B

From Table 8.1, the two LDW instructions are in parallel and are issued in cycles 1, 9, 17, ... The SUB instruction is issued in cycles 2, 10, 18, ... This is followed by the branch (B) instruction issued in cycles 3, 11, 19, ... The two parallel instructions MPY and MPYH are issued in cycles 6, 14, 22, ... The ADD instructions are issued in cycles 8, 16, 24, ...

Table 8.1 is extended to illustrate the different stages: prolog (cycles 1 through 7), loop kernel (cycle 8), and epilog (cycles 9, 10, ... not shown), as shown in Table 8.2. The instructions within the prolog stage are repeated until and including the loop kernel (cycle) stage. Instructions in the epilog stage (cycles 9, 10, ...) complete the functionality of the code.

From Table 8.2, an efficient optimized code can be obtained. Note that it is possible to start processing a new iteration before previous iterations are finished. Software pipelining allows us to determine when to start a new loop iteration.

Loop Kernel (Cycle)

Within the loop kernel, in cycle 8, each functional unit is used only once. The minimum iteration interval is the minimum number of cycles required to wait before the initiation of a successive iteration. This interval is 1. As a result, a new iteration can be initiated every cycle.

Within loop cycle 8, multiple iterations of the loop execute in parallel. In cycle 8, different iterations are processed at the same time. For example, the ADDs add data for iteration 1, while MPY and MPYH multiply data for iteration 3, LDWs load data for iteration 8, SUB decrements the counter for iteration 7, and B branches for iteration 6. Note that the values being multiplied are loaded into registers five cycles prior to the cycle when the values are multiplied. Before the first multiplication occurs, the fifth load has just completed. This software pipeline is eight iterations deep.

Example 8.11: Dot Product Using Software Pipelining for a Fixed-Point Implementation

This example implements the dot product using software pipelining for a fixed-point implementation. From Table 8.2, one can readily obtained the ASM code *dotpiped-fix.asm* shown in Figure 8.13. The loop count is 100 since two multiplies and two accumulates are calculated per iteration. The following instructions start in the following cycles:

Cycle 1: LDW, LDW (also initialization of count and accumulators A7 and B7)

Cycle 2: LDW, LDW, SUB

Cycles 3–5: LDW, LDW, SUB, B

Cycles 6–7: LDW, LDW, MPY, MPYH, SUB, B

Cycles 8–107: LDW, LDW, MPY, MPYH, ADD, ADD, SUB, B

Cycle 108: LDW, LDW, MPY, MPYH, ADD, ADD, SUB, B

The prolog section is within cycles 1 through 7; the loop kernel is in cycle 8, where all the instructions are in parallel; and the epilog section is in cycle 108. Note that SUB is made conditional to ensure that Al is no longer decremented once it reaches zero.

```
;dotpipedfix.asm   ASM code for dot product with software pipelining
;For fixed-point implementation
;cycle 1
            MVK         .S1   100,A1          ;loop count
      ||    ZERO        .L1   A7              ;init accum A7
      ||    ZERO        .L2   B7              ;init accum B7
      ||    LDW         .D1   *A4++,A2        ;32-bit data in A2
      ||    LDW         .D2   *B4++,B2        ;32-bit data in B2
;cycle 2
      ||    LDW         .D1   *A4++,A2        ;32-bit data in A2
      ||    LDW         .D2   *B4++,B2        ;32-bit data in B2
      || [A1] SUB       .S1   A1,1,A1         ;decrement count
;cycle 3
      ||    LDW         .D1   *A4++,A2        ;32-bit data in A2
      ||    LDW         .D2   *B4++,B2        ;32-bit data in B2
      || [A1] SUB       .S1   A1,1,A1         ;decrement count
      || [A1] B         .S2   LOOP            ;branch to LOOP
;cycle 4
      ||    LDW         .D1   *A4++,A2        ;32-bit data in A2
      ||    LDW         .D2   *B4++,B2        ;32-bit data in B2
      || [A1] SUB       .S1   A1,1,A1         ;decrement count
      || [A1] B         .S2   LOOP            ;branch to LOOP
;cycle 5
      ||    LDW         .D1   *A4++,A2        ;32-bit data in A2
      ||    LDW         .D2   *B4++,B2        ;32-bit data in B2
      || [A1] SUB       .S1   A1,1,A1         ;decrement count
      || [A1] B         .S2   LOOP            ;branch to LOOP
;cycle 6
      ||    LDW         .D1   *A4++,A2        ;32-bit data in A2
      ||    LDW         .D2   *B4++,B2        ;32-bit data in B2
      || [A1] SUB       .S1   A1,1,A1         ;decrement count
      || [A1] B         .S2   LOOP            ;branch to LOOP
      ||    MPY         .M1x  A2,B2,A6        ;lower 16-bit product into A6
      ||    MPYH        .M2x  A2,B2,B6        ;upper 16-bit product into B6
;cycle 7
      ||    LDW         .D1   *A4++,A2        ;32-bit data in A2
      ||    LDW         .D2   *B4++,B2        ;32-bit data in B2
      || [A1] SUB       .S1   A1,1,A1         ;decrement count
      || [A1] B         .S2   LOOP            ;branch to LOOP
      ||    MPY         .M1x  A2,B2,A6        ;lower 16-bit product into A6
      ||    MPYH        .M2x  A2,B2,B6        ;upper 16-bit product into B6
;cycles 8-107 (loop cycle)
      ||    LDW         .D1   *A4++,A2        ;32-bit data in A2
      ||    LDW         .D2   *B4++,B2        ;32-bit data in B2
      || [A1] SUB       .S1   A1,1,A1         ;decrement count
      || [A1] B         .S2   LOOP            ;branch to LOOP
      ||    MPY         .M1x  A2,B2,A6        ;lower 16-bit product into A6
      ||    MPYH        .M2x  A2,B2,B6        ;upper 16-bit product into B6
      ||    ADD         .L1   A6,A7,A7        ;accum in A7
      ||    ADD         .L2   B6,B7,B7        ;accum in B7
;branch occurs here
;cycle 108 (epilog)
            ADD         .L1x  A7,B7,A4        ;final accum of odd/even
```

FIGURE 8.13. ASM code using software pipelining for fixed-point implementation
(dotpipedfix.asm).

*Example 8.12: Dot Product Using Software Pipelining for
a Floating-Point Implementation*

This example implements the dot product using software pipelining for a floating-point implementation. Table 8.3 shows a floating-point version of Table 8.2. LDW becomes LDDW, MPY/MPYH become MPYSP, and ADD becomes ADDSP. Both MPYSP and ADDSP have three delays slots. As a result, the loop kernel starts in cycle 10 in lieu of cycle 8. The SUB and B instructions start in cycles 4 and 5, respectively, in lieu of cycles 2 and 3. ADDSP starts in cycle 10 in lieu of cycle 8. The software pipeline for a floating-point implementation is 10 deep.

Figure 8.14 shows the ASM code *dotpipedfloat.asm*, which implements the floating-point version of the dot product. Since ADDSP has three delay slots, the accumulation is staggered by four. The accumulation associated with one of the ADDSP instructions at each loop cycle follows:

Loop Cycle	Accumulator (one ADDSP)	
1	0	
2	0	
3	0	
4	0	
5	p0	;first product
6	p1	;second product
7	p3	
8	p4	
9	p0 + p4	;sum of first and fifth products
10	p1 + p5	;sum of second and sixth products
11	p2 + p6	
12	p3 + p7	
13	p0 + p4 + p8	;sum of first, fifth, and ninth products
14	p1 + p5 + p9	
15	p2 + p6 + p10	
16	p3 + p7 + p11	
17	p0 + p4 + p8 + p12	
.	.	
.	.	
.	.	
99	p2 + p6 + p10 + ... + p94	
100	p3 + p7 + p11 + ... + p95	

This accumulation is shown associated with the loop cycle. The actual cycle is shifted by 9 (by the cycles in the prolog section). Note that the first product, p0, is

TABLE 8.3 Schedule Table of Dot Product After Software Pipelining for Floating-Point Implementation

Cycle Units	Prolog									Loop Kernel
	1	2	3	4	5	6	7	8	9	10
.D1	LDDW	LDDW	LDDW	LDDW	LDDW	LDDW	LDDW	LDDW	LDDW	LDDW
.D2	LDDW	LDDW	LDDW	LDDW	LDDW	LDDW	LDDW	LDDW	LDDW	LDDW
.M1						MPYSP	MPYSP	MPYSP	MPYSP	MPYSP
.M2						MPYSP	MPYSP	MPYSP	MPYSP	MPYSP
.L1										ADDSP
.L2										ADDSP
.S1				SUB	SUB	SUB	SUB	SUB	SUB	SUB
.S2					B	B	B	B	B	B

```
;dotpipedfloat.asm   ASM code for dot product with software pipelining
;For floating-point implementation
;cycle 1
                MVK       .S1   100,A1               ;loop count
        ||      ZERO      .L1   A7               ;init accum A7
        ||      ZERO      .L2   B7               ;init accum B7
        ||      LDDW      .D1   *A4++,A3:A2   ;64-bit data in A2 and A3
        ||      LDDW      .D2   *B4++,B3:B2   ;64-bit data in B2 and B3
;cycle 2
        ||      LDDW      .D1   *A4++,A3:A2   ;64-bit data in A2 and A3
        ||      LDDW      .D2   *B4++,B3:B2   ;64-bit data in B2 and B3
;cycle 3
        ||      LDDW      .D1   *A4++,A3:A2   ;64-bit data in A2 and A3
        ||      LDDW      .D2   *B4++,B3:B2   ;64-bit data in B2 and B3
;cycle 4
        ||      LDDW      .D1   *A4++,A3:A2   ;64-bit data in A2 and A3
        ||      LDDW      .D2   *B4++,B3:B2   ;64-bit data in B2 and B3
        || [A1] SUB       .S1   A1,1,A1         ;decrement count
;cycle 5
        ||      LDDW      .D1   *A4++,A3:A2   ;64-bit data in A2 and A3
        ||      LDDW      .D2   *B4++,B3:B2   ;64-bit data in B2 and B3
        || [A1] SUB       .S1   A1,1,A1         ;decrement count
        || [A1] B         .S2   LOOP              ;branch to LOOP
;cycle 6
        ||      LDDW      .D1   *A4++,A3:A2   ;64-bit data in A2 and A3
        ||      LDDW      .D2   *B4++,B3:B2   ;64-bit data in B2 and B3
        || [A1] SUB       .S1   A1,1,A1         ;decrement count
        || [A1] B         .S2   LOOP              ;branch to LOOP
        ||      MPYSP .M1x  A2,B2,A6      ;lower 32-bit product into A6
        ||      MPYSP .M2x  A3,B3,B6      ;upper 32-bit product into B6
;cycle 7
        ||      LDDW      .D1   *A4++,A3:A2   ;32-bit data in A2 and A3
        ||      LDDW      .D2   *B4++,B3:B2   ;32-bit data in B2 and B3
        || [A1] SUB       .S1   A1,1,A1         ;decrement count
        || [A1] B         .S2   LOOP              ;branch to LOOP
        ||      MPYSP .M1x  A2,B2,A6      ;lower 32-bit product into A6
        ||      MPYSP .M2x  A3,B3,B6      ;upper 32-bit product into B6
```

FIGURE 8.14. ASM code using software pipelining for floating-point implementation (dotpipedfloat.asm).

```
;cycle 8
        ||        LDDW        .D1    *A4++,A3:A2    ;32-bit data in A2 and A3
        ||        LDDW        .D2    *B4++,B3:B2    ;32-bit data in B2 and B3
        || [A1]   SUB         .S1    A1,1,A1        ;decrement count
        || [A1]   B           .S2    LOOP           ;branch to LOOP
        ||      MPYSP .M1x    A2,B2,A6     ;lower 32-bit product into A6
        ||      MPYSP .M2x    A3,B3,B6     ;upper 32-bit product into B6
;cycle 9
        ||        LDDW        .D1    *A4++,A3:A2    ;32-bit data in A2 and A3
        ||        LDDW        .D2    *B4++,B3:B2    ;32-bit data in B2 and B3
        || [A1]   SUB         .S1    A1,1,A1        ;decrement count
        || [A1]   B           .S2    LOOP           ;branch to LOOP
        ||      MPYSP .M1x    A2,B2,A6     ;lower 32-bit product into A6
        ||      MPYSP .M2x    A3,B3,B6     ;upper 32-bit product into B6
;cycles 10-109 (loop kernel)
        ||        LDDW        .D1    *A4++,A3:A2    ;32-bit data in A2 and A3
        ||        LDDW        .D2    *B4++,B3:B2    ;32-bit data in B2 and B3
        || [A1]   SUB         .S1    A1,1,A1        ;decrement count
        || [A1]   B           .S2    LOOP           ;branch to LOOP
        ||      MPYSP .M1x    A2,B2,A6     ;lower 32-bit product into A6
        ||      MPYSP .M2x    A3,B3,B6     ;upper 32-bit product into B6
        ||      ADDSP .L1     A6,A7,A7     ;accum in A7
        ||      ADDSP .L2     B6,B7,B7     ;accum in B7
;branch occurs here
;cycles 110-124 (epilog)
                ADDSP .L1x    A7,B7,A0     ;lower/upper sum of products
                ADDSP .L2x    A7,B7,B0     ;
                ADDSP .L1x    A7,B7,A0     ;
                ADDSP .L2x    A7,B7,B0     ;
                NOP                        ;wait for 1st B0
                ADDSP .L1x    A0,B0,A5     ;1st two sum of products
                NOP                        ;wait for 2nd B0
                ADDSP .L2x    A0,B0,B5     ;last two sum of products
                NOP           3            ;3 delay slots for ADDSP
                ADDSP .L1x    A5,B5,A4     ;final sum
                NOP           3            ;3 delay slots for final sum
```

FIGURE 8.14. (*Continued*)

obtained (available) in loop cycle 5 since the first ADDSP starts in loop cycle 1 and has three delay slots. The first product, p0, is associated with the lower 32-bit term. The second ADDSP (not shown) accumulates the upper 32-bit sum of products.

A6 contains the lower 32-bit products and B6 contains the upper 32-bit products. The sums of the lower and upper 32-bit products are accumulated in A7 and B7, respectively.

The epilog section contains the following instructions associated with the actual cycle (not loop cycles), as shown in Figure 8.14.

Cycle	Instruction
110	ADDSP
111	ADDSP
112	ADDSP
113	ADDSP
114	NOP

Cycle	Instruction	
115	ADDSP	
116	NOP	
117	ADDSP	
118–120	NOP	3
121	ADDSP	
122–124	NOP	3

In cycles 113 through 116, A7 contains the lower 32-bit sum of products and B7 contains the upper 32-bit sum of products, or:

Cycle	A7 for Lower 32 Bits (B7 for Upper 32 Bits)
113	p0 + p4 + p8 + ... + p96
114	p1 + p5 + p9 + ... + p97
115	p2 + p6 + p10 + ... + p98
116	p3 + p7 + p11 + ... + p99

In cycle 114, A0 = A7 + B7 is available. A0 accumulates the lower and the upper sum of products, where

```
A7 = p0 + p4 + p8 + . . . + p96   (lower 32 bits)
B7 = p0 + p4 + p8 + . . . + p96   (upper 32 bits)
```

In cycle 115, B0 = A7 + B7 is available, where

```
A7 = p1 + p5 + p9 + . . . + p97   (lower 32 bits)
B7 = p1 + p5 + p9 + . . . + p97   (upper 32 bits)
```

Similarly, in cycles 116 and 117, A0 and B0 are obtained (available) as

```
A0 = sum of lower/upper 32 bits of (p2 + p6 + p10 + . . . + p98)
B0 = sum of lower/upper 32 bits of (p3 + p7 + p11 + . . . + p99)
```

In cycle 119, A5 = A0 + B0 (obtained from cycles 114 and 115). In cycle 121, B5 = A0 + B0 (obtained from cycles 116 and 117).

The final sum accumulates in A4 and is available after cycle 124.

8.6 EXECUTION CYCLES FOR DIFFERENT OPTIMIZATION SCHEMES

Table 8.4 shows a summary of the different optimization schemes for both fixed- and floating-point implementations, for a count of 200. The number of cycles can be

TABLE 8.4   Number of Cycles with Different Optimization Schemes for Both Fixed- and Floating-Point Implementations (Count = 200)

	Number of Cycles	
Optimization Scheme	Fixed-Point	Floating-Point
1. No optimization	2 + (16 × 200) = 3202	2 + (18 × 200) = 3602
2. With parallel instructions	1 + (8 × 200) = 1601	1 + (10 × 200) = 2001
3. Two sums per iteration	1 + (8 × 100) = 801	1 + (10 × 100) + 7 = 1008
4. With software pipelining	7 + (100) + 1 = 108	9 + (100) + 15 = 124

obtained for different array sizes, since the number of cycles in the prolog and epilog stages remain the same.

Note that for a count of 1000, the fixed- and floating-point implementations with software pipeling take:

Fixed-point: $7+(\text{count}/2)+1 = 508$ cycles
Floating-point: $9+(\text{count}/2)+15 = 524$ cycles

REFERENCES

1. *TMS320C6000 Programmer's Guide*, SPRU198G, Texas Instruments, Dallas, TX, 2002.

2. *Guidelines for Software Development Efficiency on the TMS320C6000 VelociTI Architecture*, SPRA434, Texas Instruments, Dallas, TX, 1998.

3. *TMS320C6000 CPU and Instruction Set*, SPRU189F, Texas Instruments, Dallas, TX, 2000.

4. *TMS320C6000 Assembly Language Tools User's Guide*, SPRU186K, Texas Instruments, Dallas, TX, 2002.

5. *TMS320C6000 Optimizing C Compiler User's Guide*, SPRU187G, Texas Instruments, Dallas, TX, 2000.

9

DSP/BIOS and RTDX Using MATLAB, Visual C++, Visual Basic, and LabVIEW

Three examples are included to introduce DSP/BIOS and several others to illustrate real-time data transfer (RTDX) using different links and schemes with MATLAB, Visual C++, Visual Basic, and LabVIEW.

DSP/BIOS provides CCS with the capability for analysis, scheduling, and data exchange in real time [1–5]. An application program can be analyzed while the DSP is running (the target processor need not be stopped). Many DSP/BIOS application programming interface (API) modules are available for real-time analysis, I/O, and so on. API functions are included with CCS to configure and control operation of the codec. They initialize the DSK, the McBSP, and the codec.

1. *Real-time analysis.* This may or may not be critical. For example, it is necessary to respond to input samples so that information is not lost. On the other hand, the transfer of data from the DSP to the host PC may be done between incoming samples.

2. *Real-time scheduling.* Data transfer is scheduled through DSP/BIOS software interrupts. Tasks/functions are initially assigned different priorities. Based on the results obtained from a CPU execution graph, one can reprioritize these different tasks. The CPU execution graph shows when various tasks are executed and whether or not the CPU misses real-time data. This graph is similar to the types of plots obtained with a logic analyzer. Figure 9.1 shows an execution graph associated with an audio example. This graph shows the execution of *threads*. A thread can be an independent stream of instructions executed by the DSp. It may contain an ISR, a function call, and so on.

Digital Signal Processing and Applications with the C6713 and C6416 DSK   By Rulph Chassaing
ISBN 0-471-69007-4   Copyright © 2005 by John Wiley & Sons, Inc.

Different types of threads are given different priorities. Hardware interrupts (HWIs) have the highest priorities, followed by software interrupts (SWI), which include periodic functions (PRD).

3. *Real-time data exchange* (RTDX). This allows the exchange of data between the host and the processor, via the onboard Joint Test Action Group (JTAG) interface, while the processor is running. RTDX consists of both target and host components. Data are transferred through "pipes" (for receiving and for transmitting). If the CPU starts missing real-time data, one can find out from the execution graph. Reprioritizing, if possible, could then solve this problem.

Figure 9.1a illustrates overloading the CPU with no-operation instructions (NOPs). As the number of NOPs is increased, the effects on the output can be monitored. Figure 9.1a indicates that the task of "audioSwi" has the highest priority and can interrupt the lower priority task of "loadPrd." In Figure 9.1b, "audioSwi" has a lower priority and has to wait for the higher-priority tasks of "loadPrd" and "Prd_swi". This causes data to be missed. For example, with music as input and with the number of NOPs increasing (up to a million), one can hear the gradual degradation of the output signal as the CPU starts missing execution. The execution graph can show when the CPU starts missing data.

Another consideration is the use of the LOG module LOG_printf( ) to monitor a program in real time. The C function *printf( )*, supported by real-time library support, takes too many cycles to be desirable for real-time monitoring (see Example 1.3); the LOG module LOG_printf( ) takes considerably less time. The LOG_printf( ) function can be used to record data in critical time, while the trans-

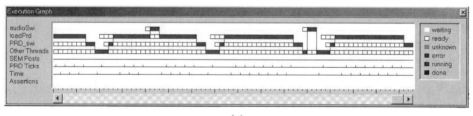

(a)

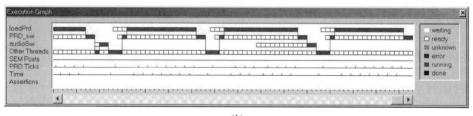

(b)

FIGURE 9.1. CCS plots of execution graphs as the CPU is being overloaded with NOPs: (a) output not degraded when setting audioSwi with the highest priority; (b) output degraded when setting audioSwi with a lower priority.

fer of data from the target processor to the host can occur in less critical time. Results of the performance of LOG_printf( ) supported with DSP/BIOS versus printf( ) supported with the runtime support library show that printf( ) can take 100 times more cycles to execute.

9.1 INTRODUCTION TO DSP/BIOS

Three examples are provided to introduce DSP/BIOS. An extensive amount of literature is available from TI on DSP/BIOS [1–5]. These examples illustrate the creation of a configuration (.cdb) file where, for example, interrupt and the execution period of a specific function can be set.

Example 9.1: Sine Generation with DIP Switch Control Through DSP/BIOS (bios_sine_ctrl)

This example illustrates the use of DSP/BIOS in controlling a generated tone with a user DIP switch. A major factor in using DSP/BIOS is the setup of a configuration file, from which scheduling, function management, hardware and software interrupt, and so on can be specified.

1. Create a new project bios_sine_ctrl.pjt. Add to the project the init and communication file C6713dskinit.c and the main C source file bios_sine_ctrl.c shown in Figure 9.2.

2. Add a configuration file to the project. Select File → New → DSP/BIOS Configuration. Select DSK6713.cdb as the configuration template.

3. Expand on Scheduling (from the configuration file). Right-click on PRD— Periodic Function Manager → Insert PRD. This inserts PRD0, which is to be renamed (right-click to rename) PRD_sinegen.

4. Right-click on PRD_sinegen and select Properties to set the period (ticks) to 5000 and function to _sinegen. Note the underscore in the function name by convention in referencing a C function. Press OK to default on the rest of the fields.

5. Repeat the previous steps 3 and 4 to set up another periodic function. Name it PRD_blinkLED0, and set its properties for a period of 200 and its function name as _blinkLED0. Save (File → Save as) this configuration file as *bios_sine_ctrl.cdb* in the folder bios_sine_ctrl. The properties of this configuration (.cdb) file are shown in Figure 9.3 within the CCS plot. Several support files are autogenerated by the configuration (.cdb) file when it is saved.

6. Add the configuration file to the project (selecting Project → Add Files to Project). Note that it is a (.cdb) type of file. Verify that it has been added to the project by expanding DSP/BIOS Config from the Projects/File View window.

```
//bios_sine_ctrl.c Sine generation with DIP Switch control
#include "bios_sine_ctrlcfg.h"              //generated support file
#include "dsk6713_led.h"
#include "dsk6713_dip.h"
#include "dsk6713_aic23.h"                  //codec-DSK support file
Uint32 fs=DSK6713_AIC23_FREQ_8KHZ;         //set sampling rate
short sine_on = 0, loop = 0, gain = 10;
short sine_table[8] = {0,707,1000,707,0,-707,-1000,-707};

void sinegen()
{
 if (DSK6713_DIP_get(2) == 0)               //if sw#2 pressed
  {
    DSK6713_LED_on(2);                      //turn on led#2
    while(++sine_on < 5000)                 //generate sine wave
     {                                      //for 5 sec
      output_sample(sine_table[loop]*gain); //output
      if (++loop > 7) loop = 0;
     }
    sine_on = 0;
  }
 DSK6713_LED_off(2);
}
void blinkLED0()
{
 DSK6713_LED_toggle(0);
 if (DSK6713_DIP_get(3) == 0) DSK6713_LED_on(3);
 else    DSK6713_LED_off(3);
}
void main()
{
 comm_poll();
 DSK6713_LED_init();
 DSK6713_DIP_init();
}
```

FIGURE 9.2. C source program for sine generation with DIP switch control through DSP/BIOS (bios_sine_ctrl.c).

7. Two support files, *bios_sine_ctrlcfg.s62* and *bios_sine_ctrlcfg_c.c*, have been autogenerated by the configuration file and added to the project. Verify that by expanding on Generated Files. A linker command file, *bios_sine_ctrlcfg.cmd*, was also generated and must be added to the project by the user.

8. A header file, *bios_sine_ctrlcfg.h*, was also autogenerated and must be included in the main C source file. Scan all files dependencies and verify that a number of chip support files have been included in the project.

Select Project → Build Options → Preprocessor. Set Define Symbols (d) to CHIP_6713 and target version to C671x. From the Linker Tab → Include Libraries

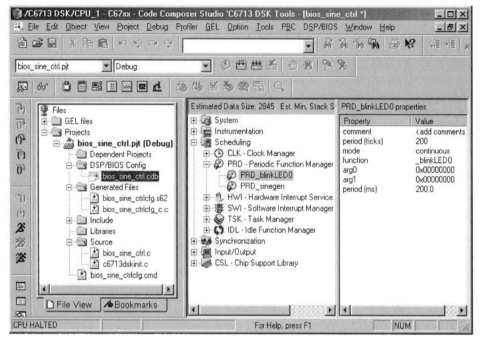

FIGURE 9.3. CCS windows displaying configuration settings (middle section) for creating the (.cdb) file bios_sine_ctrl.cdb.

(−l), include the BSL library support file: DSK6713bsl.lib. The run-time and chip support library files are already included in the autogenerated linker command file and are not added to the project.

Build the project. The necessary files are included in the folder **bios_sine_ctrl**. Verify the following:

1. LED #0 blinks.
2. When dip switch #3 is pressed, LED #3 turns on.
3. When dip switch #2 is pressed, a 1-kHz tone is generated (for about 1 second) approximately every 5 seconds (5000 ms).

In the line of code

```
while (++sine_on <5000)
```

change 5000 to 2000. Rebuild and verify that the sine wave is now being generated every 2 seconds.

Access the configuration file and change the properties of the periodic function manager *PRD_sinegen* for a period of 500 in lieu of 5000. Rebuild the project and verify that the 1-kHz tone is now generated continuously (a much faster and continuous, unnoticeable burst).

Example 9.2: Blinking of LEDs at Different Rates Using DSP/BIOS (`bios_4LED`)

This example illustrates the use of the periodic function manager within the configuration file to control the blinking rates of the four onboard LEDs. The C source program *bios_4led.c* is shown in Figure 9.4. Create the .cdb configuration file using four functions from the PRD-Periodic Function Manager: *PRD_blinkLED0,...,* *PRD_blinkLED3* with the following periods: 50, 100, 200, and 400 ms and associated functions *_blinkLED0,...,blinkLED3*, respectively. Save and add this configuration file to the project along with the BSL support library file. The linker command file, autogenerated by the configuration file, needs to be added to the project (the other autogenerated support files will be added). See Example 9.1. The necessary files for this project are in the folder bios4LED.

Build this project as **bios_4led**. Verify the different blinking rates of each of the LEDs, ranging from approximately 50 ms (LED0) to 400 ms (LED3). Change the blinking rates by changing the values of the periods set in the configuration file.

```
//Bios_4LED.c Blinking of LEDs at different rates using DSP/BIOS
#include "bios_4ledcfg.h"            //generated by .cdb file
#include "dsk6713.h"                 //BSL support file
void blinkLED0()
{
    DSK6713_LED_toggle(0);           //toggle LED0(50ms)
}
void blinkLED1()
{
    DSK6713_LED_toggle(1);           //toggle LED1(100ms)
}
void blinkLED2()
{
    DSK6713_LED_toggle(2);           //toggle LED2(200ms)
}
void blinkLED3()
{
    DSK6713_LED_toggle(3);           //toggle LED3(400ms)
}
void main()
{
    DSK6713_init();                  //init BSL
}
```

FIGURE 9.4. C program for controlling the blinking rates of the onboard LEDs using DSP/BIOS (bios_4LED.c).

```c
//bios_sine_intr.c Sine generation using .cdb in BIOS for INT11

#include <std.h>
#include <log.h>
#include "bios_sine_intrcfg.h"            //generated by .cdb
#include "dsk6713.h"
#include "DSK6713_AIC23.h"
Uint32 fs=DSK6713_AIC23_FREQ_8KHZ;        //set sampling rate
int  loop = 0, flag = 0;
short  sine_table[8]={0,707,1000,707,0,-707,-1000,-707};//sine data

interrupt void c_int11()                  //ISR defined in .cdb
{
 short out_data = 0;
 if (flag == 1)                           //if SW#3 is pressed
  {
    out_data = sine_table[loop]*10;
    if (++loop > 7) loop = 0;             //if @end of table
  }
 output_sample(out_data);                 //real-time output
}

void main()
{
 comm_intr();                             //init codec,DSK,MCBSP
 DSK6713_LED_init();                      //BSL support for LED
 DSK6713_DIP_init();                      //BSL support for SW
 LOG_printf(&trace,"Start");              //from message log
 while(1)                                 //infinite loop
  {
    if (DSK6713_DIP_get(3) == 0)          //if sw#3 is pressed
     {
       flag = 1;                          //set flag=1 if pressed
       DSK6713_LED_on(3);                 //then turn on LED#3
     }
    else
     {
       DSK6713_LED_off(3);                //if not pressed LED off
       flag = 0;                          //flag=0 if not pressed
     }
  }                                       //end of while(1)
}                                         //end of main
```

FIGURE 9.5. C program for sine generation with INT11 set in the (.cdb) configuration file (bios_sine_intr.c).

Example 9.3: Sine Generation Using BIOS to Set Up Interrupt INT11 (bios_sine_intr)

This example illustrates the generation of a sine wave when a dip switch is pressed. Figure 9.5 shows the interrupt-driven program *bios_sine_intr.c* that implements the sine generation. It uses a configuration file to set up transmit interrupt INT11 and specify the ISR. See also Examples 9.1 and 9.2.

Create a configuration file as in Example 9.1. Expand on Scheduling and HWI-Hardware Interrupt Service Routine Manager. Right-click on HWI-INT11 to select interrupt 11. Set its properties such that the interrupt source is *MCBSP_1_Transmit* and the function is *_c_int11* (the interrupt service function/routine). A vector file need not be added to the project.

A message log can be obtained with the configuration file. Expand on Instrumentation and right-click on LOG-Event Log Manager → Insert LOG, which inserts LOG0. Rename it *trace*.

Save this configuration file as *bios_sine_intr.cdb* in the folder bios_sine_intr. Add this configuration file and the autogenerated linker command file to the project, and include the BSL library support file (the run-time and CSL support library files are included in the linker command file).

Build this project as **bios_sine_intr**. Verify that a 1-kHz sine wave is generated when switch #3 is pressed. Release the switch and verify that the sine wave is no longer generated.

Select DSP/BIOS → Message Log and verify that "start" is printed in the message log window (after the program is halted).

9.2 RTDX USING MATLAB TO PROVIDE INTERFACE BETWEEN PC AND DSK

Three examples illustrate RTDX using MATLAB to provide an interface between the PC host and the DSK target. The following software tools are required:

1. The Embedded Target for TI C6000 DSP (2.0)
2. MATLAB Link for CCS

and they are available from MathWorks [6]. The required version supports the C6713 DSK (as well as platforms C6711DSK, C6416DSK, and C6701EVM). The examples and projects in this book were implemented using MATLAB's version 6.5, Revision 13.

Example 9.4: MATLAB–DSK Interface Using RTDX (rtdx_matlab_sim)

This example illustrates the interface between MATLAB and the DSK using RTDX. A buffer of data created from MATLAB (running on the host PC) is sent to the C6x processor (running on the DSK). The C source program (running on the DSK) increments each data value in the buffer and sends the buffer of data back to MATLAB. There is no real-time input or output in this simulation example. The following support files are used for this example and provided by TI: (1) c6713dsk.cmd, the linker command file; (2) *intvecs.asm*, the vector file; (3) *rtdx.lib*, the library support file; and (4) *target.h*, a header file to enable interrupt. They are included in the folder rtdx_matlab_sim.

Figure 9.6 shows the C source program *rtdx_matlab_sim.c* to illustrate the interface. It creates two channels through RTDX: an input channel to transfer data from

```
//RTDX_MATLAB_sim.c MATLAB-DSK interface using RTDX between PC & DSK

#include <rtdx.h>                       //RTDX support file
#include "target.h"                     //for init interrupt
short buffer[10] = {0};                 //init data from PC
RTDX_CreateInputChannel(ichan);         //data transfer PC-->DSK
RTDX_CreateOutputChannel(ochan);        //data transfer DSK-->PC

void main(void)
{
 int i;

 TARGET_INITIALIZE();                   //init for interrupt
 while(!RTDX_isInputEnabled(&ichan)) //for MATLAB to enable RTDX
     puts("\n\n Waiting to read "); //while waiting
 RTDX_read(&ichan,buffer,sizeof(buffer));//read data by DSK
 puts("\n\n Read Completed");
 for (i = 0; I < 10; i++)
     buffer[i]++;                       //increment by 1 data from PC
 while(!RTDX_isOutputEnabled(&ochan)) //for MATLAB to enable RTDX
     puts("\n\n Waiting to write "); //while waiting
 RTDX_write(&ochan,buffer,sizeof(buffer));//send data from DSK to PC
 puts("\n\n Write Completed");
 while(1)    {}                         // infinite loop
}
```

FIGURE 9.6. C program that runs on the DSK to illustrate RTDX with MATLAB. The buffer of data is incremented by one on the DSK and sent back to MATLAB (rtdx_matlab_sim.c).

the MATLAB on the PC to the C6x on the DSK and an output channel to transfer data from the target DSK to the PC host. When the input channel is enabled, data are *read* (received as input to the DSK) from MATLAB. After each data value in the buffer is incremented by 1, an output channel is enabled to *write* the data (sent as output from the DSK) to MATLAB. Note that the input (read) and output (write) designations are from the target DSK.

Figure 9.7 shows the MATLAB-based program *rtdx_matlab_sim.m*. This program creates a buffer of data values 1, 2, ..., 10. It requests board information, opens CCS and enables RTDX. It also loads the executable file *rtdx_matlab_sim.out* within CCS and runs the program on the DSK. Two channels are opened through RTDX: an input channel to write/send the data from MATLAB (PC) to the DSK and an output channel to read/receive the data from the DSK.

Build this project as **rtdx_matlab_sim** within CCS. The appropriate support files are included in the folder rtdx_matlab_sim. Add the necessary support files: the C source file *rtdx_matlab_sim.c*, the vector file *intvecs.asm* (from TI), *c6713dsk.cmd* (from TI), *rtdx.lib* (located in *c6713\c6000\rtdx\lib*), and the interrupt support header file *target.h* (from MATLAB). This process creates the executable file *rtdx_matlab_sim.out*.

```
%RTDX_MATLAB_sim.m MATLAB-DSK interface using RTDX. Calls CCS
%loads .out file.Data transfer from MATLAB->DSK,then DSK->MATLAB

indata(1:10) = [1:10];                          %data to send to DSK
ccsboardinfo                                    %board info
cc = ccsdsp('boardnum',0);                      %set up CCS object
reset(cc)                                       %reset board
visible(cc,1);                                  %for CCS window
enable(cc.rtdx);                                %enable RTDX
if ~isenabled(cc.rtdx)
    error('RTDX is not enabled')
end
cc.rtdx.set('timeout', 20);                     %set 20sec time out for RTDX
open(cc,'rtdx_matlab_sim.pjt');                 %open project
load(cc,'./debug/rtdx_matlab_sim.out');         %load executable file
run(cc);                                        %run
configure(cc.rtdx,1024,4);                      %configure two RTDX channels
open(cc.rtdx,'ichan','w');                      %open input channel
open(cc.rtdx,'ochan','r');                      %open output channel
pause(3)                                        %wait for RTDX channel to
open
enable(cc.rtdx,'ichan');                        %enable channel TO DSK
if isenabled(cc.rtdx,'ichan')
    writemsg(cc.rtdx,'ichan', int16(indata)) %send 16-bit data to DSK
    pause(3)
else
    error('Channel ''ichan'' is not enabled')
end
enable(cc.rtdx,'ochan');                        %enable channel FROM DSK
if isenabled(cc.rtdx,'ochan')
    outdata=readmsg(cc.rtdx,'ochan','int16') %read 16-bit data from DSK
    pause(3)
else
    error('Channel ''ochan'' is not enabled')
end
if isrunning(cc), halt(cc);                     %if DSP running halt
processor
end
disable(cc.rtdx);                               %disable RTDX
close(cc.rtdx,'ichan');                         %close input channel
close(cc.rtdx,'ochan');                         %close output channel
```

FIGURE 9.7. MATLAB program that runs on the host PC to illustrate RTDX with MATLAB. Buffer of data sent from MATLAB to the DSK (rtdx_matlab_sim.m).

Access MATLAB and make the following directory (path) active:

c6713\myprojects\rtdx_matlab_sim

Within MATLAB, run the (.m) file, typing *rtdx_matlab_sim*. Verify that the executable file is being loaded (through the CCS window) and run. Within the CCS window, the following messages should be printed: Waiting to read, Read completed,

Waiting to write, and Write completed. Then, within MATLAB, the following should be printed: *outdata* = 2 3 4 ... 11, indicating that the values (1, 2, ..., 10) in the buffer *indata* sent initially to the DSK were each incremented by 1 due to the C source program line of code: *buffer[i]++;* executed on the C6x (DSK).

Example 9.5 further illustrates RTDX through MATLAB, acquiring external real-time input data (from the DSK) and sending them to MATLAB for further processing (FFT, plotting).

Example 9.5: MATLAB-DSK Interface Using RTDX, with MATLAB for FFT and Plotting (rtdx_matlabFFT)

This example illustrates the interface between MATLAB and the DSK using RTDX. An external input signal is acquired from the DSK, and the input samples are stored in a buffer on the C6x processor. Using RTDX, data from the stored buffer are transferred from the DSK to the PC host running MATLAB. MATLAB takes the FFT of the received data from the DSK and plots it, displaying the FFT magnitude on the PC monitor. The same support tools as in Example 9.4 are required, including The Embedded Target for TI C6000 DSP (2.0) and MATLAB Link for CCS, available from MathWorks. The following support files are also used for this example and provided by TI: (1) the linker command file *c6713dsk.cmd*; (2) the vector file *intvecs.asm*; and (3) the library support file *rtdx.lib*. In the init/comm file *c6713dskinit.c*, the line of code to point at the IRQ vector table is bypassed since the support file *intvecs.asm* handles that.

Figure 9.8 shows the program *rtdx_matlabFFT.c* to illustrate the interface. It is a loop program as well as a data acquisition program, storing 256 input samples. Even though the program is polling-based, interrupt is used for RTDX. An *output* channel is created to provide the real-time data transfer from the C6x on the DSK to the PC host.

Figure 9.9 shows the MATLAB-based program *rtdx_matlabFFT.m*. This program provides board information, opens CCS and enables RTDX. It also loads the executable file (*rtdx_matlabFFT.out*) within CCS and runs the program on the DSK. Note that the output channel for RTDX is opened and data are *read* (from MATLAB running on the PC). A 256-point FFT of the acquired input data is taken, sampling at 16 kHz. The program obtains a total of 2048 buffers, and execution stops afterwards.

Build this project as **rtdx_matlabFFT** within CCS. The necessary support files are included in the folder rtdx_matlabFFT. Add the necessary support files, including *rtdx_matlabFFT.c*, *c6713dskinit.c*, *intvecs.asm* (from TI), *c6713dsk.cmd* (from TI), and *rtdx.lib* (located in *c6713\c6000\rtdx\lib*). Use the following compiler options: *-g -ml3*. The option *-ml3* (from the Advanced Category) allows for Memory Models: Far Calls and Data. This process yields the executable .out file.

```
//RTDX_MATLABFFT.c RTDX-MATLAB for data transfer PC->DSK(with loop)

#include "dsk6713_aic23.h"          //codec-DSK support file
#include <rtdx.h>                    //RTDX support file
Uint32 fs=DSK6713_AIC23_FREQ_16KHZ; //set sampling rate
RTDX_CreateOutputChannel(ochan);    //create out channel C6x-->PC

void main()
{
 short i, input_data[256]={0};      //input array size 256
 comm_poll();                       //init DSK, codec, McBSP
 IRQ_globalEnable();                //enable global intr for RTDX
 IRQ_nmiEnable();                   //enable NMI interrupt
 while(!RTDX_isOutputEnabled(&ochan)) //wait for PC to enable RTDX
    puts("\n\n Waiting... ");       //while waiting
 while(1)                           // infinite loop
 {
  i=0;
  while (i<256)                     //for 256 samples
   {
    input_data[i] = input_sample(); //defaults to left channel
    output_sample(input_data[i++]); //defaults to left channel
   }
  RTDX_write(&ochan,input_data,sizeof(input_data));//send 256 samples
 }
}
```

FIGURE 9.8. C program that runs on the DSK to illustrate RTDX with MATLAB. Input from the DSK is sent to MATLAB (`rtdx_matlabFFT.c`).

Access MATLAB and make the following directory (path) active:

c6713\myprojects\rtdx_matlabFFT

This folder contains the necessary files associated with this project. Within MATLAB, run the (.m) file *rtdx_matlabFFT*. Verify that the executable (.out) file is being loaded and run within CCS. Input a sinusoidal signal with a frequency of 2 kHz and verify that the output is the delayed (attenuated) input signal (a loop program). Within MATLAB the plot shown in Figure 9.10 is displayed on the PC monitor, which is the FFT magnitude of the input sinusoidal signal. Vary the frequency of the input signal to 3 kHz and verify the FFT magnitude displaying a spike at 3 kHz.

The FFT is executed on the PC host. As a result, on an older/slower PC, changing the input signal frequency will not yield a corresponding FFT magnitude plot immediately. *Note*: If it is desired to transfer data from the PC to the DSK, an input channel would be created using

RTDX_CreateInputChannel(ichan);
While(! RTDX_isInputEnabled(&ichan));
RTDX_read(&ichan, ...)

```
%RTDX_MATLABFFT.m MATLAB-DSK interface with loop. Calls CCS,
%loads .out file. Data from DSK→MATLAB for FFT and plotting

ccsboardinfo                                      %board info
cc=ccsdsp('boardnum',0);                          %setup CCS object
reset(cc);                                        %reset board
visible(cc,1);                                    %for CCS window
enable(cc.rtdx);                                  %enable RTDX
if ~isenabled(cc.rtdx);
    error('RTDX is not enabled')
end
cc.rtdx.set('timeout', 20);                       %set 20sec timeout for RTDX
open(cc,'rtdx_matlabFFT.pjt');                    %open project
load(cc,'./debug/rtdx_matlabFFT.out');            %load executable file
run(cc);                                          %run program
configure(cc.rtdx,1024,1);                        %configure one RTDX channel
open(cc.rtdx,'ochan','r');                        %open output channel
pause(3)                                          %wait for RTDX channel to open
fs=16e3;                                          %set sample rate in MATLAB
fftlen=256;                                       %FFT length
fp=[0:fs/fftlen:fs/2-1/fftlen];                   %for plotting within MATLAB
enable(cc.rtdx,'ochan');                          %enable channel from DSK
isenabled(cc.rtdx,'ochan');
for i=1:2048                                       %obtain 2048 buffers then stop
  outdata=readmsg(cc.rtdx,'ochan','int16'); %read 16-bit data from DSK
  outdata=double(outdata);                        %32-bit data for FFT
  FFTMag=abs(fftshift(fft(outdata)));   %FFT using MATLAB
  plot(fp,FFTMag(129:256))
  title('FFT Magnitude of data from DSK');
  xlabel('Frequency');
  ylabel('Amplitude');
  drawnow;
end
halt(cc);                                         %halt processor
close(cc.rtdx,'ochan');                           %close channel
clear cc                                          %clear object
```

FIGURE 9.9. MATLAB program that runs on the host PC to illustrate RTDX with MATLAB. MATLAB's FFT and plotting functions are used (rtdx_matlabFFT.m).

This creates an input channel, waits for the input channel to be enabled, and reads the data (input to the C6x on the DSK). In the MATLAB program, the following lines of code

> *open(cc.rtdx,'ichan','w');*
>
> *enable(cc.rtdx,'ichan');*
>
> *writemsg(…);*

open and enable an input channel and then write (send) the data from MATLAB running on the host PC to the C6x on the DSK. See Example 9.4.

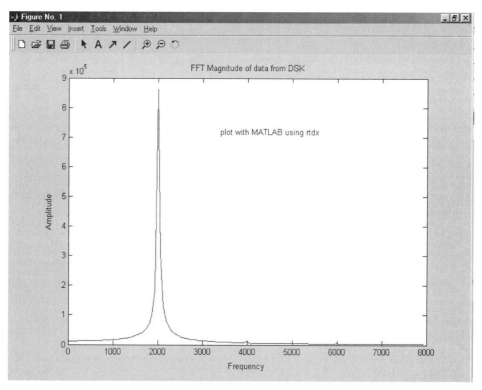

FIGURE 9.10. MATLAB's plot of the FFT magnitude of data received from the DSK.

Example 9.6: MATLAB–DSK Interface Using RTDX for FIR Filter Implementation (`rtdx_matlabFIR`)

This example further illustrates RTDX with MATLAB with the implementation of FIR filters. Figure 9.11 shows the C source program `rtdx_matlabFIR.c` that generates an input signal and implements an FIR filter on the DSK. The input signal consists of the product of random noise and a sine wave from a lookup table. This generated signal is the input to an FIR filter (see Example 4.1). The output of the filter is stored in a buffer, the address of which is transferred to MATLAB through the output RTDX channel. Initially, the implemented filter is a lowpass FIR filter with a cutoff frequency at 600 Hz. The coefficients of this filter are in the file `LP600.cof`. Two other FIR lowpass filter coefficients can also be selected in this example: LP1500.cof and LP3000.cof. These three sets of coefficients were used in Example 4.2 (FIR3LP). The address of the specific filter to be implemented is read through the RTDX input channel. All the appropriate support files for this Example are in the folder **rtdx_matlabFIR**. The CCS project is already built.

```
//FIR3LP_RTDX.c FIR-3 Lowpass with different BWs using RTDX-MATLAB
#include "lp600.cof"                         //coeff file LP @ 600 Hz
#include <rtdx.h>
#include <stdio.h>
#include "target.h"
int yn = 0;                                  //initialize filter's output
short dly[N];                                //delay samples
short h[N];                                  //filter characteristics 1xN
short loop = 0;
short sine_table[32]={0,195,383,556,707,831,924,981,1000,981,924,831,
                707,556,383,195,0,-195,-383,-556,-707,-831,-924,-981,
                -1000,-981,-924,-831,-707,-556,-383,-195};//sine values
short amplitude = 10;
#define BUFFER_SIZE 256
int buffer[BUFFER_SIZE];
int inputsample, outputsample;
short j = 0;
RTDX_CreateInputChannel(ichan);             //create input channel
RTDX_CreateOutputChannel(ochan);            //create output channel

void main()
{
 short i;
 TARGET_INITIALIZE();
 RTDX_enableInput(&ichan);                  //enable RTDX channel
 RTDX_enableOutput(&ochan);                 //enable RTDX channel
 for (i=0; i<N; i++)
  {
     dly[i] = 0;                            //init buffer
     h[i] = hlp600[i];                      //start addr of LP600 coeff
  }
 while(1)                                   //infinite loop
 {
   inputsample=rand()+amplitude*(sine_table[loop]);//generate  input
   if (loop < 31) ++loop;
   else loop = 0;
   dly[0]=inputsample;                      //FIR filter section
   yn = 0;                                  //initialize filter output
   if (!RTDX_channelBusy(&ichan))   {
       RTDX_readNB(&ichan,&h[0],N*sizeof(short));} //input coeff
   for (i = 0; i< N; i++)
       yn +=(h[i]*dly[i]);                  //y(n) += h(LP#,i)*x(n-i)
   for (i = N-1; i > 0; i--)                //starting @ bottom of buffer
       dly[i] = dly[i-1];                   //update delays
   outputsample = (yn >> 15);               //filter output
   buffer[j] = outputsample;                //store output -> buffer
   j++;
   if (j==BUFFER_SIZE) {
      j = 0;
      while (RTDX_writing != NULL) {}       //wait rtdx write to complete
      RTDX_write( &ochan, &buffer[0], BUFFER_SIZE*sizeof(int) );
   }
 }
}
```

FIGURE 9.11. C program that implements FIR filters and runs on the DSK. It illustrates RTDX with MATLAB.

1. Access MATLAB, and set the path to c:\C6713\myprojects\ RTDX_MATLABFIR. Open the MATLAB program (mwslider.m) and set the appropriate path (within the program). Within MATLAB, type mwslider. This MATLAB program mwslider.m displays a slider to select among the three sets of filter coefficients, and plots both the filtered signal and its spectrum. You should obtain Figure 9.12 (without the plots). The slider is initially set to implement the lowpass filter with a cutoff frequency of 600 Hz.

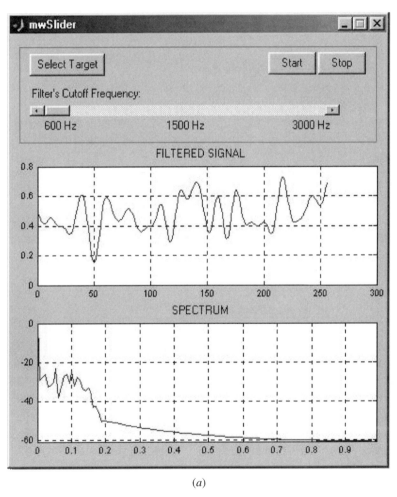

(a)

FIGURE 9.12. MATLAB plots with slider used to select one of three FIR lowpass filter coefficients. The upper and lower graphs show the filtered signal and its spectrum, respectively: (a) selecting BW of 600 Hz; (b) selecting BW of 1500 Hz; (c) selecting BW of 3000 Hz.

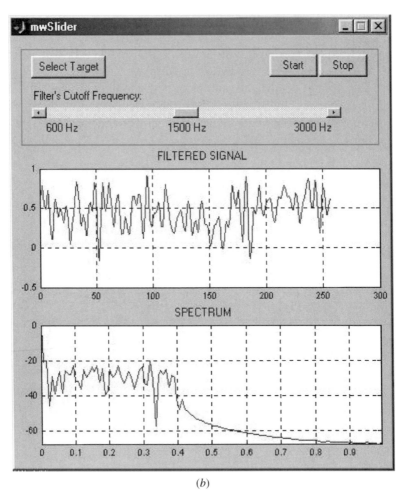

(*b*)

FIGURE 9.12. (*Continued*)

2. Select the target (this must be done first), and press OK to select the C6713 DSK board. Press Start to run. This opens CCS, and loads and runs the executable file `rtdx_matlabFIR.out`. Verify the results in Figure 9.12*a* that shows the filtered signal (upper graph) as well as its spectrum (lower graph). From the lower graph, the bandwidth is at approximately 0.15, which represents the normalized frequency v, where $v = f/F_N$ and F_N is the Nyquist frequency, 4 kHz. This corresponds to a cutoff frequency $f = 0.15 F_N = 600$ Hz. Change the slider to the middle position to select the 1500-Hz lowpass filter for implementation and verify the results in Figure 9.12*b*. Figure 9.12*c* shows that the 3000-Hz filter was selected and implemented. Note that the normalized frequency is approximately 0.75, which corresponds to a cutoff frequency, $f = 0.75 F_N = 3000$ Hz.

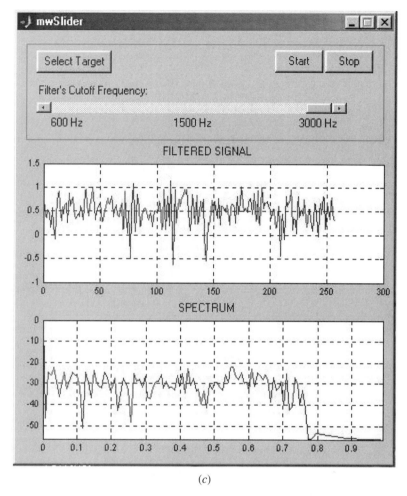

(c)

FIGURE 9.12. (*Continued*)

9.3 RTDX USING VISUAL C++ TO INTERFACE WITH DSK

Two examples are provided to illustrate the use of RTDX with Microsoft's Visual C++, one of which makes use of MATLAB's functions for finding and plotting the FFT magnitude (not for the RTDX interface). Three projects in Chapter 10 (DTMF, FIR, and Radix-4 FFT) make use of RTDX with Visual C++ to obtain a PC-DSK interface.

Example 9.7: Visual C++–DSK Interface Using RTDX for Amplitude Control of the Sine Wave (rtdx_vc_sine)

This example illustrates the use of RTDX with Microsoft Visual C++. The application running on the target DSK generates a sine wave. A procedure follows to illus-

```
//RTDX_vc_sine.c Sine generation.RTDX using Visual C++(or VB/LABVIEW)

#include "rtdx_vc_sinecfg.h"            //generated by .cdb file
#include "dsk6713_aic23.h"             //codec-dsk support file
#include <rtdx.h>                       // for rtdx support
Uint32 fs=DSK6713_AIC23_FREQ_16KHZ;     //set sampling rate
short loop = 0;
short sin_table[8] = {0,707,1000,707,0,-707,-1000,-707};
int gain = 1;
RTDX_CreateInputChannel(control_channel); //create input channel

interrupt void c_int11()                //ISR set in .cdb
{
 output_sample(sin_table[loop]*gain);
 if (++loop > 7) loop = 0;
}

void main()
{
 comm_intr();                           //init codec,dsk,MCBSP
 RTDX_enableInput(&control_channel);    //enable input channel
 while(1)                               //infinite loop
 {
  if(!RTDX_channelBusy(&control_channel)) //if channel not busy
     RTDX_read(&control_channel,&gain,sizeof(gain));//read from PC
 }
}
```

FIGURE 9.13. C program that runs on the DSK to illustrate RTDX with Visual C++. It generates a sine wave (rtdx_vc_sine.c).

trate the development of the host application with RTDX support—in particular, the development of a Visual C++ application with a slider control for adjusting the amplitude of the generated sine wave running on the C6x DSK. All the Visual C++ application files are on the CD in the folder *rtdx_vc_sine*.

CCS Component

Figure 9.13 shows the C source program *rtdx_vc_sine.c* that implements the sine generation with amplitude control. This is the same C source program used to illustrate RTDX with Visual Basic in Example 9.9 as well as with LabVIEW in Example 9.13. An RTDX input channel is created and enabled in order to read the slider data from the PC host.

Create, save, and add the configuration file *rtdx_vc_sine.cdb* to the project. Select INT11, *MCSP_1_Transmit* as the interrupt source and *_c_int11* as the function. See Examples 9.1–9.3. Add the autogenerated linker command file and the BSL library support file. The run-time and the CSL library support files are included in the autogenerated linker command file. Add also the init and communication files, but not the vector file. The necessary files are included in the folder rtdx_vc_sine.

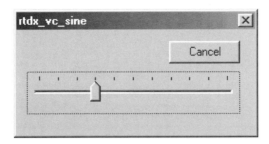

FIGURE 9.14. Gain slider obtained with Visual C++ for the project `rtdx_vc_sine`.

Build this project as **rtdx_vc_sine**. Within CCS, load and run the executable file *rtdx_vc_sine.out*. Verify that a 2-kHz sine wave is generated and outputted through the codec on the DSK.

Enable RTDX within CCS. Select Tools → RTDX → Configuration Control → Enable RTDX (activate/check it).

Visual C++ Component

Run the Visual C++ application (executable file on the CD). The gain slider in Figure 9.14 should pop up. Vary the gain slider position and verify a corresponding change in the amplitude of the generated sine wave with the DSK output connected to a speaker or a scope.

Procedure to Develop the Visual C++ Executable File

This procedure is used to develop the necessary Visual C++ support files to create the executable (`.exe`) file (already on the CD in the folder *rtdx_vc_sine*).

1. Launch Microsoft Visual C++ and select File → New to create a new project. Various types of C++ projects will be displayed in the new project dialog.

2. Select *MFCAPPWizard (exe)*, and specify *rtdx_vc_sine* as the project name and *c:\c6713\myprojects\rtdx_vc_sine* as the location. Click OK.

3. This brings out the *MFCAPPWizard* dialog. Select the application type *dialog based*, then select *next*. Click on *next* twice to accept the default settings. Then, click on *Finish* and OK. Three classes will be automatically generated and added to the project.

4. A dialog resource editor will be opened. Click on "*TODO: Place dialog controls here*" and delete it from the main dialog window by pressing the delete key. Resize the main dialog window to an appropriate size (use the lower-right corner with the mouse). Select the *slider control* from the Control Toolbox (on the right). Draw the slider control in the main dialog window by holding it down with the left mouse button and moving it to the dialog window. Release the button when the control is of the appropriate size.

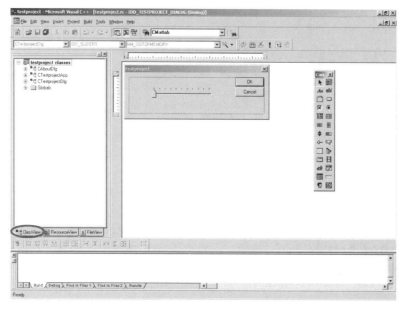

FIGURE 9.15. Visual C++ windows displaying the *classview* pane and the gain slider control for the project `rtdx_vc_sine`.

5. Right click on the *slider control* in the main dialog window, and select the *properties* menu item. Click on the *styles tab* and select the *Tick Marks* and the *Auto Ticks* options. From the *Point* list, select the *Top/Left* option. Close the slider control property dialog.

6. Click on the *ClassView* pane (bottom-left window) to expose the three classes that constitute the project, as shown in Figure 9.15, along with the slider control. These classes are:

   · *CaboutDlg*
   · *CtestprojectApp*
   · *CtestprojectDlg*

   where t*estproject* is the project name specified initially in step 2 (*rtdx_vc_sine*). The class of interest is *CTestprojectDlg* since it is the class that controls the main dialog window. The *CTestprojectApp* class is a standard class included in most projects to handle application startup, since there is no *main* function, as in a typical C++ console application. The *CAboutDlg* class is responsible for displaying an *About* message dialog, as in most window-based applications.

7. From the main menu, select *View* and select the *ClassWizard* menu item. This pops up the *MFCClassWizard* dialog window. (Make sure to select

CTestprojectDlg from the class name.) Select the *Member Variables* tab, and then select *IDC_SLIDER1* from the list of control IDs.

8. Click the *Add Variable* button to display the *Add Member Variable* dialog. Choose an appropriate member variable name, such as *m_slider*, and make sure that the *Category* field is *Value* and the *Variable type* field is *int*. Click OK to return to the *ClassWizard* dialog window.

9. Create a new class for RTDX. Click on the *Add Class* button and select *from a type library*. Browse in the folder *c:\c6713\cc\bin* and select (or type) the file *Rtdxint.dll*. This pops up the *Confirm classes* dialog. Click OK to return to the *ClassWizard* dialog. Click OK again to dismiss the *ClassWizard* dialog. The new class *IRtdxExp* has been added for the functionality of RTDX.

10. From the *ClassView* pane (lower-left window):

    (a) Select the class *CTestprojectDlg*. Right-click on the class and select *Add member variable*. For variable type, use *IRtdxExp** (note the pointer notation), and for variable name use *pRTDX* (or another name). Click OK to dismiss the dialog. This creates a pointer that represents and manipulates the class *IRtdxExp* created in the previous step.

    (b) Right-click on the class *CTestprojectDlg* and select *Add Windows Message Handler*. This will bring up the *New Windows Message* dialog. From the list, find and select the message *WM_DESTROY*. Click on the *Add and Edit* button to insert the new windows message. Add the following lines of code just after the function

    > *CDialog::OnDestroy( ).*
    > *if(pRTDX->Close( ))*
    >   *MessageBox("Could not close the channel!", "Error");*

    (c) Right-click on the class *CTestprojectDlg* and choose the *Add Windows Message Handler* to bring up again the *New Windows Message* dialog. Select the *WM_HSCROLL* message and click on the *Add and Edit* button. Add the following lines of code just above the function *CDialog::OnHScroll(nSBCode, nPos, pScrollBar)*. This is shown in Figure 9.16.

    > *long buffer;*
    > *UpdateData(TRUE);*
    > *pRTDX->WriteI4((long)m_slider, &buffer);*
    > *UpdateData(FALSE);*

    (d) Select the class *CTestprojectDlg* and expand it. Locate the function *OnInitDialog( )* and double-click on it. Add the following lines of code just above the *return* instruction:

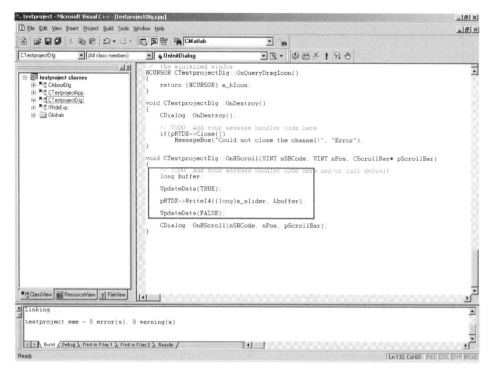

FIGURE 9.16. Visual C++ windows handler for the message *WM_HSCROLL*.

> *CSliderCrtl* pSliderCrtl = (CSliderCrtl*)*
>   *GetDlgItem(IDC_SLIDER1);*
> *pSliderCrtl->SetRange(1,10);*
> *pRTDX = new IRtdxExp;*
> *pRTDX->CreateDispatch(_T("RTDX"));*
> *if(pRTDX->SetProcessor(_T("C6713 DSK"),_T("CPU_1")))*
>   *MessageBox("Could not set the processor!", "Error");*
> *if(pRTDX->Open("control_channel","W"))*
>   *MessageBox("Could not open the channel!","Error");*

(e) Double-click on the class *CTestprojectDlg* and add the following line of code just before the class definition statement:

> *#include "rtdxint.h"*

(f) Select the class *CTestprojectApp* and expand it. Double-click on the function *InitInstance( )* and add the following line of code:

> *AfxOleInit( );*

just above the line *CTestAppDlg dlg;.*

The added lines of code can be verified from the file *rtdx_vc_sineDlg.cpp* (on the CD). Select Build (menu item from the main project window) → Rebuild All to create the application (executable) file.

Example 9.8: Visual C++–DSK Interface Using RTDX with MATLAB Functions for FFT and Plotting (rtdx_vc_FFTmatlab)

This example illustrates real-time data communication using RTDX with Microsoft Visual C++, invoking MATLAB's FFT and plotting functions. MATLAB is not used in this example to provide the RTDX communication link between the PC and the DSK, as in Examples 9.4–9.6. Instead, only the MATLAB's functions for FFT and plotting are invoked.

The folder **rtdx_vc_FFTmatlab** contains the Visual C++ support files, including the application/executable file rtdx_vc_FFTmatlab.exe (already built). See also Example 9.7.

Running Executable from CCS
The folder *rtdx_MatlabFFT* for Example 9.5 includes the main C source program (Figure 9.8) rtdx_MatlabFFT.c, which implements a loop program. It also creates and enables an output channel to write/send data acquired from the DSK to the PC. It illustrated RTDX with MATLAB in Example 9.5, and it can be used in this example to illustrate this Microsoft Visual C++ application. The (.m) MATLAB file that provides the RTDX communication link between the DSK and the PC in Example 9.5 is *not* used in this example. Only, the MATLAB's FFT and plotting functions are used.

Input into the DSK a 2-kHz sine wave with an approximate amplitude of 1 V p-p. Within the CCS window, select Tools → RTDX → Enable RTDX (check it). Load and run rtdx_matlabFFT.out. The RTDX communication link is not yet produced, and "waiting" is printed continuously within the CCS window.

Running Visual C++ Application
Run the Visual C++ application rtdx_vc_FFTMatlab.exe located in the folder *rtdx_vc_FFTMatlab\debug* (double-click on it).

Verify a loop program with the DSK output to a scope, and an FFT plot of the 2-kHz sine wave as shown in Figure 9.17, obtained using MATLAB's FFT and plotting functions (see also Example 9.5). Change the input sine wave frequency to 3 kHz and verify that the MATLAB plots 3-kHz sine wave.

You can readily add the labels for the x and y axes in Figure 9.17 by modifying the file rtdx_vc_FFTMatlabDlg.cpp. Find the section of code where the MATLAB functions are invoked for FFT and plotting. After the line of code for the figure's title, insert the appropriate xlabel and ylabel functions. Launch Microsoft Visual C++. Select File and open the workspace (.dsw) file located in the folder **rtdx_vc_FFTmatlab**. Select Build → Rebuild All to recreate a

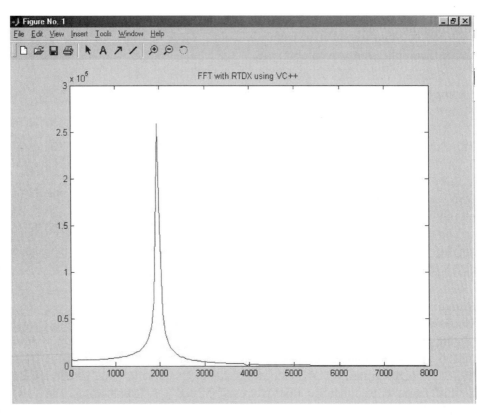

FIGURE 9.17. Plot of FFT magnitude (with MATLAB) to illustrate RTDX using Visual C++ for the project `rtdx_vc_matlabFFT`.

new application (.exe) file. Verify that the FFT plot now contains the x and y axis labels.

Creation of Visual C++ Application and Support Files

1. Repeat steps 1–3 in Example 9.7. The Resource Dialog editor should be opened. Resize the main dialog window. Right-click on the *TODO:Place dialog control here* and select the *Properties* menu item. From the resulting property dialog in the *Caption* field, enter any messages that you want displayed in the dialog window (such as RTDX with Visual C++ to . . .), and then close the property dialog window.

2. Click on the *ClassView* pane to expose the three classes: *CaboutDlg*, *CRtdx_vc_fftMatlabApp*, and *CRtdx_vc_fftMatlabDlg*. Figure 9.18*a* shows the *ClassView* pane displaying these classes and a message inserted by the user in step 3. You can adjust the size of the dialog window so that it looks like Figure 9.18*b*, which will pop up when you run the executable file. (Delete the cancel button and change the text of the OK button to Exit, which is already done.)

3. Select *View* from the main menu, then *ClassWizard*. This pops up the *MFC ClassWizard* dialog window. Repeat step 9 in Example 9.7.

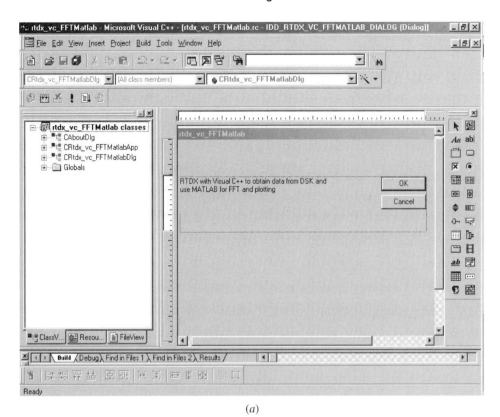

(a)

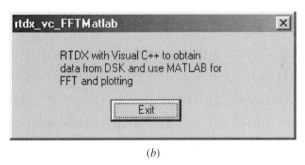

(b)

FIGURE 9.18. (*a*) Classview pane displaying the three classes, and a message inserted by the user, for the project `rtdx_vc_FFTmatlab`; (*b*) message when application file is executed.

4. Repeat step 3, but select *New* (instead of *from a type library*). For the class name, enter *CRTDXThread*. Click on the *Base Class* list, select *CWinThread*, and click OK. The newly created *CRTDXThread* class can be used to run a separate window thread that continuously polls the open RTDX channel for incoming real-time data. This is more efficient than having the main program poll the RTDX channel.

5. From the *ClassView* pane, right-click on the class *CRtdx_vc_fftMatlabDlg* and select *Add member variable*. For the type, use *CRTDXThread** (note the pointer notation) and for the name, use *pRTDXThread* (or another name) and click OK to dismiss the dialog. Double-click on *CRtdx_vc_fftMatlabDlg* to open its class definition file, and add the following line of code (just before the class definition):

   #include "RTDXThread.h"

6. Create a class for the functionality with MATLAB:

   (a) Click on *Insert* from the main menu and select *New Class*. For the class type, select *Generic class* and for the name, type *CMatlabClass*. Then click OK to close the dialog.

   (b) Select and double-click on the *CMatlabClass* (from the *ClassView* pane) to open its class definition file. Add the following lines of code (just above the class definition):

   #include "Engine.h"
   #pragma comment(lib, "libeng.lib")
   #pragma comment(lib, "libmx.lib")

   Right-click on *CMatlabClass* and select *Add member variable*. For the type, use *Engine** (note the pointer notation) and for the name, use *pEngine* (or another name), and then click OK.

   (c) Double-click on *CMatlabClass* to reveal its class definition. Add the following lines of code below the definition for *pEngine (below Engine* pEngine)*:

   public:
   void OpenMatlab(LPCTSTR lpCommand);
   ...   //already added
   int CreateBuffer(char* pOutputBuffer, int nLength);

   (d) Click on the *File View* pane (next to the *ClassView* pane), and expand *Rtdx_vc_fftMatlab* to expose three folders. Expand on *Source Files*, and double-click on *MatlabClass.cpp*. Add this section of code at the end of this file (after the pair of brackets):

   void CMatlabClass::OpenMatlab(LPCTSTR lpCommand)
   {
   pEngine = engOpen(lpCommand);
   ...
   return engOutputBuffer(pEngine, pOutputBuffer, nLength);
   }

(e) Right click on the class *CRtdx_vc_fftMatlabDlg* and select *Add Windows Message Handler*. Find and select the message *WM_DESTROY*, and click on *Add and Edit* to insert the new windows message. Add the following lines of code beneath the function *CDialog::OnDestroy( )*:

> *nFlag = 0;*
> *WaitForSingleObject(pRTDXThread-> m_hThread, INFINITE);*

(f) Right-click on the class *CRtdx_vc_fftMatlabDlg* and click on *Add member function*. For the type, use UINT and for the declaration, type *static RTDXThreadFunction(LPVOID lpVoid)* and then click OK.

(g) Expand the class *CRtdx_vc_fftMatlabDlg*, double-click on the member function *RTDXThreadFunction(LPVOID lpVoid)*, and add the following lines of code in the function body (between the pair of brackets):

> *CMatlabClass* pMatlab;*
> *IRtdxExp *pRtdx;*
> . . .
> *pMatlab->ExecuteLine(_T("fs = 16e3;"));*
> . . .
> *pMatlab->ExecuteLine(_T("plot(fp, fftMag(129: 256))"));*
> . . .
> *return 0;*

Scroll to the top of the file and add the following two *include* files and the global variable *nflag*:

> *#include "MatlabClass.h"*
> *#include "Rtdxint.h"*
> *int nFlag = 1;*

(h) With the class *CRtdx_vc_fftMatlabDlg* expanded, double-click on the member function *OnInitDialog( )* and add the following line of code just before the return instruction:

> *pRTDXThread = (CRTDXThread*)AfxBeginThread*
> *(RTDXThreadFunction, m_hWnd);*

7. The path of MATLAB libraries and *include* files need to be added before building the project. Select Tools → Options to display the *Options* dialog, and click on the *Directories* tab. Select the *Include Files* item from *Show directo-*

ries for. Click twice on the rectangle below the list of *Directories*, then click on the "..." displayed on the right. Browse in your MATLAB installation directory for the include path *c:\Matlab_folder\extern\include* (for example, *matlabR13* as the *Matlab_folder*). From the *Show directories for* list, select the *library file* item. Click twice on the rectangle below the list of *Directories* and select the "..." (as before). Browse in your MATLAB folder for the path *c:\Matlab_folder\extern\lib\win32\microsoft\msvc60*, and click on OK to save the changes.

Build the Visual C++ application project. Select Build → Rebuild All to create `rtdx_vc_FFTMatlab.exe`.

9.4 RTDX USING VISUAL BASIC TO PROVIDE INTERFACE BETWEEN PC AND DSK

Two examples are provided to illustrate the interface between the PC host and the DSK with RTDX using Visual Basic.

Example 9.9: Visual Basic–DSK Interface Using RTDX for Amplitude Control of a Sine Wave (`rtdx_vbsine`)

This example generates a sine wave outputted through the codec on the DSK. It illustrates RTDX using Visual Basic (VB) to create a slider and control the amplitude of the generated sine wave.

CCS Component

Figure 9.19 shows the C source program *rtdx_vbsine.c* that implements the sine generation with amplitude control. This is the same C source program used to illustrate RTDX with Visual C++ in Example 9.4 and LabVIEW in Example 9.13. An RTDX input channel is created and enabled in order to read the slider data from the PC host. This example is not meant to teach the reader VB, but rather to use it.

Create, save, and add a configuration file *rtdx_vbsine.cdb* to the project. Select INT11, *MCSP_1_Transmit* as the interrupt source, and *_c_int11* as the function (see Examples 9.1–9.3). Add the autogenerated linker command file and the BSL library support file. The run-time and the CSL library support files are included in the autogenerated linker command file. Add also the init and communication file, but not the vector file. The necessary files are included in the folder `rtdx_vbsine`.

Build this project as **rtdx_vbsine**. Within CCS, load and run the executable file *rtdx_vbsine.out*. Verify that a 2-kHz sine wave is generated and outputted through the codec on the DSK.

Enable RTDX within CCS. Select Tools → RTDX → Configuration Control → Enable RTDX (activate/check it).

```
//rtdx_vbsine.c Sine generation.RTDX with Visual Basic(VC++/LABVIEW)

#include "rtdx_vbsinecfg.h"              //generated by .cdb file
#include "dsk6713_aic23.h"              //codec-dsk support file
#include <rtdx.h>                       // for rtdx support
Uint32 fs=DSK6713_AIC23_FREQ_16KHZ;     //set sampling rate
short loop = 0;
short sin_table[8] = {0,707,1000,707,0,-707,-1000,-707};
int gain = 1;
RTDX_CreateInputChannel(control_channel); //create input channel

interrupt void c_int11()                //ISR set in .cdb
{
 output_sample(sin_table[loop]*gain);
 if (++loop > 7) loop = 0;
}

void main()
{
 comm_intr();                           //init codec,dsk,MCBSP
 RTDX_enableInput(&control_channel);    //enable input channel
 while(1)                               //infinite loop
 {
  if(!RTDX_channelBusy(&control_channel)) //if channel not busy
     RTDX_read(&control_channel,&gain,sizeof(gain));//read from PC
 }
}
```

FIGURE 9.19. C program that generates a sine wave. It illustrates RTDX using VB to control the amplitude of the generated sine wave (rtdx_vbsine.c).

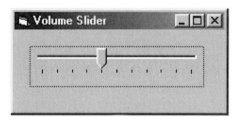

FIGURE 9.20. Volume slider to control the amplitude of the DSK output signal. Object created with VB for the project rtdx_vbsine.

VB Component

The folder *rtdx_vbsine* contains a subfolder PC that contains the support files associated with VB. Click on the (.vbp) VB project file to open VB. The project consists of the file *slider.frm* that describes the slider and the file *boardproc_frm.frm* that describes the board information. These two files are included with CCS. The slider is the same as that used in an example *(hostio1)* included with CCS. Within VB, select Run → Start. Press OK for the board information and the slider box shown in Figure 9.20 should pop up. Connect the DSK output to a scope. Vary the slider position and verify the change in the amplitude of the generated output sine wave (keep the mouse cursor on the slider button to change the slider value). Note that the Application (.exe) file, included on the CD, also can be used to run the VB project directly. This application file can be re-created within VB after loading the project file and selecting File → Make `rtdx_vbsine.exe`.

The next example implements a loop using RTDX with VB, where the amplitude of the output signal is changed using a gain value sent by the PC host to the C6x processor.

Example 9.10: Visual Basic–DSK Interface Using RTDX for Amplitude Control of Output in a Loop Program (rtdx_vbloop)

This example extends the previous example with a loop program using VB and RTDX to control the amplitude of an output signal. A window where the user can enter a gain value is built in VB. That gain value is sent from the PC host to the C6x processor. Figure 9.21 shows the C source program *rtdx_vbloop.c* that implements this project example. See also the previous example.

An RTDX input channel is created and enabled. When the RTDX channel is not busy, the C6x processor reads the data from the PC. Create and add a configuration file to set the interrupt service function, and add similar support files to the project, as in the previous example.

Build this project as **rtdx_vbloop**. Input a sine wave with an approximate amplitude and frequency of 0.5 V p-p and 2 kHz, respectively. Verify that the DSK

```
//rtdx_vbloop.c RTDX with Visual basic(or VC++)for loop gain control
#include "rtdx_vbloopcfg.h"              //generated by .cdb file
#include "dsk6713_aic23.h"
#include <rtdx.h>                        //codec-DSK support file
Uint32 fs=DSK6713_AIC23_FREQ_16KHZ;     //set sampling rate
int gain = 1;                           //initial gain value
RTDX_CreateInputChannel(control_channel); //create input channel

interrupt void c_int11()                //ISR
{
 output_sample(gain*input_sample());    //output = scaled input
}

void main()
{
 comm_intr();                           //init codec,DSK,MCBSP
 RTDX_enableInput(&control_channel);    //enable RTDX channel
 while(1)                               //infinite loop
 {
 if(!RTDX_channelBusy(&control_channel)) //if channel not busy
   RTDX_read(&control_channel,&gain,sizeof(gain));//read gain from PC
 }
}
```

FIGURE 9.21. C program that implements a loop. It illustrates RTDX using VB to control the amplitude of an output signal from the DSK (`rtdx_vbloop`).

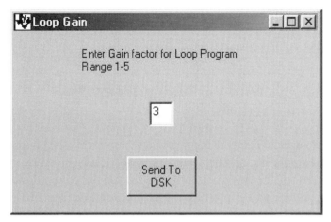

FIGURE 9.22. Gain slider to control the amplitude of the DSK output signal. Loop gain object created with VB for the project `rtdx_vbloop`.

output exhibits the characteristics of a loop program, as in Examples 2.1 and 2.2. Enable RTDX within CCS as in the previous example.

The subfolder *PC* within the folder *rtdx_vbloop* contains the support files associated with VB. The VB project includes the board information file, as in the previous example, and *gain.frm*, a block where the user can enter a gain value to control the amplitude of the output sine wave. The object *gain.frm* was created with VB. Run the application (`.exe`) file. Enter a gain value of 3 (see Figure 9.22) and verify the increase in amplitude of the output sine wave.

Note that instead of using *gain.frm* in the project, you can use *slider.frm* from the previous example to obtain the slider.

9.5 RTDX USING LABVIEW TO PROVIDE INTERFACE BETWEEN PC AND DSK

Three examples are provided to illustrate RTDX with LabVIEW for filter design and for adjusting the gain of a generated sinusoid. These examples are not intended to teach LabVIEW, but rather to illustrate the interface between the DSK and LabVIEW. The source files (LabVIEW Instrument `.vi`) are included on the CD. You can test these examples even if you do not have the LabVIEW tools. If you do, you can further open the source as a block diagram of a virtual instrument (VI) consisting of individual block components (as smaller VIs). VIs are available for signal generation, plotting, and so on.

The following tools are required:

1. LabVIEW Full Development System, V. 7.0
2. LabVIEW DSP Test Integration Toolkit for TI DSP, V. 2.0

and are available from National Instruments [7]. The DSP test integration toolkit provides the RTDX link between LabVIEW and the DSK. To create the executable (application) file, the professional version is required.

Example 9.11: LabVIEW–DSK Interface Using RTDX for FIR Filtering (`rtdx_lv_filter`)

This example illustrates RTDX using LabVIEW to provide the communication link between the C6x running on the DSK and LabVIEW running on the host PC. LabVIEW is used for the design of an FIR filter, for the generation of a sine wave as input to the filter, and for plotting the filtered output. The FIR filter is implemented on the DSK. All the necessary files for this example are included in the folder **rtdx_lv_filter**.

1. Click on the LabVIEW Instrument (`.vi`) file `rtdx_lv_filter` to open the (`.vi`) window shown in Figure 9.23. The initial filter settings are for an FIR

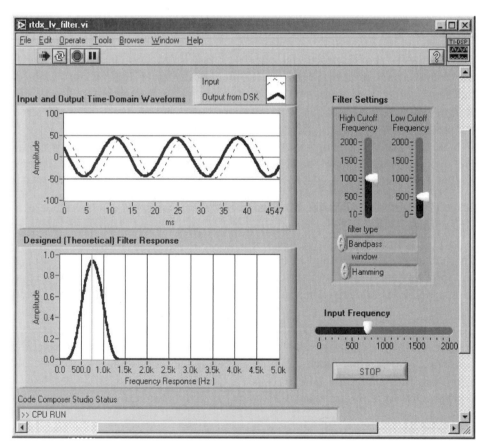

FIGURE 9.23. LabVIEW Instrument window for FIR filter design and plotting to illustrate RTDX for the project `rtdx_lv_filter`.

bandpass filter design using a Hamming window, and with low and high cutoff frequencies of 500 and 1000 Hz, respectively. Select Operate → Run. In Figure 9.23, the upper graphs show both the input sine wave generated with LabVIEW and the output of the filter implemented on the DSK. The theoretical frequency response of the designed filter is also plotted showing a center frequency at 750 Hz. Vary the input signal frequency between 300 and 1200 Hz and verify that the filter's output amplitude starts with zero, reaches a maximum at 750 Hz, and then decreases again towards zero.

Change the filter settings for a lowpass with a bandwidth (low cutoff frequency) of 1500 Hz. Vary the frequency of the input signal between 0 and 1600 Hz. Verify that the amplitude and frequency of the filtered output signal are the same as those of the input signal for frequencies between 0 and 1300 Hz. The output signal's amplitude decreases towards zero for input frequencies beyond 1300 Hz.

Various windows for the filter design are available, such as Hamming, Hanning, Blackman, and so on. Experiment with different filter characteristics.

2. From Figure 9.23, select Window → Show Block Diagram. The LabVIEW tools are required to view the block diagram (the source). Figure 9.24 shows a section of the block diagram that contains various components (smaller

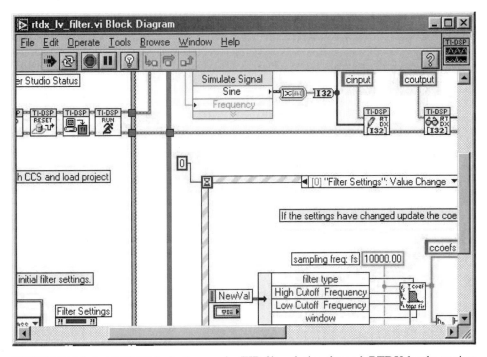

FIGURE 9.24. LabVIEW block diagram for FIR filter design through RTDX for the project `rtdx_lv_filter`.

blocks). A full description and the function of different blocks can be readily obtained by highlighting each block.

CCS is invoked from LabVIEW to build the project and to load and run the (.out) file (from the current directory) on the DSK. (See the CPU status within CCS in Figure 9.23.) Input and output arrays of data, specified as 32-bit integers (cinput, coutput), are transferred to the DSK through RTDX (Figure 9.24).

3. Figure 9.25 shows the C source program rtdx_lv_filter.c that runs on the DSK. It creates two input channels (for the sine wave data and the filter coefficients generated by LabVIEW) and one output channel for the filtered output data (coutput). Inputs to the DSK are obtained using RTDX_read() or RTDX_readNB() to read/input the sine data (cinput) and the coefficients

```
//rtdx_lv_filter.c RTDX with LABVIEW->filter design/plot DSK output
#include <rtdx.h>                        //RTDX support
#include "target.h"                      //init target
#define kBUFFER_SIZE 48                  //RTDX read/write buffers
#define kTAPS 51
double gFIRHistory [kTAPS+1];
double gFIRCoefficients [kTAPS];
int input[kBUFFER_SIZE],output[kBUFFER_SIZE];
int gain;
double FIRFilter(double val,int nTaps,double* history,double* coefs);
int  ProcessData (int* output, int* input, int gain);
RTDX_CreateInputChannel(cinput);        //create RTDX input data channel
RTDX_CreateInputChannel(ccoefs);        //input channel for coefficients
RTDX_CreateOutputChannel(coutput);      //output channel DSK->PC(Labview)
void main()
{
int i;
TARGET_INITIALIZE();                    //init target for RTDX
RTDX_enableInput(&cinput);              //enable RTDX channels
RTDX_enableInput(&ccoefs);              //for input, coefficients, output
RTDX_enableOutput(&coutput);
gFIRCoefficients[0] = 1.0;
for (i = 1; i<kTAPS; i++)
  gFIRCoefficients[i] = 0.0;
for (;;)                                //infinite loop
 {
  while(!RTDX_read(&cinput,input,sizeof(input)));//wait for new buffer
  if (!RTDX_channelBusy(&ccoefs))       //if new set of coefficients
    RTDX_readNB(&ccoefs,&gFIRCoefficients,sizeof(gFIRCoefficients));
  ProcessData (output, input, 1);       //filtering on DSK
  RTDX_write(&coutput,&output,szeof(output));//output from DSK->LABVIEW
 }
}
```

FIGURE 9.25. C program running on the DSK that implements an FIR filter and illustrates RTDX with LabVIEW (rtdx_lv_filter.c).

```
int ProcessData (int *output,int *input,int gain) //calls FIR filter
{
int i;
double filtered;
for(i=0; i<kBUFFER_SIZE; i++) {
  filtered=FIRFilter(input[i]*gain,kTAPS,gFIRHistory,gFIRCoefficients);
  output[i] = (int)(filtered + 0.5);}              //scale output
return 0;
}


double FIRFilter (double val,int nTaps,double* history,double* coefs)
{                                                  //FIR Filter
 double temp, filtered_val, hist_elt;
 int  i;
 hist_elt = val;
 filtered_val = 0.0;
 for (i = 0; i <  nTaps; i++)
  {
   temp = history[i];
   filtered_val += hist_elt * coefs[i];
   history[i] = hist_elt;
   hist_elt = temp;
  }
 return filtered_val;
}
```

FIGURE 9.25. (*Continued*)

(ccoefs). The filter is implemented on the DSK by the function FIRFilter, and the filtered output (coutput) is sent to Labview for plotting using RTDX_write(). If the filter characteristics are changed, a new set of coefficients (ccoefs) is calculated within Labview and sent to the DSK through RTDX.

Example 9.12: LabVIEW–DSK Interface Using RTDX for Controlling the Gain of a Generated Sinusoid (rtdx_lv_gain)

In this example, LabVIEW is used to control the amplitude of a generated sine wave and to plot the scaled output sine wave. An array of data representing the generated sine wave and a gain value are sent from LabVIEW to the DSK. Through RTDX, the C6x on the DSK scales the received sine wave input data and sends the resulting scaled output waveform to LabVIEW for plotting. The necessary files for this example are in the folder **rtdx_lv_gain**.

1. Click on the LabVIEW Instrument (.vi) file rtdx_lv_gain to obtain Figure 9.26. Run it as in Example 9.11. The project rtdx_lv_gain.pjt is

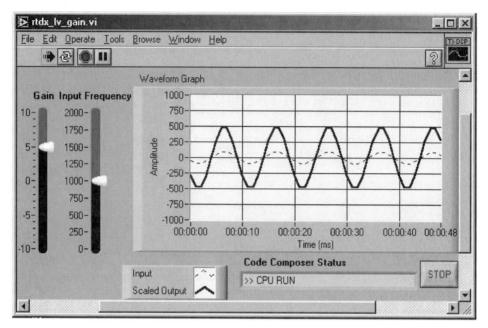

FIGURE 9.26. LabVIEW Instrument window to control the gain of a generated sine wave through RTDX for the project rtdx_lv_gain.

opened within CCS, and loaded and run on the DSK. See the Code Composer Status in Figure 9.26. Verify that the amplitude of the output sine wave is five times that of the input. You can vary the input signal frequency as well as the gain settings to control the scaled output amplitude waveform. The output frequency is the same as the input frequency. You can readily change the input signal type to a square wave, a triangle, or a sawtooth.

From the block diagram, one can verify that the input and output data are transferred through RTDX as two arrays (using [I32]), whereas the gain is transferred as a single value (using I32). The brackets represent the array notation (using 32-bit integer format).

2. Figure 9.27 shows the C source program rtdx_lv_gain.c that runs on the DSK. Through RTDX, the input and output channels are enabled and opened for the C6x on the DSK to read the generated sine wave data and the user set gain value and to write the scaled sine wave data to LabVIEW for plotting.

```
//rtdx_lv_gain.c RTDX with LABVIEW to control gain of generated sine

#include <rtdx.h>                            //RTDX support
#include "target.h"                          //init target
#define kBUFFER_SIZE 49
RTDX_CreateInputChannel(cinput);            //create RTDX input channel
RTDX_CreateInputChannel(cgain);            //input channel for gain
RTDX_CreateOutputChannel(coutput);         //channel for scaled output

void Gain(int *output,int *input,int gain) //scale array of input array
{
 int i;
 for(i=0; i<kBUFFER_SIZE; i++)
    output[i]=input[i]*gain;                //scaled output
}

void main()
{
 int input[kBUFFER_SIZE];
 int output[kBUFFER_SIZE];
 int gain = 5;                              //initial gain setting
 TARGET_INITIALIZE();                       //init target for RTDX
 RTDX_enableInput(&cgain);                  //enable RTDX channels
 RTDX_enableInput(&cinput);                 //for input array
 RTDX_enableOutput(&coutput);               //for output array
 for (;;)                                   //infinite loop
  {
   if (!RTDX_channelBusy(&cgain))           //if new gain value
      RTDX_readNB(&cgain, &gain, sizeof(gain));   //read it
   while(!RTDX_read(&cinput,input,sizeof(input)));//wait for input
   Gain (output, input, gain);                   //function to scale
   RTDX_write(&coutput,&output,sizeof(input));   //output DSK-->host
  }
}
```

FIGURE 9.27. C program running on the DSK that generates a sine wave and illustrates RTDX with LabVIEW (rtdx_lv_gain.c).

Example 9.13: LabVIEW–DSK Interface Using RTDX for Controlling the Amplitude of a Generated Sinusoid with Real-Time Output from the DSK (rtdx_lv_sine)

This example illustrates the use of LabVIEW to control the amplitude of a sine wave generated on the DSK. See also Examples 9.11 and 9.12. The sine wave is generated using the same C source program that illustrates RTDX with Visual C++ (Figure 9.13) and VB (Figure 9.19).

Figure 9.28 shows the LabVIEW Instrument file rtdx_lv_sine. Run it. Connect the output of the DSK to a scope and verify the change in the output sine wave by varying the Volume slider within LabVIEW.

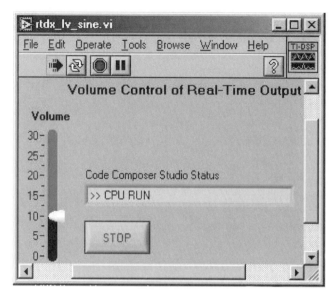

FIGURE 9.28. LabVIEW Instrument window for controlling the DSK output amplitude of a sine wave through RTDX for the project `rtdx_lv_sine`.

Acknowledgments

A special thanks for the contributions on RTDX by: Aghogho Obi, from WPI, with the examples using Visual C++, Mary Ann Nazario, from the MathWorks, with Example 9.6, and Mike Triborn, from National Instruments, with the Examples using LabVIEW.

REFERENCES

1. *TMS320C6000 DSP/BIOS User's Guide*, SPRU303B, Texas Instruments, Dallas, TX, 2000.

2. *An Audio Example Using DSP/BIOS*, SPRA598, Texas Instruments, Dallas, TX, 1999.

3. *TMS320C6000 DSP/BIOS Application Programming Interface (API) Reference Guide*, SPRU403A, Texas Instruments, Dallas, TX, 2000.

4. *DSP/BIOS by Degrees: Using DSP/BIOS Features in an Existing Application*, SPRA591, Texas Instruments, Dallas, TX, 1999.

5. *Real-Time Data Exchange*, SPRY012, Texas Instruments, Dallas, TX, 1998.

6. The MathWorks, Inc. Available at www.mathworks.com

7. National Instruments. Available at www.ni.com

10

DSP Applications and Student Projects

This chapter can be used as a source of experiments, projects, and applications along with Chapters 4–9—for example, using one of the four RTDX schemes in Chapter 9 in conjunction with an adaptive filter example from Chapter 7 or an IIR filter from Chapter 5. The filter can be designed using a software package, the coefficients sent to the DSK through RTDX for implementation in real time. The RTDX scheme provides a real-time interaction between the host PC and the C6x on the DSK.

Additional ideas for projects can be found in Refs. 1–6. A wide range of projects has been implemented on the floating-point C30 and C31 processors [7–21] as well as on the fixed-point TMS320C25 [22–28]. They range in topic from communications and controls to neural networks and also can be used as a source of ideas to implement other projects.

10.1 DTMF SIGNAL DETECTION USING CORRELATION, FFT, AND GOERTZEL ALGORITHM

This project implements the detection of a dual-tone multifrequency (DTMF) tone and is decomposed into four smaller projects. The first mini-project uses a correlation scheme and displays the detected DTMF signals with the onboard LEDs. The second mini-project expands on the first one and uses RTDX that provides a PC-DSK interface to display on the PC monitor the detected DTMF signals by the C6x on the DSK. The third mini-project uses the FFT to estimate the DTMF signals. The fourth mini-project uses Goertzel's algorithm and implements the DTMF detection

Digital Signal Processing and Applications with the C6713 and C6416 DSK   By Rulph Chassaing
ISBN 0-471-69007-4   Copyright © 2005 by John Wiley & Sons, Inc.

TABLE 10.1   DTMF Encoding

Frequencies	1209 Hz	1336 Hz	1477 Hz
697 Hz	1	2	3
770 Hz	4	5	6
852 Hz	7	8	9
941 Hz	*	0	#

on the C6416 DSK (can be transported readily to the C6713 DSK). The complete executable files for all four subprojects are included on the CD.

A DTMF signal consists of two sinusoidal signals: one from a group (row) of four low frequencies and the other from a group (column) of three high frequencies. This is illustrated in Table 10.1. When a key is pressed from a telephone, a DTMF signal is generated. For example, pressing button 6 generates a tone consisting of the summation of the two tones with frequencies of 770 and 1477 Hz, as shown in Table 10.1. For easier detection, these frequencies are chosen so that the sum or difference of any two frequencies does not equal that of any of the other frequencies.

Various schemes can be used to decode DTMF signals:

1. A correlation scheme, as described in this first mini-project. An RTDX option in the second mini-project provides a PC-DSK interface displaying the dialed (received) numbers on the PC screen.
2. The FFT (or the DFT) to detect the signals corresponding to the DTMF tones. The FFT is used in the third mini-project to estimate the weights associated with the seven frequencies.
3. Use of a bank of FIR filters so that each filter passes only one of the frequencies. The average power at the output of two of these filters should be larger than that at the other outputs, yielding the corresponding DTMF tone (not used in this project).
4. Use of Goertzel's algorithm [2,22,28,29] in lieu of the FFT or DFT since only two frequencies need be detected/selected. This method (see Appendix F) can be more efficient than the FFT when a "small" number of spectrum points are required rather than the entire spectrum (implemented in Appendix H with the C6416 DSK and can be readily transported to the C6713 DSK).

Each DTMF signal can be represented as

$$u(t) = A(\sin(\omega_1 t + \varphi_1) + \sin(\omega_2 t + \varphi_2))$$

where ω_1 and ω_2 are the two frequencies that need to be determined, and φ_1 and φ_2 are unknown phases. Frequency f_1 is one of the following frequencies: 697, 770, 852, or 941 Hz; and frequency f_2 is one of the following frequencies: 1209, 1336, or 1477 Hz [30,31].

10.1.1 Using a Correlation Scheme and Onboard LEDs for Verifying Detection

The correlation scheme is as follows. Let the input signal be $u(t) = A(\sin(2\pi 697t + \varphi_1) + \sin(2\pi 1209t + \varphi_2))$. Since the input signal includes $\sin(2\pi 697t + \varphi_1)$, the correlation of the input signal with $\sin(2\pi 697t + \varphi_1)$ must be higher than the correlations with $\sin(2\pi 770t + \varphi_1)$, $\sin(2\pi 852t + \varphi_1)$, and $\sin(2\pi 941t + \varphi_1)$. The Fourier transform $\int u(t)e^{-j\omega t}dt$ has a peak at 697 Hz. Using Euler's formula for the exponential function, it becomes a correlation of $u(t)$ with sine and cosine functions. As a result, the input frequency can be determined by correlating the input signal with the sine and cosine for each possible frequency. The algorithm is as follows:

1. For each frequency, find the following correlations:

$$W_{\sin 697} = \sum_{n=1}^{N} u(t_n)\cdot\sin(2\pi 697t_n), \quad W_{\cos 697} = \sum_{n=1}^{N} u(t_n)\cdot\cos(2\pi 697t_n),$$

. . .

$$W_{\sin 1477} = \sum_{n=1}^{N} u(t_n)\cdot\sin(2\pi 1477t_n), \quad W_{\cos 1477} = \sum_{n=1}^{N} u(t_n)\cdot\cos(2\pi 1477t_n)$$

2. For each frequency, find the maximum between sine weight and cosine weight:

$$W_{697} = \max(|W_{\sin 697}|, |W_{\cos 697}|),$$

. . .

$$W_{1477} = \max(|W_{\sin 1477}|, |W_{\cos 1477}|)$$

3. Among the first four weights, choose the largest one; and among the last three weights, choose the largest one:

$$W_1 = \max(|W_{697}|, |W_{770}|, |W_{852}|, |W_{941}|),$$
$$W_2 = \max(|W_{1209}|, |W_{1336}|, |W_{1477}|)$$

4. The frequencies present in the input signal can then be obtained. If both W_1 and W_2, are larger than a threshold, turn on the appropriate LEDs corresponding to each character, as shown in Table 10.2.

Figure 10.1 shows the C source program `DTMF.c` that can be completed readily. Build this project as DTMF. You can test this project first since the complete executable file `DTMF.out` is included on the CD in the folder DTMF. It can be tested using one of the following:

TABLE 10.2   Characters and Corresponding LEDs

1	0001
2	0010
3	0011
4	0100
5	0101
6	0110
7	0111
8	1000
9	1001
*	1010
0	1011
#	1100

```
//DTMF.c Core program to decode DTMF signals and turn on LEDs
#define N    100
#define thresh 40000
short i;short buffer[N]; short sin697[N],cos697[N],sin770[N],cos770[N];
...
long weight697,weight697_sin,weight697_cos; long ...weight1477_cos;
long weight1,weight2,choice1,choice2;
interrupt void c_int11()
{
  for (i = N-1; i > 0; i--)
      buffer[i]=buffer[i-1];                  // initialize buffer
  buffer[0] = input_sample();                 //input into buffer
  output_sample(buffer[0]*10);                //output from buffer
  weight697_sin=0;  weight697_cos=0;          //weight @ each freq
  ...
  weight1477_sin = 0;  weight1477_cos =  0;
  for (i = 0; i < N; i++)
  {
   weight697_sin = weight697_sin + buffer[i]*sin697[i];
   weight697_cos = weight697_cos + buffer[i]*cos697[i];
   ...
   weight1477_cos= weight1477_cos + buffer[i]*cos1477[i];
  }
//for each freq compare sine and cosine weights and choose largest
  if(abs(weight697_sin)>abs(weight697_cos))   weight697=abs(weight697_sin);
  else weight697 = abs(weight697_cos);
  ...
  if(abs(weight1477_sin)>abs(weight1477_cos)) weight1477 = abs(weight1477_sin);
  else weight1477 = abs(weight1477_cos);
  weight1=weight697; choice1=1;//among weight697,..weight941->largest
  if(weight770 > weight1) {weight1 = weight770; choice1=2;} //...
  if(weight941 > weight1) {weight1 = weight941; choice1=4;}
  weight2=weight1209; choice2=1;//among weight1209,..weight1477->largest
  if(weight1336> weight2) {weight2 = weight1336; choice2=2;}
```

FIGURE 10.1. Core C program using correlation to detect DTMF tones (dtmf.c).

```
if(weight1477> weight2) {weight2 = weight1477; choice2=3;}
if((weight1>thresh)&&(weight2>thresh)) //set threshhold
{  // depending on choices1 and 2 turn on corresponding LEDs
if((choice1 == 1)&&(choice2 == 1)) { //button "1" -> 0001
  DSK6713_LED_off(0);DSK6713_LED_off(1);DSK6713_LED_off(2);DSK6713_LED_on(3);}
... //for button "2","3",..,"*","0"
if((choice1 == 4)&&(choice2 == 3))   //button "#" -> 1100
  {DSK6713_LED_on(0);DSK6713_LED_on(1);DSK6713_LED_off(2);DSK6713_LED_off(3);}
}            //end of if > threshold
else { //weights below threshold, turn LEDs off
  DSK6713_LED_off(0);DSK6713_LED_off(1);DSK6713_LED_off(2);DSK6713_LED_off(3);}
 return;
}
void main()
{
DSK6713_LED_init();
DSK6713_LED_off(0);DSK6713_LED_off(1);DSK6713_LED_off(2);DSK6713_LED_off(3);
for (i = 0; i < N; i++)    //define sine/cosine for all 7 frequencies
 {
  buffer[i]=0;
  sin697[i]=1000*sin(2*3.14159*i/8000.*697);
  cos697[i]=1000*cos(2*3.14159*i/8000.*697);
  ...
  cos1477[i]=1000*cos(2*3.14159*i/8000.*1477);
 }
 comm_intr();  while(1); //init, infinite loop
}
```

FIGURE 10.1. (*Continued*)

1. A phone to create the DTMF signals and a microphone to capture these signals as input to the DSK's mic input. See Chapter 1 on how to modify the header file c6713dskinit.h to select the mic (in lieu of line) input as active by changing the value of register 4 from 0x0011 to 0x0015 in the header file. (An appropriate microphone with the necessary pre-amp can be used and connected directly to the line input on the DSK.) For the threshold value set in the program, use 1000000 with the microphone input option. Dial a few numbers and verify the corresponding LEDs turning on based on the number detected.

2. Figure 10.2 shows the core of the MATLAB program *DTMF*.m that generates/plays DTMF signals as input to the DSK. This program can be completed readily. Verify that all 12 DTMF signals 0, 1, ..., # are consecutively generated by the MATLAB program, each lasting approximately 1.5 second. Also verify that the corresponding LEDs on the DSK are turned on for each detected DTMF signal. For the line input, use a threshold value of 40000 in the program.

3. A tone generator using DialpadChameleon (can be downloaded from the web). This provides a pad with keys to generate short DTMF signals that can be used as input to the DSK.

```
%DTMF.m Core MATLAB file to generate DTMF signals

clear all
t = 1:8000;
t = t/8000;
num_1 = zeros(8000,1);
num_2 = zeros(8000,1);
...                                        ;also num_0, num_star
num_pound = zeros(8000,1);

for n = 1:8000
        num_1(n) = sin(2*pi*697*t(n)) + sin(2*pi*1209*t(n));
        num_2(n) = sin(2*pi*697*t(n)) + sin(2*pi*1336*t(n));
        ...
        num_pound(n)=sin(2*pi*941*t(n))+sin(2*pi*1477*t(n));
end

for i = 1:100000000
        soundsc(num_1);
        pause(1.5);
        soundsc(num_2);
        pause(1.5);
        ...
        soundsc(num_pound);
        pause(1.5);
end
```

FIGURE 10.2. Core MATLAB program to generate DTMF tones (dtmf.m).

The length of the signal affects the reliability of detection. If the buffer size is too small, the probability of turning on the wrong LEDs increases because of the uncertainty in frequency associated with short signals. If the buffer is too long, it complicates the detection near the transmission points. The Dialpad signals have the shortest duration.

10.1.2 Using RTDX with Visual C++ to Display Detected DTMF Signals on the PC

Figure 10.3a shows the core of the C source program DTMF_BIOS_RTDX.c for the RTDX version to provide a PC-DSK interface for displaying the DTMF signals on the PC monitor. These signals are detected by the C6x on the DSK and transferred to the PC for display. Figure 10.3a can be completed readily. The complete RTDX with Visual C++ support files are included on the CD. Examples 9.7 and 9.8 and Sections 10.3 and 10.5 illustrate RTDX using Visual C++.

Build this project as **DTMF_BIOS_RTDX**. Examples 9.1–9.3 introduce the use of the configuration (.cdb) file. The interrupt is set within this configuration file. The complete executable (.out) file is also on the CD. Load/run the executable (.out) file within CCS. Select Tools → Configuration Control → Enable RTDX (check it). Use one of the three options (as in the non-RTDX version) to input the DTMF signals.

```
//DTMF_BIOS_RTDX.c Addtl. code to DTMF.c for RTDX version using VC++

#include <rtdx.h>                            //RTDX support file
RTDX_CreateOutputChannel(ochan);            //output channel for DSK->PC
#define thresh 80000                        //defines a threshold
short value = 0; short w = 0;               //used for RTDX version
.... see DTMF.c
if((weight1>thresh)&&(weight2>thresh))       //set threshold
 if((choice1 == 1)&&(choice2 == 1)) {        //button "1" -> 0001
   DSK6713_LED_off(0);DSK6713_LED_off(1);DSK6713_LED_off(2);DSK6713_LED_on(3);
   value = 1;
 }
 . . . //for button "2", "3",..., "*", "0"
 if((choice1 == 4)&&(choice2 == 3)) { //button "#" -> 1100
   DSK6713_LED_on(0);DSK6713_LED_on(1);DSK6713_LED_off(2);DSK6713_LED_off(3);
    value = 12;
 }
}                       //end of if > than the threshold value (see DTM
else { //weights below threshold, turn LEDs off
 DSK6713_LED_off(0);DSK6713_LED_off(1);DSK6713_LED_off(2);DSK6713_LED_off(3);
 value = 0;
}
w = w + 1;
if w > 50;
{
 w = 0;
 RTDX_write(&ochan,&value,sizeof(value));//send value to PC
}
return;
}                                   //end of interrupt service routine
void main()
{
. . . as in DTMF.c
comm_intr();
while(!RTDX_isOutputEnabled(&ochan))
     puts("\n\n Waiting . . . ");  //wait for output channel->enabled
while(1);                             //infinite loop
}
```

(*a*)

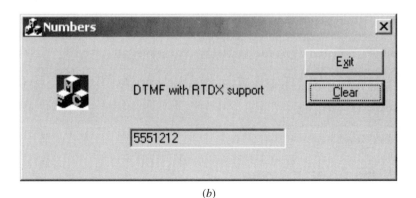

(*b*)

FIGURE 10.3. (*a*) Core C program to detect DTMF signals with RTDX for PC-DSK interface (`dtmf_BIOS_RTDX.c`); (*b*) PC screen displaying detected DTMF signals with RTDX for PC-DSK interface.

Run the application Visual C++ file `DTMF_BIOS_RTDX.EXE`. Verify the corresponding detected DTMF signals on the LEDs also displayed on the PC monitor, as shown in Figure 10.3*b*.

Implementation Issues

1. A number is sent to the PC (through RTDX) every 50th time and can be changed.

2. The threshold value can be adjusted.

3. A "`length`" of 15 is set in the file `numbersDlg.cpp`. This is used to analyze the last 15 numbers and determine if a button was pressed. A smaller value can cause false detection due to noise, whereas it can be more difficult to recognize a short DTMF signal with a larger value of `length`.

If the number 1 is pressed using a Dialpad, dozens of 1s are transmitted through RTDX and appear in the data stream. With no button pressed, a stream of 0s is transmitted. The algorithm distinguishes the actual buttons that are pressed. An array of size `length` stores the last `length` numbers. The number of 1s in the array goes into `Weight1`, the number of 2s in the array goes into `Weight2`, and so on. If any of the weights is greater than 70% of `length`, then it is decided that the number corresponding to that weight was pressed. The character corresponding to this number is then added to the string shown in Figure 10.3*b*. Note that each weight should be followed by `Weight0` (except `Weight0`).

10.1.3 Using FFT and Onboard LEDs for Verifying Detection

Figure 10.4 shows the core of the C source program that implements this mini-project using an FFT scheme to detect the DTMF signals. Examples 6.5 and 6.6 and Sections 10.4 and 10.5 illustrate the radix-4 FFT. The FFT is used to estimate the weights associated with the seven frequencies. For example, the 697-Hz signal corresponds to a weight of 697(256/8000) $\simeq$ 22, and we would use the 22nd value of the FFT array. A 256-point FFT is used with a sampling frequency of 8000 Hz. Similarly, the 770-Hz signal corresponds to a weight of 770(256/8000) $\simeq$ 25, and we would use the 25th value of the FFT array, and so on for the other weights (28, 31, 39, 43, and 47). We then find the largest weights associated with the first four frequencies to determine the row frequency signal and the largest weights associated with the last three frequencies to determine the column frequency signal. For the largest weights, the corresponding LEDs are turned on (as in Section 10.1.1). As with the previous schemes, the same input (MATLAB, Dialpad, or microphone) can be used. Verify similar results.

10.1.4 Using Goertzel Algorithm

Example H.6 (Appendix H) implements the DTMF detection on the C6416 DSK using the Goertzel algorithm. The complete support files are included on the CD. Transport this method to the C6713 DSK.

```
//DTMF_Bios_FFT.c Core program using radix-4 FFT and onboard LEDs
. . .                          //see radix-4 example in Chapter 6
short input_buffer[N] = {0};  //to store input samples...same as x
float output_buffer[7] = {0}; //to store magnitude of FFT
short buffer_count, i, J;
short nFlag;                   //indicator to begin FFT
short nRow, nColumn;
double delta;
float tempvalue;
interrupt void c_int11()
{
 input_buffer[buffer_count] = input_sample();
 output_sample((short)input_buffer[buffer_count++]);
 if(buffer_count >= N)          //if accum more than N points->begin FFT
   {
    buffer_count = 0;           //reset buffer_count
    nFlag = 0;                  //flag to signal completion
    for(i = 0; i < N; i++)
       {
         x[2*i] = (float)input_buffer[i];   //real part of input
         x[2*i+1] = 0;                       //imaginary part of input
       }
   }
}
void main(void)
{
 nFlag = 1;
 buffer_count = 0;
 . . . //generate twiddle constants, then index for digit reversal
 comm_intr();
 while(1)                        //infinite loop
 {
  while(nFlag);    //wait for ISR to finish buffer accum samples
  nFlag = 1;
  //call radix-4 FFT, then digit reverse function
  output_buffer[0]=(float) sqrt(x[2*22]*x[2*22]+x[2*22+1]*x[2*22+1]);
  . . . //for weigths 25,28,31,39,43
  output_buffer[6]=(float) sqrt(x[2*47]*x[2*47]+x[2*47+1]*x[2*47+1]);
  tempvalue = 0;                      //choose largest row frequency
  nRow = 0;
  for(j = 0; j < 4; j++)
    {
     if(tempvalue < output_buffer[j])
       {
        if(output_buffer[j] > 0.5e4)
          {
             nRow = j + 1;
             tempvalue = output_buffer[j];
          }
       }
    }                               //end of for loop
  tempvalue = 0;                    //choose largest column frequency
  nColumn = 0;
  for(j = 4; j < 7; j++)
   . . .                            //as with the rows
```

FIGURE 10.4. Core C program using FFT to detect DTMF tones (dtmf_bios_FFT.c).

```
        nColumn = j - 3;
  . . .                               //as with the rows
  } //end of for loop
if((nRow != 0) && (nColumn != 0))
{
  if((nRow==1)&&(nColumn==1))        //for button 0001 ("1")
{DSK6713_LED_off(0);DSK6713_LED_off(1);DSK6713_LED_off(2);DSK6713_LED_on(3);}
  if((nRow==1)&&(nColumn==2))        //for button 0010
{DSK6713_LED_off(0);DSK6713_LED_off(1);DSK6713_LED_on(2);DSK6713_LED_off(3);}
  //for button "3", "4", ..., "#"
}
else
{DSK6713_LED_off(0);DSK6713_LED_off(1);DSK6713_LED_off(2);DSK6713_LED_off(3);}
};                                    //end of while (1) infinite loop
}                                     //end of main
```

FIGURE 10.4. (*Continued*)

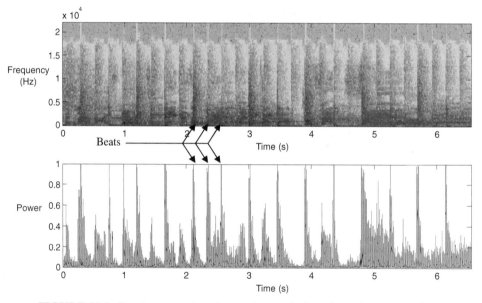

FIGURE 10.5. Spectrogram plot of a music sample for a beat detector project.

10.2 BEAT DETECTION USING ONBOARD LEDs

This mini-project implements a beat detection scheme using the onboard LEDs [32]. Music visualization is a continuously progressing area in audio processing, not only for analysis of music but also for entertainment visualization purposes. The scheme is based on the idea that the drum is the most energy-rich component of the music. In this project, the beat of the music is the drum pattern or bass line of the piece of music. Figure 10.5, obtained with MATLAB's capability for plotting the spectrogram of an input .wav file, shows a representative sample section of a piece of music featuring a live drum, a voice, and other instruments. The beat pattern is visible in the spectrogram of the music, and the energy plot shows that the beat of the drum can be the most energy-rich portion of the music.

Furthermore, it is advantageous to filter out any higher-frequency portions of the music that may also have high energy. This has the added advantage that the parts of the music containing no bass line will not "confuse" the algorithm.

Implementation

Figure 10.6 shows the partial C source program beatdetector.c that can be completed readily. The project can be tested first using the executable (.out) file on the CD in the folder beatdetector. The incoming music signal is continuously sampled at 8 kHz (with a 4-kHz antialiasing filter on the codec) and stored in a buffer. The buffer has 4000 points and is decomposed into 20 chunks, each chunk consisting of 200 points. The signal energy of a smaller portion of the buffer—a "chunk" of the larger buffer—consisting of the most recently collected samples is compared to the signal energy of the entire buffer. When this portion of the signal has a significantly higher energy than the rest of the signal, it is considered to be a beat. The average algorithm is described by the following equations:

$$\langle E \rangle = \frac{1}{N} \sum_{k=0}^{N} B[k]^2$$

$$\langle e \rangle = \frac{1}{n} \sum_{k=i_0}^{i_0+n} B[k]^2$$

$$beat = \begin{cases} true & \langle e \rangle > \langle E \rangle \cdot C \\ false & otherwise \end{cases}$$

$\langle E \rangle$ and $\langle e \rangle$ represent the average energy of the buffer and of each chunk, respectively. C is the comparison factor (sensitivity), B is the buffer, and i_0 is the start position in the chunk buffer. N and n represent the number of points in the buffer and in the chunk, respectively. The first two equations represent the average for the entire buffer and for a chunk, respectively, and the third equation describes the actual beat detection logic.

To fine-tune this method, the following can be adjusted: (1) the length N of the larger buffer (the total signal being compared against), (2) the length n of the chunks (the "instantaneous" signal), and (3) the sensitivity C of the energy comparison. Values for C ranging from 0.5 to 2 were tested, and a value of 1.3 seems to be optimal for most types of music.

A larger buffer size can give a better energy average; however, this has several drawbacks:

1. A larger chunk size means lower accuracy since the beat status can only be updated as often as a single chunk is filled and processed.

```
//Beatdetector.c Core program for beat detection project

const int chunks = 20;              //number of frames in buffer
const int instant_length = 200;     //length of 1 buffer
#define average_length 4000         //length of buffer
const float c = 1.3;                //confidence multiplier
double ae = 0, ie = 0;
short buffer[average_length];       //Buffer
void main()
{
  comm_poll();                      //init DSK, codec, McBSP
  while(average_counter < average_length){ //sample entire buffer
      buffer[average_counter] = input_sample();
      average_counter++;
  }
  while(1) {                        //infinite loop
    instant_counter = 0;
    while(instant_counter<instant_length){  //sample one frame and
     buffer[chunk_counter*instant_length+instant_counter]=input_sample();
     instant_counter++;             //move it to circular buffer
    }
  for (average_counter=0;average_counter<average_length;average_counter++) {
    ae=ae+buffer[average_counter]*buffer[average_counter];//av energy
  }                                              //in entire buffer
  ae = ae / average_length;
  for (instant_counter=0;instant_counter<instant_length;instant_counter++) {
    ie=ie+buffer[chunk_counter*instant_length+instant_counter]
       *buffer[chunk_counter*instant_length+instant_counter];
  }                                 //average energy in last few msec
  ie = ie / instant_length;
  if (ie > ae*c){//if energy in short buffer>whole buffer,turn on LEDs
  ..
  else {                           //if not, turn off LEDs
  ..
  }
  chunk_counter++;                 //incr position in chunk counter
  if(chunk_counter>=chunks) chunk_counter=0; //right point in buffer
  }                                //end of while(1) infinite loop
}                                  //end of main
```

FIGURE 10.6. Core C program for beat detection (beatdetector.c).

2. The larger the buffer, the longer the processing time for calculating the average energy, so the buffer size is limited by the processing speed of the board.

3. A larger buffer requires the use of external memory, which can mean a reduction in speed.

A buffer stored in internal memory with a length of half a second (4000 points) decomposed into 20 chunks seems to work best. The LEDs onboard the DSK are flashed whenever a beat is detected. To expand on this project, the beat information can be fed back (from the DSK output) as data or as an audio signal to control, for example, external light effects. Alternatively, it can be fed back to the host PC for further processing, such as calculating beats per minute. RTDX can then be used to provide an interface between the PC host and the DSK (see Chapter 9).

Build this project as **beatdetector** and verify that this detection scheme (with several different types of music) recognizes the drum in most cases, with very few false positives.

10.3 FIR WITH RTDX USING VISUAL C++ FOR TRANSFER OF FILTER COEFFICIENTS

This project implements an FIR filter using VC++ with RTDX to transfer the coefficients. Chapters 4 and 9 discuss FIR filters and RTDX with VC++, respectively. All the appropriate files for this project are on the CD in the folder rtdx_vc_FIR. Figure 10.7 shows the C source program rtdx_vc_FIR.c that runs on the DSK. It implements the FIR filter, and creates and enables an input channel through RTDX to read a new set of coefficients. These coefficients are transferred through RTDX from the PC host to the C6x running on the DSK.

1. Build this project as **rtdx_vc_FIR**. A configuration (.cdb) file is created to set INT11. Note that the project includes several autogenerated support files including the linker command file. The init/comm. file is included in the project for real-time input and output. The vector file is not included since INT11 is set within the configuration file. See Example 9.2.

   Within CCS, load and run the executable file. Select Tools → RTDX → Configuration control and enable RTDX (check it).

2. Run the Visual C++ application file included in the folder rtdx_vc_FIR\VC_FIR_RTDX\Debug. A message for the user to load a coefficient file pops up, as shown in Figure 10.8. Load the coefficient file LP600.cof, looking in the folder rtdx_vc_FIR. This coefficient file was designed with MATLAB and used in Example 4.2 to implement a lowpass FIR filter with a cutoff frequency at 600 Hz. Verify this result.

   Load LP1500.cof and LP3000.cof, which represent FIR lowpass filter with 81 coefficients and with cutoff frequencies at 1500 and 3000 Hz, respectively. Verify that these FIR filters can be implemented readily.

The coefficient files are transferred in real time to the C program running on the DSK, using the function RTDX_read( ) in Figure 10.7. The coefficients are stored in the buffer RtdxBuffer, along with N that represents the number of coefficients (81) as the first value in the coefficient file (the lowpass coefficient files in the example FIR3LP have been modified for this project). Experiment with different sets of coefficients.

```
//rtdx_vc_FIR.c FIR with RTDX using VC++ to transfer coefficients file
#include "dsk6713_aic23.h"
#include <rtdx.h>
#define RTDX_BUFFER_SIZE 256                    //change for higher order
Uint32 fs = DSK6713_AIC23_FREQ_8KHZ;
RTDX_CreateInputChannel(control_channel);       //create input channel
short* pFir;                                    //->filter's Impulse response
short RtdxBuffer[RTDX_BUFFER_SIZE]={0};         //buffer for RTDX
short dly[RTDX_BUFFER_SIZE] = {0};              //buffer for input samples
short i;
short N;                                        //order of filter
int yn;
interrupt void c_int11()
{
  dly[0] = input_sample();
  yn = 0;
  for(i = 0; i < N; i++)
      yn += pFir[i]*dly[i];
  for(i = N - 1; i > 0; i--)
      dly[i] = dly[i-1];
  output_sample(yn >> 15);
}
void main()
{
  N = 0;                                        //initial filter order
  pFir = &RtdxBuffer[1];                        //-> 2nd element in buffer
  comm_intr();
  RTDX_enableInput(&control_channel);           //enable RTDX input channel
  while(1)                                       //infinite loop
  {
   if(!RTDX_channelBusy(&control_channel))      //if free, read->buffer
   {                                            //read N and coefficients
    RTDX_read(&control_channel,&RtdxBuffer,sizeof(RtdxBuffer));
    N = RtdxBuffer[0];                          //extract filter order
   }
  }
}
```

FIGURE 10.7. C source program that runs on the DSK to implement an FIR filter using RTDX with Visual C++ to transfer the coefficients from the PC to the DSK (rtdx_vc_FIR.c).

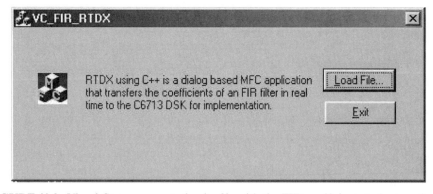

FIGURE 10.8. Visual C++ message to load a file with the FIR coefficients to be transferred through RTDX from the PC to the DSK.

10.4 RADIX-4 FFT WITH FREQUENCY DOMAIN FILTERING

This mini-project extends the radix-4-based Examples 6.5 and 6.6 (see *FFTr4_sim.c* and *FFTr4.c*), using the same optimized FFT support functions. It illustrates real-time radix-4 FFT with filtering implemented in the frequency domain. Figure 10.9 shows the core of the C source program `FFTr4_filter.c` for this project. From the comments in Figure 10.9, and Examples 6.5 and 6.6, the reader can readily complete the full program and verify the results of this project.

A 256-point FFT is implemented. The calculation of the twiddle constants and the function calls to (1) the index for digit reversal, (2) the radix-4 optimized FFT function, and (3) the digit reversal are as in Example 6.5 (*FFTr4_sim.c*). To obtain the inverse FFT (IFFT), (1) take the conjugate of the input, (2) invoke the FFT optimized function, and (3) invoke the digit reversal function, as in Example 6.5.

A gel file (on the CD) uses three sliders in this project: the first one to test whether to turn on any filter, the second to determine whether to turn on a lowpass or a highpass filter, and the third to control the bandwidth of the lowpass or the highpass filter.

Build this project as **FFTr4_filter**. Input a sinusoidal wave with an approximate amplitude of 1 V p-p and a frequency of 2 kHz. Load/run the program and verify that the output is the original input signal. Vary slightly the input signal frequency and verify the corresponding variation in the output signal frequency. Taking the FFT of the input signal and the IFFT of the result yields the original input signal. Use the sliders to turn on the filters and verify that the bandwidth of these two filters can be adjusted.

Note: This project can be tested first since the associated executable file (`real-time.out`) is on the CD in the folder `FFTr4_filter`.

10.5 RADIX-4 FFT WITH RTDX USING VISUAL C++ AND MATLAB FOR PLOTTING

This project implements a radix-4 FFT using TI's optimized functions. The resulting FFT magnitude of a real-time input is sent to MATLAB for plotting. In real time, the output data are sent to the PC host using RTDX with Visual C++. Chapter 9 includes two examples using RTDX with Visual C++, Chapter 6 includes two examples (one in real time) to implement a radix-4 FFT, and Section 10.4 contains a mini-project using radix-4 FFT with frequency-domain filtering. The necessary files are in the folder `rtdx_vc_FFTr4`. This includes the Visual C++ support and executable files in the folder `rtdx_vc_FFTr4\rtdxFFT`.

CCS Component

The C source program `rtdx_vc_FFTr4.c` runs on the DSK and is shown in Figure 10.10a. An output RTDX channel is created and enabled to write (send) the resulting FFT magnitude data in the buffer `output_buffer` to MATLAB running on

```
//FFTr4_filter.c Core program radix-4 FFT with freq domain filtering

float input_buf[2*N];              //to store input-same as x
short input_short[N];              //used for debugging
short filterfrequency = 1;         //slider for BW control
short filteron = 0;                //slider for filter on/off
short filter_type = 0;             //slider for LP or HP filter
short n2 = 2*N;                    //to save processor time
short nover2 = N/2;                //to save processor time

void main(void)
{        //initialize DSK, call index digit reversal, calculate W
 ...
 while (1)                          //infinite loop
 {
  for (i=(n2-2); i > 0; i -= 2)
  {                                  //shift input buffer
   input_buf[i-2] = input_buf[i];
   input_buf[i-1] = input_buf[i+1];
  }
  for (i=0;i < n2;i += 2)
  {
   output_sample((short)(x[i]/(nover2/8)));//out most recent samples
   input_short[i/2] = (input_sample());
   input_buf[i] = (float)input_short[i/2];
   input_buf[i+1] = 0;
  }
  for (i=0; i < n2;i++)
     x[i] = input_buf[i];          //copy input_buf to x
  //call FFT cfftr4_dif(x, w, N);  then digit reverse
  if (filteron)
  {                                  //LP/HP adjustable BW
   if (filter_type == 0)
   {                                 //lowpass filter
    for(i = (filterfrequency); i<(n2-filterfrequency) ; i++)
      { x[i] = 0; }
   }
   else if (filter_type == 1)
   {                                 //highpass filter
    for(i = 0; i<(filterfrequency) ; i++)
     {
      x[i] = 0;
      x[n2 - i - 1] = 0;
     }
   }
  }
  //for IFFT:conjugate input, call ASM FFT and digit reverse functions
 }                                  //for infinite while loop
}                                   //for main
```

FIGURE 10.9. Core C program for radix-4 FFT and filtering in the frequency domain (FFTr4_filter.c).

```
//rtdx_vc_FFTr4.c Core r4-FFT using RTDX with VC++(MATLAB for plotting)
. . . N=256,16kHz rate,align x&w,... see Examples in Chapter 6
#include <rtdx.h>
short input_buffer[N] = {0};          //store input samples(same as x)
float output_buffer[N] = {0};         //store magnitude FFT
short buffer_count=0;
short nFlag=1;                        //when to begin the FFT
short i, j;
RTDX_CreateOutputChannel(ochan);      //output channel C6x->PC transfer
interrupt void c_int11()              //ISR
{
 input_buffer[buffer_count] = input_sample(); //input -->buffer
 output_sample(input_buffer[buffer_count++]); //loop
 if(buffer_count >= N)
  {                                   //if more than N pts, begin FFT
   buffer_count = 0;                  //reset buffer_count
   nFlag = 0;                         //flag to signal completion
   for(i = 0; i < N; i++)
    {
     x[2*i]=(float)input_buffer[i]; //real component of input
     x[2*i+1] = 0;                    //imaginary component of input
    }
  }
}
void main(void)
{
 . . . //generate twiddle constants and digit reversal index
 comm_intr();                         //init DSK
 while(!RTDX_isOutputEnabled(&ochan));//wait for PC to enable RTDX
 while(1)                             //infinite loop
  {
   while(nFlag);                      //wait to finish accum samples
   nFlag = 1;
   cfftr4_dif(x, w, N);               //call radix-4 FFT function
   digit_reverse((double *)x, IIndex, JIndex, count);
   for(j = 0; j < N; j++)
     output_buffer[j]=(float)sqrt(x[2*j]*x[2*j]+x[2*j+1]*x[2*j+1]);
   RTDX_write(&ochan,output_buffer,sizeof(output_buffer));//Send DSK>PC
  };
}
```

(*a*)

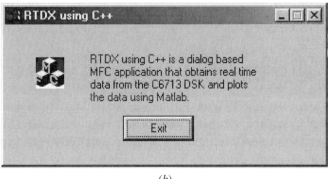

(*b*)

FIGURE 10.10. (*a*) C program to implement radix-4 FFT and illustrate RTDX with Visual C++, using MATLAB for FFT and plotting (rtdx_vc_FFTr4.c); (*b*) message when the VC++ application file is executed.

the PC host for plotting (only). RTDX is achieved using Visual C++. The radix-4 FFT support functions for generating the index for digit reversal, and for digit reversal, were used in Chapter 6. The complex radix-4 FFT function `cfftr4_dif.c` is also on the CD (the ASM version was used in Chapter 6). Note that the real and imaginary components of the input are consecutively arranged in memory (as required by the FFT function). Digit reversal is performed on the resulting FFT since it is scrambled and needs to be resequenced. After the FFT magnitude is calculated and stored in `output_buffer`, it is sent to MATLAB through an output RTDX channel.

The project uses DSP/BIOS only to set interrupt INT 11 using the (`.cdb`) configuration file (see Example 9.3). As a result, a vector file is not required. The BSL file needs to be added (the support files for RTDX and CSL are included in the autogenerated linker command file, which must be added to the project by the user).

Build this project within CCS as **rtdx_vc_FFTr4**. Within CCS, select Tools → RTDX and configure the buffer size to 2048 (not 1024), and then enable RTDX (check it). From the configuration (`.cdb`) file, select Input/Output → RTDX. Right-click for properties to increase the buffer size from 1024 to 2056. Load and run the (`.out`) file. Input a 2-kHz sine wave with an approximate amplitude of 1/2 V p-p. The output from the DSK is like a loop program.

Visual C++ Component

Execute/run the application file `rtdxFFT.exe` located in the VC++ folder `rtdx_vc_FFTr4\rtdxFFT` (within debug). Figure 10.10*b* will pop up, followed by the FFT magnitude plot from MATLAB. Verify that the FFT of the 2-kHz sine wave output is plotted within MATLAB, as in Example 9.5.

The Visual C++ file `rtdxFFTDlg.cpp` includes the code section for MATLAB to set the sampling rate and plot the received data. It is located in the dialog class within the thread

$$UINT\ CRtdxFFTDlg::RTDXThreadFunction(LPVOID\ lpvoid)$$

Re-create the executable (application) file. Launch Microsoft Visual C++ and select File → Open Workspace to open `rtdxFFT.dsw`. Build and Rebuild All.

10.6 SPECTRUM DISPLAY THROUGH EMIF USING A BANK OF 32 LEDs

This mini-project takes the FFT of an input analog audio signal and displays the spectrum of the input signal through a bank of 32 LEDs. The specific LED that turns on depends on the frequency content of the input signal. The bank of LEDs is controlled through the external memory interface (EMIF) bus on the DSK. This EMIF bus is a 32-bit data bus available through the 80-pin connector J4 on board the DSK.

The FFT program in Chapter 6 using TI's optimized ASM-coded FFT function is extended for this project. Figure 10.11 shows the core of the program that imple-

```
//graphic_FFT.c Core program.Displays spectrum to LEDs through EMIF

#include "output.h"          //contains EMIF address
int *output = (int *)OUTPUT; //EMIF address in header file
. . .
while (1)                    //infinite loop
 {
  .                          //same as in FFTr2.c
  .
  for(i = 0; i < N/2; i++)
   {
    if (Xmag[i] > 20000.0)   //if mag FFT >20000
     {
       out = out + 1 << i;   //shifts one to appropriate bit location
     }
   }
  *output = out;             //output to EMIF bus
  out = 0;                   //reset out variable for next iteration
 }
```

FIGURE 10.11. Core C program to implement radix-2 FFT using TI's optimized FFT support functions. It displays the spectrum to 32 LEDs through EMIF `graphic_FFT.c`).

ments this project—using a 64-point radix-2 FFT, sampling at 32 kHz—and does not output the negative spike (32000) for reference. The executable (`.out`) file is on the CD in the folder **graphic_FFT**. and can be used first to test this project. See also the project used to display the spectrum through EMIF using LCDs in Section 10.7.

EMIF Consideration

To determine whether the data is being outputted through the EMIF bus, the following program is used:

```
# define OUTPUT 0xA0000000   //output address (EMIF)
int *output = (int*)OUTPUT;  //map memory location to variable
void main( )
{
*output = 0x00000001;        //ouput 0x1 to the bus
}
```

This program defines the output EMIF address and gives the capability to read and write to the EMIF bus. Test the EMIF by writing different values lighting different LEDs. The final version of the program includes a header file to define the output EMIF address.

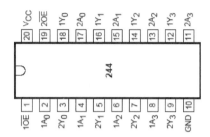

FIGURE 10.12. Line driver used with external LEDs to display the spectrum in project `graphic_FFT`.

EMIF-LEDs

A total of 32 LEDs connect through four line drivers (74LS244). Current limiting resistors of 300 ohms are connected between each LED and ground. The line drivers allow for the needed current to light up the LEDs. The current drawn by the LED is limited to 10 mA so that the line drivers are not overloaded. Figure 10.12 shows one of the line drivers. Pin 20 is connected to +5 V and pin 10 to ground. Pins 1 and 19 are also connected to ground to enable the output of the line driver. Each line driver supports eight inputs and eight outputs. The pins labeled with "Y" are output pins. Each of the output pins (on a line driver) is connected to pins 33–40, which correspond to data pins 31–24 on the EMIF bus. The arrangement is the same with the other three line drivers connecting to pins 43–50 (data pins 23–16), pins 53–60 (data pins 15–8), and pins 63–70 (data pins 7–0), respectively. Pin 79 on the EMIF bus is used for universal ground. See also the schematics of connectors J3 and J4 shown in the file *c6713_dsk_schem.pdf*, included with CCS. Table 10.3 shows the EMIF signals.

Note: Pin 75 on J3 (not J4), the 80-pin connector for the external peripheral interface, is to be connected to ground since it is an enable pin for the EMIF interface and enables the output voltages on these pins.

Implementation

The real-time radix-2 FFT program example in Chapter 6 is slightly modified to check the amplitude of a specific frequency and determine whether or not it is above a set threshold value of 20000. If so, the value of that specific frequency is sent to the EMIF output port to light the appropriate LED(s). From Figure 10.11, when a value of the FFT magnitude is larger than the set threshold, the variable *out* is output. This output corresponds to a bit that is shifted by the value of the index i that is the corresponding frequency location in the FFT array. This bit shift moves a binary 1 to the appropriate bit location corresponding to the specific LED to be lit. This process is repeated for every value in the magnitude FFT array. If multiple values in the FFT array are larger than the set threshold of 20000, then the appropriate bit-shifted value is accumulated. This process lights up all the LEDs that have

TABLE 10.3 EMIF Signals

Pin	Signal	I/O	Description	Pin	Signal	I/O	Description
1	5V	Vcc	5V voltage supply pin	2	5V	Vcc	5V voltage supply pin
3	EA21	O	EMIF address pin 21	4	EA20	O	EMIF address pin 20
5	EA19	O	EMIF address pin 19	6	EA18	O	EMIF address pin 18
7	EA17	O	EMIF address pin 17	8	EA16	O	EMIF address pin 16
9	EA15	O	EMIF address pin 15	10	EA14	O	EMIF address pin 14
11	GND	Vss	System ground	12	GND	Vss	System ground
13	EA13	O	EMIF address pin 13	14	EA12	O	EMIF address pin 12
15	EA11	O	EMIF address pin 11	16	EA10	O	EMIF address pin 10
17	EA9	O	EMIF address pin 9	18	EA8	O	EMIF address pin 8
19	EA7	O	EMIF address pin 7	20	EA6	O	EMIF address pin 6
21	5V	Vcc	5V voltage supply pin	22	5V	Vcc	5V voltage supply pin
23	EA5	O	EMIF address pin 5	24	EA4	O	EMIF address pin 4
25	EA3	O	EMIF address pin 3	26	EA2	O	EMIF address pin 2
27	BE3#	O	EMIF byte enable 3	28	BE2#	O	EMIF byte enable 2
29	BE1#	O	EMIF byte enable 1	30	BE0#	O	EMIF byte enable 0
31	GND	Vss	System ground	32	GND	Vss	System ground
33	ED31	I/O	EMIF data pin 31	34	ED30	I/O	EMIF data pin 30
35	ED29	I/O	EMIF data pin 29	36	ED28	I/O	EMIF data pin 28
37	ED27	I/O	EMIF data pin 27	38	ED26	I/O	EMIF data pin 26
39	ED25	I/O	EMIF data pin 25	40	ED24	I/O	EMIF data pin 24
41	3.3V	Vcc	3.3V voltage supply pin	42	3.3V	Vcc	3.3V voltage supply pin
43	ED23	I/O	EMIF data pin 23	44	ED22	I/O	EMIF data pin 22
45	ED21	I/O	EMIF data pin 21	46	ED20	I/O	EMIF data pin 20
47	ED19	I/O	EMIF data pin 19	48	ED18	I/O	EMIF data pin 18
49	ED17	I/O	EMIF data pin 17	50	ED16	I/O	EMIF data pin 16
51	GND	Vss	System ground	52	GND	Vss	System ground
53	ED15	I/O	EMIF data pin 15	54	ED14	I/O	EMIF data pin 14
55	ED13	I/O	EMIF data pin 13	56	ED12	I/O	EMIF data pin 12
57	ED11	I/O	EMIF data pin 11	58	ED10	I/O	EMIF data pin 10
59	ED9	I/O	EMIF data pin 9	60	ED8	I/O	EMIF data pin 8
61	GND	Vss	System ground	62	GND	Vss	System ground
63	ED7	I/O	EMIF data pin 7	64	ED6	I/O	EMIF data pin 6
65	ED5	I/O	EMIF data pin 5	66	ED4	I/O	EMIF data pin 4
67	ED3	I/O	EMIF data pin 3	68	ED2	I/O	EMIF data pin 2
69	ED1	I/O	EMIF data pin 1	70	ED0	I/O	EMIF data pin 0
71	GND	Vss	System ground	72	GND	Vss	System ground
73	ARE#	O	EMIF async read enable	74	AWE#	O	EMIF async write enable
75	AOE#	O	EMIF async output enable	76	ARDY	I	EMIF asynchronous ready
77	N/C	—	No connect	78	CE1#	O	Chip enable 1
79	GND	Vss	System ground	80	GND	Vss	System ground

frequencies with corresponding amplitudes above the set threshold value. Setting the threshold value at 20000 creates a range of frequencies from about 150 Hz to 15 kHz.

Build this project as **graphic_FFT** and verify that the lights adapt to the input audio signal in real time. You can also test this program with a signal generator as input to the DSK. Increase the frequency of the input signal and verify the sequence associated with the LEDs that turn on.

10.7 SPECTRUM DISPLAY THROUGH EMIF USING LCDS

This project implements a graphical frequency display through the use of a 2×16 character liquid-crystal display (LCD) (LCM-S01602DTR/M from Lumex). Each LCD character is decomposed into two separate states to form a bar graph displaying the spectrum of an input signal. See also the previous project, which displays a spectrum through EMIF using a bank of 32 LEDs. Figure 10.13 shows the core of the program, $EMIF\_LCD.c$, that implements this project. It uses the C-coded FFT function called from $FFT256c.c$ in Chapter 6 to obtain the spectrum (for the section of code that is excluded without outputting the negative spike for reference).

FFT Component
One component of the program is based on the FFT program example in Chapter 6 that calls a C-coded FFT function (see *FFT256c.c*). The FFT component uses 256 points and samples at 32 kHz to allow a frequency display range from 0 to 16 kHz. The second component of the program is associated with the EMIF-LCD.

LCD Component
Since the LCD is 16 characters wide, each character is chosen to correspond to one band. The FFT range then can be decomposed linearly into sixteen 1-kHz bands, with each band being determined in a nested "for loop." The 256-point FFT is then decomposed into 16 bands with eight samples per band. The average of the samples is taken and placed into an array of size 16. Using thresholds, this array is then parsed to determine which character (blank or filled) is to be displayed on the LCD.

Each LCD character has two different states, either fully on or fully off (four states total). These characters are then placed in arrays, one array for the top row of the LCD and one for the bottom row. These arrays are accessed by the function that writes data to the appropriate LCD. Two functions are used to transfer data to the LCD:

1. The first function, *LCD_PUT_CMD*, is used primarily by an initialization function (*init_LCD*). It masks the proper data bits and configures the control lines. The LCD has setup and hold times that must be achieved for proper operation. The *LCD_PUT_CMD* function sets the control lines, with delays to ensure that there are no timing glitches, and then pulses the enable control line. Clocking the data into the LCD occurs during the falling edge of the enable line.

2. The second function, *LCD_PUT_CHAR*, sends the characters to the LCD and requires different control signals. The cursor address is autoincremented so that a character is sent to the proper position on the LCD.

```
//EMIF.LCD.c Core C program. Displays spectrum to LCDs through EMIF
#define IOPORT 0xA1111111          //EMIF address
int *ioport = (int *)IOPORT;       //pointer to get data out
int input, output;                 //temp storage
void set_LCD_characters();         //prototypes
void send_LCD_characters();
void init_LCD();
void LCD_PUT_CMD(int data);
void LCD_PUT_CHAR(int data);
void delay();
float bandage[16];                 //holds FFT array after downsizing
short k=0, j=0;
int toprow[16] = {0, 0, 0, 0, 0, 0, 0, 0, 0, 0, 0, 0, 0, 0, 0, 0};
int botrow[16] = {0, 0, 0, 0, 0, 0, 0, 0, 0, 0, 0, 0, 0, 0, 0, 0};
short rowselect = 1;               //start on top row
short colselect = 0;               //start on left of LCD
#define LCD_CTRL_INIT 0x38         //initialization for LCD
#define LCD_CTRL_OFF 0x08
#define LCD_CTRL_ON 0x0C
#define LCD_AUTOINC 0x06
#define LCD_ON 0x0C
#define LCD_FIRST_LINE 0x80
#define LCD_SECOND_LINE 0xC0       //address of second line
main()
{
 ..
 init_LCD();                       //init LCD
 while(1)                          //infinite loop
  {
   for(k=0; k<16; k++){            //for 16 bands
           float sum = 0;          //temp storage
           for(j=0; j<8; j++)      //for 8 samples per band
               sum += x1[8*k+j];   //sum up samples
           bandage[k] = (sum/8);   //take average
   }
   set_LCD_characters();           //set up character arrays
   send_LCD_characters();          //put them on LCD
  }                                //end of infinite loop
}                                  //end of main
interrupt void c_int11()           //ISR
{
 output_sample(bandage[buffercount/16]); //out from iobuffer
 ..
}
void set_LCD_characters()          //to fill arrays with characters
{
 int n = 0;                        //temp index variable
 for (n=0; n<16; n++)
  {
   if(bandage[n] > 40000)          //first threshold
    {
     toprow[n] = 0xFF;             //block character
     botrow[n] = 0xFF;
    }
```

FIGURE 10.13. Core C program using a C-coded FFT function to display the spectrum to LCDs through EMIF (EMIF_LCD.c).

```
      else if(bandage[n] > 20000)        //second threshold
        {
         toprow[n] = 0x20;               //blank space
         botrow[n] = 0xFF;
        }
      else                               //below second threshold
        {
         toprow[n] = 0x20;
         botrow[n] = 0x20;
        }
    }
}
void send_LCD_characters()
{
 int m=0;
 LCD_PUT_CMD(LCD_FIRST_LINE);           //start address
 for (m=0; m<16; m++)                   //display top row
   LCD_PUT_CHAR(toprow[m]);
 LCD_PUT_CMD(LCD_SECOND_LINE);          //second line
 for (m=0; m<16; m++)                   //display bottom row
   LCD_PUT_CHAR(botrow[m]);
}
void init_LCD()
{
 LCD_PUT_CMD(LCD_CTRL_INIT);            //put command
 LCD_PUT_CMD(LCD_CTRL_OFF);             //off display
 LCD_PUT_CMD(LCD_CTRL_ON);              //turn on
 LCD_PUT_CMD(0x01);                     //clear display
 LCD_PUT_CMD(LCD_AUTOINC);              //set address mode
 LCD_PUT_CMD(LCD_CTRL_ON);              //set it
}
void LCD_PUT_CMD(int data)
{
 *ioport = (data & 0x000000FF);         //RS=0, RW=0
 delay();
 *ioport = (data | 0x20000000);         //bring enable line high
 delay();
 *ioport = (data & 0x000000FF);         //bring enable line low
 delay();
}
void LCD_PUT_CHAR(int data)
{
 *ioport = ((data & 0x000000FF)| 0x80000000);   //RS=1, RW=0
 *ioport = ((data & 0x000000FF)| 0xA0000000);   //enable high
 *ioport = ((data & 0x000000FF)| 0x80000000);   //enable Low
 delay();
}
void delay()                            //create 1 ms delay
{
 int q=0, junk=2;
 for (q=0; q<8000; q++)
     junk = junk*junk;
}
```

FIGURE 10.13. (*Continued*)

TABLE 10.4   EMIF-LCD Pin Connections

LCD Pin Number	Name	Function	DSK (EMIF) Pin Connection J4
1	Vss	Ground	Gnd
2	Vdd	Supply	+5V
3	Vee	Contrast	Gnd
4	RS	Register select	ED31
5	R/W	Read/write	ED30
6	E	Enable	ED29
7	D0	Data bit 0	ED0
8	D1	Data bit 1	ED1
9	D2	Data bit 2	ED2
10	D3	Data bit 3	ED3
11	D4	Data bit 4	ED4
12	D5	Data bit 5	ED5
13	D6	Data bit 6	ED6
14	D7	Data bit 7	ED7

With only one port to use, the two functions *LCD_PUT_CHAR* and *LCD_PUT_CMD* include bitwise AND and OR operations to mask and set only certain bits.

The delay function creates a 1-ms delay to meet the timing requirements (setup and hold times) of the LCD for proper operation.

EMIF-LCD Pins Description

Table 10.4 displays information of the LCD pins and the EMIF connector. EMIF pins information on connector J4, is shown in Table 10.3 (associated with the previous project) and contained in the file *c6713_dsk_schem.pdf*, included with CCS. The least significant data pins (ED0–ED7) for the characters are selected, and the three most significant data pins (ED29–ED31) for the control lines are selected. The first six pins on the LCD are used for power and control signals. To enable the data for output through the EMIF bus, pin 75 of the External Peripheral Interface connector J3 (not J4) is to be connected to ground (see also the previous project).

Build this project as **EMIF_LCD**. Use either an input signal from a signal generator or an input audio signal. Verify the graphical frequency display on the LCDs.

Some possible improvements to this project include:

1. More thresholds so that more levels of frequency intensities can be represented. More than four thresholds would better illustrate the frequency intensity.

2. The bands can be displayed logarithmically instead of linearly. A logarithmic display would allow for a wider range of frequencies. An up-sampling scheme would then be used.

10.8 TIME-FREQUENCY ANALYSIS OF SIGNALS WITH SPECTROGRAM

This project makes use of the short time Fourier transform (STFT) for the analysis of signals, resulting in a spectrogram plot [33,34]. A spectrogram is a plot of the frequencies that make up a particular signal. The magnitude of the frequency at a particular time is represented by the colors in the graph. This plot of frequency versus time provides information on the changing frequency content of a signal over time.

The spectrogram is the square of the absolute value of the STFT of a signal. The STFT looks at a nonstationary signal as small blocks in time and takes the Fourier transform of each block to obtain the frequency content of the signal at that time. This involves multiplying the signal with a moving window to observe smaller segments of the signal and taking the Fourier transform of the product. The use of a sliding window and its size needs to be determined. A large window size (length) can be chosen to enhance the frequency resolution, but at the expense of the time resolution, and vice versa. The window increment, which represents the distance between successive windows, also needs to be determined.

A spectrogram can be more useful than a plot of the spectrum since there can be a different spectrum for each time. The spectrogram is plotted as frequency versus time as a three-dimensional plot. Consider a musical scale consisting of eight musical notes representing the C scale major: C, D, E, F, G, A, B, C with the following sinusoidal frequencies: 262, 294, 330, . . . , 523, respectively, starting with the middle C at a frequency of 262 Hz. The subsequent C is one octave higher at 523 Hz, which represents a doubling in frequency. A spectrogram plot of frequency versus time would identify each note as it is played.

Time-frequency analysis techniques include the STFT, Gabor expansion, and energy distribution based techniques such as the Wigner–Ville distribution. These techniques are used to study the behavior of nonstationary signals such as music and speech signals.

The files for this project are in the folder **spectrogram** (with separate subfolders). The spectrogram project is decomposed into three separate sections (versions), all of which make use of MATLAB's function `imagesc` to plot the spectrogram:

1. Simulation using MATLAB to read a.wav file and plot its spectrogram
2. RTDX with MATLAB and use of a C-coded FFT function
3. RTDX with Visual C++ and a radix-4 optimized FFT function

10.8.1 Simulation Using MATLAB

This is a simulated version using MATLAB. Figure 10.14a shows the MATLAB file `spectrogram.m` that plots a spectrogram, using the function *wavread* to read a .wav file `chirp.wav` that is a swept sinusoidal signal. MATLAB's FFT function is also used, as well as the function `imagesc`, to find the spectrogram of the input .wav file.

```
%Spectrogram.m Reads .wav file,plots spectrogram using STFT with MATLAB

[x,fs,bits] = wavread('chirp.wav'); %read .wav file
N = length(x);
t=(0:N-1)/fs;
set(0,'DefaultAxesColorOrder',[0 0 0],...
     'DefaultAxesLineStyleOrder','-|-.|--|:');
figure(1); plot(t,x);                    %plots time-domain signal
xlabel('Time (sec)'); ylabel('Amplitude'); title('Waveform of signal');
M=256;  B=floor(N/M);                    %divide signal->blocks of M samples
x_mat=reshape(x(1:M*B),[M B]);           %reshape vector into MxB matrix
win=hamming(M);                          %Hamming window before FFT
win_mat=repmat(win,[1 B]);
x_fft=fft(x_mat.*win_mat);               %perform FFT
y=abs(x_fft(1:M/2,:));                    %want positive freq and mag info
t=(1:B)*(M/fs);                          %values for time and freq axes
f=((0:M-1)/(M-1))*(fs/2);
figure(2);
imagesc(t,f,dB(y));                      %plot spectrogram
colormap(jet);    colorbar;    set(gca,'ydir','normal');
xlabel('Time (sec)');  ylabel('Frequency (Hz)');  title('Spectrogram');
```
(*a*)

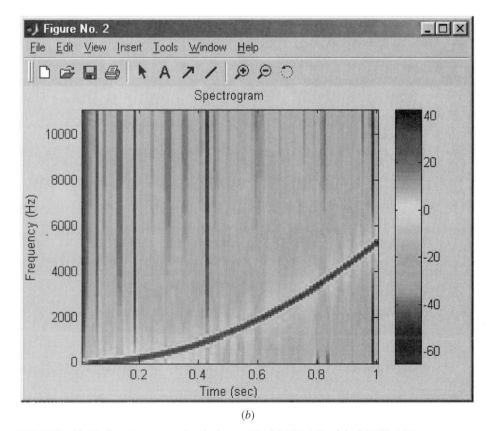
(*b*)

FIGURE 10.14. Spectrogram simulation with MATLAB: (*a*) MATLAB program to read and find the spectrogram of an input .wav file; (*b*) spectrogram plot of an input chirp signal.

Run the MATLAB program and verify Figure 10.14*b* as the spectrogram of a chirp signal. It illustrates the increase in frequency of the swept sinusoidal signal over time. You can readily test other .wav files on the CD, such as speech_RTDSP.wav (on the CD) with the spoken words "Real-Time Digital Signal Processing."

10.8.2 Spectrogram with RTDX Using MATLAB

This version of the project makes use of RTDX with MATLAB for transferring data from the DSK to the PC host. Section 9.1 introduces the use of a configuration (.cdb) file and Section 9.2 illustrates RTDX with MATLAB.

Figure 10.15*a* shows the core source program spectrogram_rtdx_mtl.c that runs on the DSK and can be readily completed using the program FFT256c.c in Chapter 6 (the complete executable file is on the CD). It calls the C-coded FFT function used in Chapter 6 and enables an RTDX output channel to write/send the

```
//Partial_Spectrogram_rtdx_mtl.c Core program for Time-Frequency
//analysis with spectrogram using RTDX-MATLAB
. . . See FFT256c.c
#include <rtdx.h>                          //RTDX support file
#include "hamming.cof"                     //Hamming window coefficients
RTDX_CreateOutputChannel(ochan);          //create output channel C6x->PC

main()
{
//. . . calculate twiddle constants
 comm_intr();                             //init DSK, codec, McBSP
 while(!RTDX_isOutputEnabled(&ochan))     //wait for PC to enable RTDX
     puts("\n\n Waiting . . . ");         //while waiting
 while(1)                                 //infinite loop
  {
    . . .
    for (i = 0 ; i < PTS ; i++)           //swap buffers
     {
       samples[i].real=h[i]*iobuffer[i];  //multiply by Hamming coeffs
       iobuffer[i] = x1[i];               //process frame to iobuffer
     }
    . . . use FFT magnitude squared
    RTDX_write(&ochan,x1,sizeof(x1)/2);   //send 128 samples to PC
  }                                       //end of infinite loop
}                                         //end of main
interrupt void c_int11()                  //ISR
{. . . as in FFT256c.c }
```

(*a*)

FIGURE 10.15. Spectrogram using RTDX with MATLAB: (*a*) core program to calculate FFT and transfer FFT data from the DSK to the PC; (*b*) spectrogram plot of an external chirp input signal; (*c*) spectrogram plot of a 500-Hz square wave input signal.

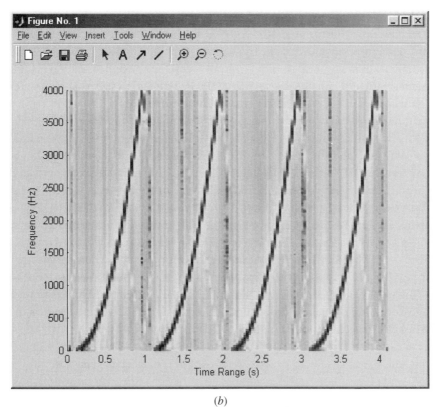

(b)

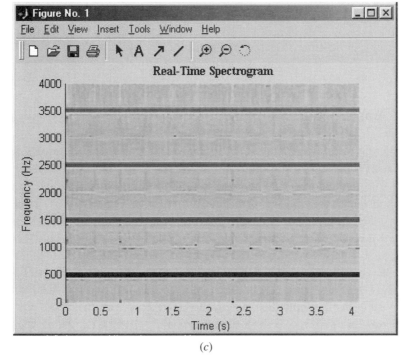

(c)

FIGURE 10.15. (*Continued*)

resulting FFT data to the PC running MATLAB for finding the spectrogram. A total of $N/2$ (128 points) are sent (in lieu of 256) for better resolution (continuity). The (.cdb) configuration file is used to set interrupt INT11, as in Section 9.1. From this configuration file, select Input/Output → RTDX. Right-click on properties and change the RTDX buffer size to 8200. Within CCS, select Tools → RTDX → Configure to set the host buffer size to 2048 (from 1024).

An input signal is read in blocks of 256 samples. Each block of data is then multiplied with a Hamming window of length 256 points. The FFT of the windowed data is calculated and squared. Half of the resulting FFT of each block of 256 points is then transferred to the PC running MATLAB to find the spectrogram. Build this project as **spectrogram_rtdx_mtl**. Within CCS, select Tools → RTDX → Configure.

Open MATLAB, select the appropriate path, and run spectrogram_rtdx.m (on the CD). Within MATLAB, CCS will enable RTDX, and will load and run the COFF (.out) executable file. Then MATLAB will plot the resulting spectrogram of an input signal. Input/play Chirp.wav (output of a sound card as input to the DSK) and verify the spectrogram of this input signal plotted by MATLAB, as shown in Figure 10.15b. For a chirp input signal, the transfer of 128 points (in lieu of 256) yields a better spectrogram.

For a faster and accurate plot, delete the commands within the MATLAB file that include the labels (x and y axes, and title) in the spectrogram plot.

Use a 500-Hz square wave as input and verify the spectrogram plot shown in Figure 10.15c. A darker red strip is formed at the 500-Hz fundamental frequency, and lighter red strips at the other harmonics of 1500, 2500, and 3500 Hz. For this type of input, you may choose to transfer the entire block of 256-point FFT data at each time.

You can extend this project version using TI's optimized FFT function (see Chapter 6).

10.8.3 Spectrogram with RTDX Using Visual C++

This project is also tested using RTDX with Visual C++ for data transfer from the DSK to the PC host. The program spectrogram_rtdx_r4.c (on the CD) implements a 256-point radix-4 FFT using TI's optimized FFT function and the associated support files for digit reversal. See also the two radix-4 FFT examples in Chapter 7 and Section 10.5. As with the MATLAB version for RTDX, only 128 points are transferred at a time.

Change the buffer size to 8200 within the (.cdb) file, as with the previous MATLAB version. Within CCS, change the host buffer size from 1024 to 2048. Enable RTDX (there is no MATLAB file for doing so). Load/run the .out file.

The Visual C++ support files are on the CD. Access/run the VC++ application file vc_spectrogram.exe. You should get the Visual C++ dialog message in Figure 10.16 until MATLAB plots the spectrogram of a real-time input signal. Input/play the (.wav) chirp signal and verify that the results are identical to those

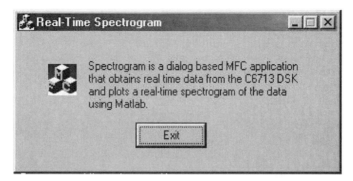

FIGURE 10.16. Visual C++ dialog message for a spectrogram.

achieved with the spectrogram in Figure 10.15*b*, being continuously updated within MATLAB. The file `vc_spectrogramdlg.cpp` contains the MATLAB commands for plotting the spectrogram. However, MATLAB is not used in this version to provide the RTDX link.

As in Section 10.8.2, you can obtain a fast and accurate plot by deleting the commands for including the title and the labels within the spectrogram plot. These commands are in the file `vc_spectrogramdlg.cpp`.

You can extend this project version using the radix-2 FFT (in lieu of the radix-4). Chapter 6 includes several examples based on the radix-2 FFT.

10.9   AUDIO EFFECTS (ECHO AND REVERB, HARMONICS, AND DISTORTION)

This project illustrates various audio effects such as distortion, echo and reverb, and harmonics [35]. Figure 10.17 shows the core program `soundboard.c` (virtually complete) that implements this project. The overall program flow consists of pre-amplification, distortion, echo/reverb, harmonics, and postamplification. Preamp and postamp are included to avoid overdriving the output. A sampling rate of 16 kHz is chosen, and a total of 10 sliders are used for the overall control. The slider gel file is on the CD in the folder `soundboard`.

The distortion effect is the simplest to implement. It requires overamplifying each sample and clipping it at maximum and minimum values. The acquired input sample is amplified based on whether it is positive or negative. The amplification polynomial used for the distortion component is used to amplify the signal in a nonlinear fashion. The result is scaled by a distortion magnitude controlled by a slider, then clipped so as not to overdrive the output.

The resulting output is processed for an echo/reverb effect (see Examples 2.6 and 2.7 on echo effects). The length of the echo is controlled by changing the buffer size where the samples are stored. A dynamic change of the echo length leads to a reverb effect. A fading effect with a decaying echo is obtained with a slider.

```
//Soundboard.c  Core C program for sound effects
union {Uint32 uint; short channel[2];} AIC23_data;
union {Uint32 uint; short channel[2];} AIC23_input;
short EchoLengthB = 8000;          //echo delay
short EchoBuffer[8000];            //create buffer
short echo_type = 1;               //to select echo or delay
short Direction = 1;               //1->longer echo,-1->shorter
short EchoMin=0,EchoMax=0;         //shortest/longest echo time
short DistMag=0,DistortionVar=0,VolSlider=100,PreAmp=100,DistAmp=10;
short HarmBuffer[3001];            //buffer
short HarmLength=3000;             //delay of harmonics
float output2;
short DrumOn=0,iDrum=0,sDrum=0;    //turn drum sound when = 1
int   DrumDelay=0,tempo=40000;     //delay counter/drum tempo
short ampDrum=40;                  //volume of drum sound
..                                 //addtl casting
interrupt void c_int11()           //ISR
{
AIC23_input.uint = input_sample(); //newest input data
input=(short)(AIC23_input.channel[RIGHT]+AIC23_input.channel[LEFT])/2;
input = input*.0001*PreAmp*PreAmp;
output=input;
output2=input;                     //distortion section
if (output2>0)
output2=0.0035*DistMag*DistMag*DistMag*((12.35975*(float)input)
      - (0.359375*(float)input*(float)input));
else  output2 =0.0035*DistMag*DistMag*DistMag*(12.35975*(float)input
      + 0.359375*(float)input*(float)input);
output2/=(DistMag+1)*(DistMag+1)*(DistMag+1);
if (output2 > 32000.0)  output2 = 32000.0 ;
else if (output2 < -32000.0 )  output2 = -32000.0;
output= (output*(1/(DistMag+1))+output2); //overall volume slider
input = output;                    //echo/reverb section
iEcho++;                           //increment buffer count
if (iEcho >= EchoLengthB) iEcho = 0;      //if end of buffer reinit
output=input + 0.025*EchoAmplitude*EchoBuffer[iEcho];//newest+oldest
if(echo_type==1) EchoBuffer[iEcho] = output; //for decaying echo
else EchoBuffer[iEcho]=input;      //for single echo (delay)
EchoLengthB += Direction;          //alter the echo length
if(EchoLengthB<EchoMin+100){Direction=1;} //echo delay is shortest->
if(EchoLengthB>EchoMax){Direction=-1;}    //longer,if longest->shorter
input=output;                      //output echo->harmonics gen
if(HarmBool==1) {                  //everyother sample...
 HarmBool=0;                       //switch the count
 HarmBuffer[iHarm]=input;          //store sample in buffer
 if(HarmBool2==1){                 //everyother sample...
  HarmBool2=0;                     //switch the count
  HarmBuffer[uHarm] += SecHarmAmp*.025*input;//store sample in buffer
 }
 else{HarmBool2=1; uHarm++;        //or just switch the count,
     if(uHarm>HarmLength) uHarm=0; //and increment the pointer
 }
}
```

FIGURE 10.17. Core C program to obtain various audio effects (`soundboard.c`).

```
else{HarmBool=1; iHarm++;                    //or just switch the count
if(iHarm>HarmLength) iHarm=0;}               //and increment the pointer
output=input+HarmAmp*0.0125*HarmBuffer[jHarm];//add harmonics to output
jHarm++;                                     //and increment the pointer
if(jHarm>HarmLength) jHarm=0;                //reinit when maxed out
DrumDelay--;                                 //decrement delay counter
if(DrumDelay<1) {                            //drum section
     DrumDelay=50000-Tempo;                  //if time for drumbeat
     DrumOn=1;                               //turn it on
}
if(0){                                       //if drum is on
 output=output+(kick[iDrum])*.05*(ampDrum);//play next sample
 if((sDrum%2)==1) {iDrum++;}                 //but play at Fs/2
 sDrum++;                                    //incr sample number
 if(iDrum>2500){iDrum=0; DrumOn=0;}          //drum off if last sample
}
output = output*.0001*VolSlider*VolSlider;
AIC23_data.channel[LEFT]=output;
AIC23_data.channel[RIGHT]=AIC23_data.channel[LEFT];
output_sample(AIC23_data.uint);             //output to both channels
}
main()          //init DSK,codec,McBSP and while(1) infinite loop
```

FIGURE 10.17. (*Continued*)

The third effect is harmonics boost. A harmonics buffer is used for this effect. Two main loop sections are created to produce two separate sets of harmonics. The larger (outer) loop combines the input with samples from the harmonics buffer at twice the input frequency. The smaller (inner) loop produces the next harmonics at four times the input frequency. The magnitudes of the harmonics are controlled with a slider.

These effects were tested successfully using the input from a keyboard with the keyboard output to a speaker. The audio output is sent to both channels of the codec (see Example 2.3), using the stereo capability of the onboard codec. The executable and gel files are included in the folder **soundboard**.

A drum effect section is included in the program for expanding the project. The use of external memory must be considered when applying many effects.

10.10 VOICE DETECTION AND REVERSE PLAYBACK

This project detects a voice signal from a microphone, then plays it back in the reverse direction. Figure 10.18 shows the block diagram that implements this project. All the necessary files are in the folder detect_play. Two circular buffers are used: an input buffer to hold 80,000 samples (10 seconds of data) continuously being updated and an output buffer to play back the input voice signal in the reverse direction. The signal level is monitored, and its envelope is tracked to determine whether or not a voice signal is present.

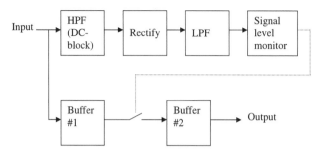

FIGURE 10.18. Block diagram for the detection of a voice signal from a microphone and playback of that signal in the reverse direction.

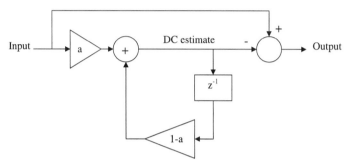

FIGURE 10.19. DC blocking first-order IIR highpass filter for voice signal detection and reverse playback.

When a voice signal appears and subsequently dies out, the signal-level monitor sends a command to start the playback of the accumulated voice signal, specifying the duration of the signal in samples. The stored data are transferred from the input buffer to the output buffer for playback. Playback stops when one reaches the end of the entire signal detected.

The signal-level monitoring scheme includes rectification and filtering (using a simple first-order IIR filter). An indicator specifies when the signal reaches an upper threshold. When the signal drops below a low threshold, the time difference between the start and end is calculated. If this time difference is less than a specified duration, the program continues into a no-signal state (if noise only). Otherwise, if it is more than a specified duration, a signal-detected mode is activated.

Figure 10.19 shows the DC blocking filter as a first-order IIR highpass filter. The coefficient a is much smaller than 1 (for a long time constant). The estimate of the DC filter is stored as a 32-bit integer.

The lowpass filter for the envelope detection is also implemented as a first-order IIR filter, similar to the DC blocking filter except that the output is returned directly rather than being subtracted from the input. The filter coefficient a is larger for this filter to achieve a short time contant.

Build and test this project as **detect_play**. You may need to change the header file c6713dskinit.h for a microphone input (see Chapter 1).

10.11 PHASE SHIFT KEYING—BPSK ENCODING AND DECODING WITH PLL

See also the two projects on binary phase shift keying (BPSK) and modulation schemes in Sections 10.12 and 10.13. This project is decomposed into smaller mini-projects as background for the final project. The final project is the transmission of an encoded BPSK signal with voice as input and the reception (demodulation) of this signal with phase-locked loop (PLL) support on a second DSK. All the files associated with these projects are located in separate subfolders within the folder **PSK**.

10.11.1 BPSK Single-Board Transmitter/Receiver Simulation

BPSK is a digital modulation technique that separates bits by shifting the carrier 180 degrees. A carrier frequency signal is chosen that is known by both the transmitter and the receiver. Each bit is encoded as a phase shift in the carrier at some predetermined period. When a 0 is sent, the carrier is transmitted with no phase shift, and when a 1 is sent, the carrier is phase-shifted by 180 degrees [36–39].

CCS Component
The necessary files for this project are on the CD in **BPSK_sim** within the folder **PSK**. Figure 10.20 shows the C source program BPSK_sim.c that modulates a bit stream of 10 bits set in the program. Since there is no carrier synchronization, demodulation is performed by the same program on the same DSK board.

Build this project as **BPSK_sim**. Connect the DSK output to the input to verify the demodulation of the transmitted sequence. Run the program. The demodulator program prints the demodulated sequence within CCS. Verify that it is the same as the sequence set in the array encodeSeq to be encoded.

The array buffer stores the entire received vector that can be plotted within CCS. Select View → graph → Time/Frequency. Use buffer as the address, 190 as the acquisition and display size, 8000 as the sample rate, and a 16-bit signed integer format. Figure 10.21a shows the CCS plot of the received sequence: {1, 0, 1, 1, 0, 0, 0, 1, 0, 1} as set in the program. Note that when the received sequence changes from a 0 to a 1 or from a 1 to a 0, a change of phase is indicated in the positive and negative y axis, respectively. Change the sequence to be encoded in the program to {0, 1, 0, 0, 1, 1, 1, 0, 1, 0} and verify the CCS plot in Figure 10.21b.

MATLAB Component
The MATLAB program BPSK_sim.m is also included on the CD. It simulates the modulation and demodulation of a random bit stream. Run this MATLAB file and verify the plots in Figures 10.22a and 10.22b for signal-to-noise ratios (SNR) of 0.5 and 5.0, respectively. They display the transmitted and received waveforms of a random bit stream. The SNR can be changed in the program. The MATLAB program also displays the decision regions and detection, as shown in Figures 10.23a and 10.23b, for SNRs of 0.5 and 5.0, respectively. With small values of SNR, the

```
//BPSK.c BPSK Modulator/Demod. DSK Output sequence --> Input
#include "dsk6713_aic23.h"          //codec-DSK support file
#include <math.h>
#include <stdio.h>
Uint32 fs=DSK6713_AIC23_FREQ_16KHZ; //set sampling rate
#define PI 3.1415926
#define N 16                        //# samples per symbol
#define MAX_DATA_LENGTH 10          //size of mod/demod vector
#define STABILIZE_LEN 10000         //# samples for stabilization
float phi_1[N];                     //basis function
short r[N] = {0};                   //received signal
int rNum=0,  beginDemod=0;          //# of received samples/demod flag
short encSeqNum=0, decSeqNum=0;     //# encoded/decoded bits
short encSymbolVal=0,decSymbolVal=0;//encoder/decoder symbol index
short encodeSeq[MAX_DATA_LENGTH]={1,0,1,1,0,0,0,1,0,1};//encoded seq
short decodeSeq[MAX_DATA_LENGTH];   //decoded sequence
short sigAmp[2] = {-10000, 10000};  //signal amplitude
short buffer[N*(MAX_DATA_LENGTH+3)];//received vector for debugging
short buflen=0,  stabilizeOutput=0;
interrupt void c_int11()            //interrupt service routine
{
 int i,  outval= 0;
 short X = 0;
 if(stabilizeOutput++ < STABILIZE_LEN) //delay start to Stabilize
 {
  r[0] = input_sample();
  output_sample(0);
  return;
 }
 if(encSeqNum < MAX_DATA_LENGTH)    //modulate data sequence
 {
  outval = (int) sigAmp[encodeSeq[encSeqNum]]*phi_1[encSymbolVal++];
  if(encSymbolVal>=N) {encSeqNum++;  encSymbolVal=0; }
  output_sample(outval);
 }
 else output_sample(0);            //0 if MAX_DATA_LENGTH exceeded
 r[rNum++] = (short) input_sample();//input signal
 buffer[buflen++] = r[rNum - 1];
 if(beginDemod)                    //demod received signal
 {
  if(decSeqNum<2 && rNum==N)  {    //account for delay in signal
     decSeqNum ++;    rNum = 0; }
  if(rNum == N)                    //synchronize to symbol length
  {
   rNum = 0;
   for(i=0; i<N; i++)             //correlate with basis function
         X += r[i]*phi_1[i];
   decodeSeq[decSeqNum-2] = (X >= 0) ? 1: 0;   //do detection
   if(++decSeqNum == MAX_DATA_LENGTH+2) //print received sequence
   {
     for(i=0; i<decSeqNum-2; i++)
           printf("Received Value: %d\n", decodeSeq[i]);
```

FIGURE 10.20. C program that modulates a sequence of 10 numbers to illustrate BPSK, using a single DSK for modulation and demodulation (BPSK.c).

```
      exit(0);
     }
    }
   }
  else   { beginDemod = 1; rNum = 0; }
 }
void main()
{
   int i; comm_intr();                      //init DSK, codec, McBSP
   for(i=0; i<=N; i++)
     phi_1[i] = sin(2*PI*i/N);              //basis function
   while(1);                                //infinite loop
 }
```

FIGURE 10.20. (*Continued*)

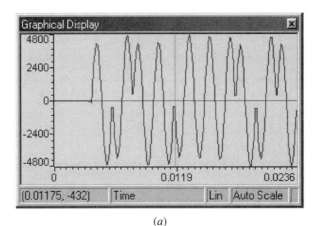

(*a*)

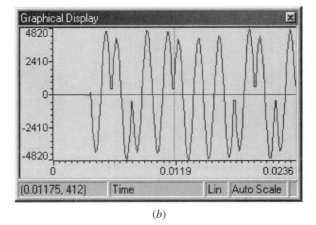

(*b*)

FIGURE 10.21. CCS plot of a received sequence, representing a BPSK modulated signal:
(*a*) sequence of {1, 0, 1, 1, 0, 0, 0, 1, 0, 1}; (*b*) sequence of {0, 1, 0, 0, 1, 1, 1, 0, 1, 0}.

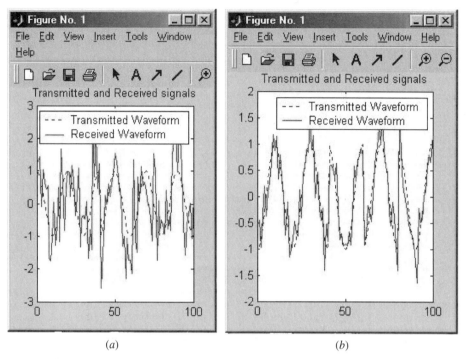

(a)   (b)

FIGURE 10.22. MATLAB plots simulating the modulation of a random bit stream showing the transmitted and received waveforms for (*a*) SNR = 0.5; (*b*) SNR = 5.0.

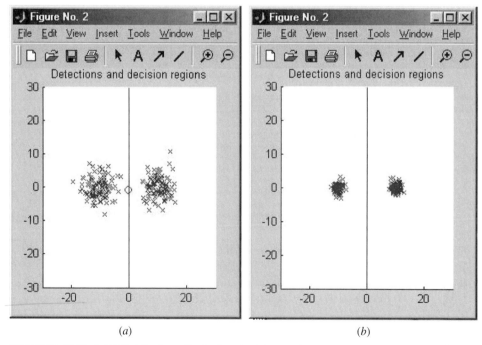

(a)   (b)

FIGURE 10.23. MATLAB plots displaying decision regions and detection for (*a*) SNR = 0.5; (*b*) SNR = 5.0.

```
//BPSK_ReIn.c Illustrates transmitter/voice encoder with Real IN
#include "dsk6713_aic23.h"            //codec-DSK support file
#include <math.h>
Uint32 fs=DSK6713_AIC23_FREQ_32KHZ; //set sampling rate
#define NUMSAMP 4                     //# samples per symbol
#define MAX_DATA_LENGTH 10            //size of mod/demod vector
short encSeqNum=0, encSymbolVal=0;   //# encoded bits/symbol index
short sin_table[NUMSAMP]={0,10000,0,-10000};
short sample_data;  short bits[16]={0};  short outval=1;

interrupt void c_int11()                  //interrupt service routine
{
 int i;
 short j=0;
 sample_data=(short)input_sample(); //input sample
 if(encSeqNum == 32)                    //decimate 32kHz to 1kHZ
 {
  encSeqNum = 0;
  if((sample_data>1000)||(sample_data<-1000)) {//above noise threshold
  for(i=0;i<8;i++) bits[i]=(sample_data&(1<<i))?1:-1;} //8sig bits
  else {for(i=0;i<8;i++) bits[i]=0;} //get next bit
 }
 outval = (short) bits[j];
 output_sample(outval*sin_table[encSymbolVal++]);//output next sample
 if(encSymbolVal>=NUMSAMP) {encSymbolVal=0; j++;} //reset encSymbolVal
 encSeqNum++;
 if (j==8)    j=0;                      //start next sample
}
void main()
{comm_intr();     while(1);}            //init DSK/infinite loop
```

FIGURE 10.24. C program to illustrate a transmitter/voice encoder using a real-time input signal (`bpsk_ReIn.c`).

received signals fall outside of the appropriate decision regions, resulting in errors in detection. The received signal is noisier, resulting in some false detection. This occurs when the correlator produces an incorrect phase for the incoming symbol. Correct detections are marked with blue ×'s and incorrect detections with red circles. For larger values of SNR, there are no false detections and the correlated signals lie well within the detection region.

10.11.2 BPSK Transmitter/Voice Encoder with Real-Time Input

CCS Component

Figure 10.24 shows the C source program `bpsk_ReIn.c` that implements a transmitter/voice encoder with a real-time input signal. You can use your voice as input from a microphone connected to the mic input (unless you have a microphone with the appropriate preamplification to connect directly to the DSK line input). Note

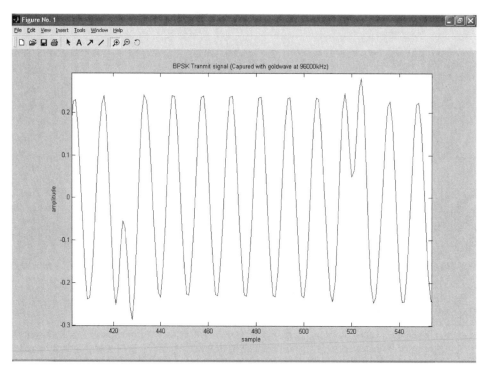

FIGURE 10.25. Plot of encoded DSK output using voice as input to the DSK.

that if you do this, you need to change the value of register 4 in the init/comm. header file (see Chapter 1). Or you can use `TheForce.wav` as input.

Build this project as **BPSK_ReIN**. All the necessary files for this project are on the CD in **BPSK_ReIn** within the folder **PSK**. Use `TheForce.wav` or voice as input to the DSK, with the DSK output to a scope. Verify that a representative segment of the encoded BPSK output signal from the DSK is as shown in Figure 10.25.

MATLAB Component

The corresponding MATLAB file for this project `bpsk_ReIn.m` is on the CD. Verify the resulting MATLAB plots in Figure 10.26. The upper graph shows the received waveform signal segment of (`TheForce.wav`). A `.wav` file is used to model the input signal being encoded as a BPSK signal. The plots show successive samples being encoded and decoded. The `.wav` sample is decimated to 1 kHz, converted to a bit stream, and then modulated to a BPSK signal that is then plotted. The upper graph shows which amplitude of the voice signal is being modulated into a BPSK signal. Note that as the circle moves along the received waveform in the upper graph, the corresponding BPSK signal and transmitted bits are displayed in the lower graph and are continuously encoded (updated).

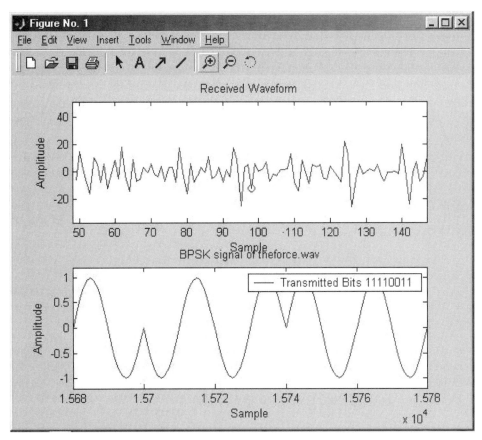

FIGURE 10.26. MATLAB plots of an encoded voice signal (lower graph) and received segment of `TheForce.wav` (upper graph).

10.11.3 Phase-Locked Loop

This project is a PLL receiver. In BPSK, the receiver must be able to lock onto the phase of a received signal in order to distinguish between 1s and 0s. A sinusoid of 1 kHz, with varying phase, is used as the real-time input to the DSK. This input signal has eight unique phase shifts. The real-time output signal is the phase of the received signal. Two DSKs are required to implement this project.

To determine the phase of an incoming sinusoid, the maximum of the correlation coefficient is calculated between the received sinusoid and a sinusoid offset by a phase estimate. The correlation coefficient, Y, between two sinusoids is given by

$$Y = \int_{0}^{2\pi} \sin(t \cdot \omega + \phi_{\text{carrier}}) \cdot \sin(t \cdot \omega + \phi_{\text{est}})$$

The received sine wave has a phase of ϕ_{carrier}, and an estimate of the phase is ϕ_{est}. The correlation coefficient has a maximum value when ϕ_{carrier} and ϕ_{est} are equal.

To determine this maximum, begin with an initial estimate of ϕ_{est}. For every period of the incoming signal that is received, that signal is correlated with a sine wave that has a phase slightly larger and slightly smaller than ϕ_{est}. This yields two values for the correlation coefficient, one at $\phi_{est} + \varepsilon$ and the other at $\phi_{est} - \varepsilon$. The difference between these two values gives an approximation of the derivative of the correlation coefficient. Using the difference between the correlation coefficients at $\phi_{est} + \varepsilon$ and $\phi_{est} - \varepsilon$ as an estimate of the derivative, a new value for ϕ_{est} is calculated using

$$\phi_{est} = \phi_{est} + (Y_{+\varepsilon} - Y_{-\varepsilon})$$

where

$$Y_{+\varepsilon} = \int_0^{2\pi} \sin(t \cdot \omega + \phi_{carrier}) \cdot \sin(t \cdot \omega + \phi_{est} + \varepsilon)$$

$$Y_{-\varepsilon} = \int_0^{2\pi} \sin(t \cdot \omega + \phi_{carrier}) \cdot \sin(t \cdot \omega + \phi_{est} - \varepsilon)$$

This process is repeated every time a full period of the incoming sine wave is received. Eventually, $\phi_{carrier}$ and ϕ_{est} will be equal and the derivative estimated by the difference in the correlation coefficient $\phi_{est} + \varepsilon$ and $\phi_{est} - \varepsilon$ will be 0. When this occurs, the receiver is considered locked onto the signal.

Implementation

1. Figure 10.27 shows the C source program `sine8_phase_shift.c` used to generate a 1-kHz sine wave with eight unique phase shifts as the output of the first DSK. This output sine wave has varying phases but a constant frequency. Build this project as **sine8_phase_shift**. Verify that the DSK output connected to a scope is as shown in Figure 10.28. Every 50 periods of the sine wave, the loop index in the program is incremented by 1 to skip one of the lookup values set in `sine_table`. This results in a transmitted sine wave with eight different phase values. Connect the output of the DSK into the input of the second DSK.

2. Figure 10.29 shows the C source program `bpsk_demod.c` (on the CD) that implements a PLL demodulator on the second DSK. Note that the first DSK is still running even though the USB port is unplugged and reconnected to the second DSK. See also the example `scrambler` in Chapter 4. Figure 10.30a shows a CCS plot of the demodulator output. Note that eight different amplitude values are shown for each period of the received input sinusoid. This plot is obtained within CCS using `phiBuf` as the starting address, with 500 points as the acquisition and display size. You can readily change the demodulator program so that the phase shift is every five periods of the sine wave. You can

```
//sin8_phase_shift.c Sine generation. Illustrates phase shift
#include "dsk6713_aic23.h"              //codec-DSK support file
Uint32 fs=DSK6713_AIC23_FREQ_8KHZ;     //set sampling rate
short loop = 0;
short sine_table[8]={0,707,1000,707,0,-707,-1000,-707};//sine values
short phase_change_idx = 0;
interrupt void c_int11()                //interrupt service routine
{
  output_sample(sine_table[loop]);
  if (loop < 7)  ++loop;                //reinit index loop
  else  loop = 0;
  if (phase_change_idx++ >= 50*8)    //phase shift every 50 periods
  {
   if (loop == 7)   loop = 0;         //skip a value
   else             loop++;
   phase_change_idx = 0;
  }
  return;
}
void main()
{
  comm_intr();  while(1);              //init DSK/infinite loop
}
```

FIGURE 10.27. C program that generates a sine wave with eight unique phase shifts (sine8_phase_shift.c).

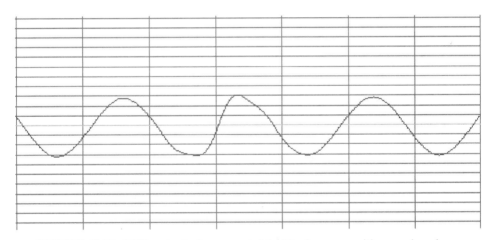

FIGURE 10.28. DSK output of a generated 1-kHz sine wave with a varying phase.

further adjust the indexing through the sine values to create a phase shift showing four (or two) different amplitude values.

Figure 10.30*b* shows a CCS plot of the PLL output buffer that receives only one period of the sine wave. Use a starting address of r_symbol, an acquisition and display size of 16, and a 16-bit signed integer (not a 32-bit float, as for phiBuf).

```
//BPSK_demod.c PLL demodulator. Input from 1st DSK
#include "dsk6713_aic23.h"          //codec-DSK support file
#include <math.h>
Uint32 fs=DSK6713_AIC23_FREQ_16KHZ; //set sampling rate
#define NUMSAMP 16                   //# samples per symbol
#define PI   3.1415926
short sample_data;                   //input sample
short ri=0,   r[10000]={0};          //buffer index/received data
short r_symbol[NUMSAMP];             //buffer to receive one period
short SBind=0,   phiBind=0;          //symbol/phi buffer index
float phiBuf[1000] = {0};            //buffer to view phi estimates
float y1, y2,   damp=1;              //correlation vectors,damping
float phi = PI;                      //phase estimate

interrupt void c_int11()             //interrupt service routine
{
 int i,   max=1;
 sample_data=(short)input_sample();  //receive sample
 r[ri++] = sample_data;
 r_symbol[SBind++] = sample_data;    //put sample in symbol buffer
 if(ri >= 10000)   ri = 0;           //reset buffer index
 if(SBind == NUMSAMP)                //after one period is received
 {                                   //then perform phi estimate
  SBind = 0;                         //reset buffer index
  y1 = 0, y2 = 0;
  for(i=0; i<NUMSAMP; i++)           //correlate received symbol
   {
    y1 += r_symbol[i]*sin(2*PI*i/NUMSAMP + phi - 0.1);
    y2 += r_symbol[i]*sin(2*PI*i/NUMSAMP + phi + 0.1);
    if(r_symbol[i] > max)            max = r_symbol[i];
   }
  y1=y1/max;       y2=y2/max;        //normalize correlation coefs
  phi = phi + 0.4*(y2 - y1)*phi;     //determine new estimate for phi
  if(phi < 1)              phi=phi+2*PI; //normalize phi
  if(phi >(2*PI+1))    phi=phi-2*PI;
  phiBuf[phiBind++]=phi;             //put phi in buffer for viewing
  if(phiBind >= 1000)   phiBind = 0; //reset buffer index
 }
output_sample(phi);
}
void main()
{
  comm_intr();   while(1);           //init DSK/infinite loop
}
```

FIGURE 10.29. C program implementing a PLL demodulator (bpsk_demod.c).

10.11.4 BPSK Transmitter and Receiver with PLL

The support files for this project are in the subfolders **transmitter** and **receiver**. This project is the final product and includes the demodulation of a transmitted BPSK signal. It uses two DSKs: one to transmit a BPSK signal and the other to demodulate it. The transmitter.c program shown in Figure 10.31 uses the stereo capability of the AIC23 codec to transmit a 12-kHz carrier signal through the right

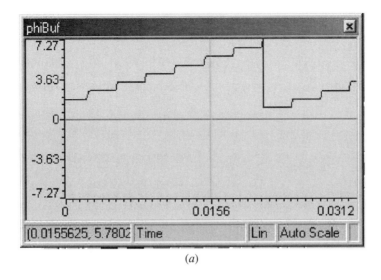

(a)

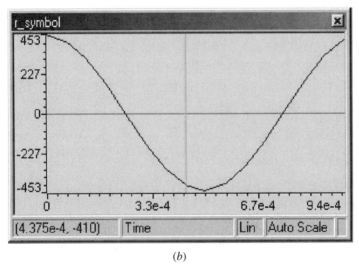

(b)

FIGURE 10.30. CCS plot of a PLL demodulator: (a) output showing eight different amplitudes; (b) output buffer that receives only one period.

channel and the BPSK encoded voice signal through the left channel. See Example 2.3 for the use of an adapter or a stereo cable for obtaining stereo input and output. In this case, you can use a stereo cable that connects the output of the first DSK running the transmitter program to the input of the second DSK running `receiver.c`. Use voice or `TheForce.wav` as input. Verify the successful reception (demodulation) of the transmitted BPSK signal (`TheForce.wav`), with the receiver output connected to a speaker.

See Example 4.13 for the use of an FIR filter function implemented in ASM code. For this project, $N = 8$, so that the size of the circular buffer is 512 bytes (a 16-bit value occupies two memory locations).

```c
//transmitter.c Transmits voice as a BPSK signal
#include "dsk6713_aic23.h"          //codec-DSK support file
#include <math.h>
#include "lp1500.cof"               //1500 Hz coeff lowpass filter
Uint32 fs=DSK6713_AIC23_FREQ_48KHZ; //set sampling rate
#define NUMSAMP   4                 //# samples per Symbol
#define MAX_DATA_LENGTH 10          //size of Mod/Demod vector
#define NUM_BITS 8                  //number of bits per sample
#define SYNC_INTERVAL 100           //interval between sync bits
short encSeqNum   = 8;              //number of encoded bits
short encSymbolVal = 0;             //encoder symbol index
short sin_table[NUMSAMP]={0,1000,0,-1000}; //for carrier
short bits[8];                      //holds encoded sample
short sampleBuffer[2000];           //to view sample
short sIndex = 0;                   //index sampleBuffer
short syncSequence[8]={1,1,1,-1,1,-1,-1,1};//synchronization sequence
short outval=1;                     //bit value to be encoded
short encodeVal = 0;                //filtered input value
int yn = 0;                         //init filter's output
short gain=10;                      //gain on output
short syncTimer = 0;               //tracks time between syncs
#define LEFT   0                    //setup left/right channel
#define RIGHT 1
union {Uint32 uint; short channel[2];} AIC23_data;

interrupt void c_int11()            //interrupt service routine
{
 int i;
 short sample_data;
 sample_data = input_sample();
 yn = fircircfunc(sample_data,h,N); //asm func passing to A4,B4,A6
 if(encSymbolVal >= NUMSAMP)        //increment through waveform
 {
  encSymbolVal = 0;
  encSeqNum++;
 }
 if(encSeqNum == NUM_BITS)          //when all 8 bits sent
 {                                  //get a new sample
  encSeqNum = 0;
  if(syncTimer++ >= SYNC_INTERVAL)  //determine whether
  {                                 //to send sync sequence
   syncTimer = 0;
   for(i=0; i<8; i++)               //put sync sequence in bit
      bits[i] = syncSequence[i];
  }
  else
  {                                 //get the bits
   encodeVal = (short) (yn >> 15);
   for(i=8; i<16; i++)              //encode input sequence
      bits[i-8]=(encodeVal&(1<<i)) ? 1 : -1; //shift
  }
}
```

FIGURE 10.31. C program for BPSK transmission (transmitter.c).

```
  sampleBuffer[sIndex++] = encodeVal;
  if(sIndex >= 2000)    sIndex = 0;
}
outval = (short) bits[encSeqNum];
AIC23_data.channel[RIGHT]=gain*sin_table[encSymbolVal];//carrier
AIC23_data.channel[LEFT]=gain*outval*sin_table[encSymbolVal++];//data
output_sample(AIC23_data.uint);    //output to both channels
}
void main(){
 comm_intr();    while(1); }          //init,infinite loop
```

FIGURE 10.31. (*Continued*)

The input is lowpass-filtered, decimated, and converted to an 8-bit stream. The bit stream is then modulated as a BPSK signal, and four output samples are generated for each bit. Each sample of the voice is a 16-bit integer. Because of sampling rate limitations, only the most significant 8 bits are used for transmission. This yields a resolution of 256 sample levels for the amplitude of the voice, which results in some degradation in the fidelity of the received signal.

The procedure is to sample the voice, get the most significant 8 bits, then transmit one period of a sine wave for each bit. Each period of a sine wave is constructed by outputting to the D/A converter four values of the sine wave. Therefore, for one voice sample, 30 output samples are necessary. This is a severe limitation since the maximum sampling rate is 96 kHz. The maximum sampling rate of the voice that we can implement is then 96 kHz/32, or 3 kHz.

The receiver uses eight samples to determine the phase of the phase-locked loop component allowing for a 48-kHz sampling rate by the transmitter. It can be verified that the receiver's voice bandwidth is approximately 3 kHz. To reconstruct a byte, the receiver must know where the frame starts for each byte. The transmitter periodically sends a synchronization sequence that is 1 byte long. This occurs once every 100 bytes.

To achieve frame synchronization, a synchronization sequence is sent periodically by the transmitter. This sequence is 8 bits long and is detected by the receiver by correlating the incoming bits with the expected sequence. A trigger variable looks over the previously received 8 bits and counts the number of bits that match the synchronization sequence. If the trigger variable is equal to 8, then the synchronization sequence was detected. With 8 bits in the synchronization sequence, there are 256 possible values, so that there is a 1/256 possibility that the sequence will occur randomly. This is too high a probability, and since we are receiving bits at 12 kHz (96 kHz/eight samples per bit), we would expect the sequence to occur randomly about 47 times a second (12 kHz/256). To lower this rate, we make sure that successive synchronization sequences are separated by the expected interval before declaring that the sequence has actually been received. When a correlation is detected, the frame index is reset to zero.

Since the receiver is reconstructing voice samples at a rate of 64 kHz, it needs to interpolate received voice samples to provide the DAC with a sample every time the interrupt routine is invoked. The receiver uses Newton's Forward interpolation with a third-degree polynomial to interpolate the sample values [39]. The generic expansion follows for points f_0 through f_n:

$$p(x) = f_0 + u\Delta_1 f_1 + [u(u-1)/2!]\Delta_2 f_2 + [u(u-1)(u-2)/3!]\Delta_3 f_3 + \ldots$$
$$+ [u(u-1)(u-2)\ldots(u-n+1)/n!]\Delta_n f_n + \ldots$$

where $u = [(x - x_i)/(x_{i+1} - x_i)]$ and f_i is the value of the function $f(n)$ at x_i. To interpolate, based on three points, this equation becomes

$$p(x) = f_0 + u(f_1 - f_0) + [u(u-1)/2](f_2 - 2f_1 + f_0)$$

Interpolating the output values significantly increases the quality of the output voice.
   Possible improvements include the following:

1. At least a quadrature phase-shift keying (QPSK) scheme can be used for the transmitter/receiver to allow much higher data rates across the channel.
2. Noise can be added to the system to increase the practicality of the project.
3. In addition to a phase estimator, a frequency estimator can be added to the receiver. Channels can sometimes introduce frequency distortion into a signal, and this would help the correlator to decode the modulated sequence.

10.12 BINARY PHASE SHIFT KEYING

This mini-project implements BPSK (see also Section 10.11). Two separate boards are used, one to modulate a signal simulating the transmitter component and the other to demodulate the received signal, simulating the receiver component.

Modulation

The modulation scheme transmits binary data using the polar nonreturn to zero (NRZ), ±1 V for the input data. The input is multiplied by a carrier signal with a frequency of $f_c = 8\,\text{kHz}$. For input data with values of ±1 V, the amplitude of the carrier remains the same, but not the phase. An input of +1 V yields a carrier output with a zero-phase shift, while an input of −1 V yields a carrier output that has been shifted by 180°.

A 100-Hz square wave with an amplitude of ±1 V is chosen as the input data. Using a threshold detector at 0 V, it is determined from the input whether the output signal carrier is a positive or a negative cosine. An 8-kHz cosine as the carrier is generated using a 4-point lookup table, sampling at 32 kHz. If the sampled data are greater than zero, then the output carrier is the generated cosine multiplied by +1; if the sampled data are less than zero, then the output carrier is the generated cosine multiplied by −1. Whenever the input signal switches from +1 to −1, or vice versa, the phase of the cosine wave is scaled by 180°. This change in phase looks like an *M* or a *W* on an oscilloscope. Figure 10.32 shows the core of the C

```
//BPSK_modulate.c  Core program for BPSK modulation
. . .
short cos_table[4] = {1000,0,-1000,0};
interrupt void c_int11()
{
  input_data  = ((short)input_sample());
  if(input_data>0) bpsk_signal = cos_table[i++];
  else bpsk_signal = -1*cos_table[i++];
  output_sample(bpsk_signal);
  if(i > 3) i=0;
}
void main()
{ comm_intr(); while(1); }
```

FIGURE 10.32. Core C program for BPSK modulation (`bpsk_modulate.c`).

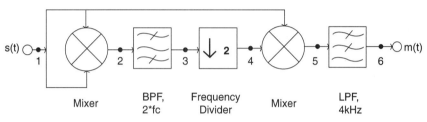

s(t) 1   Mixer   2   BPF,   3   Frequency   4   Mixer   5   LPF,   6   m(t)
                        2*fc   Divider   4kHz

FIGURE 10.33. Carrier recovery block diagram for BPSK demodulation.

source code *bpsk_modulate.c* for the modulation scheme. Build the modulation component of the project. Verify that the output is an 8-kHz sinusoidal waveform, which becomes the input to the second DSK.

Demodulation

The second DSK simulates a pozar as a carrier recovery to demodulate the received signal. Demodulation can occur regardless of the input phase. The carrier recovery scheme is shown in Figure 10.33 and consists of a mixer, a bandpass filter centered at 16 kHz, a frequency divider by 2, a second mixer, and a lowpass filter with a cutoff frequency of 4 kHz. The output at each node is (with an input $m(t) = \pm 1$ V, $f_m = 100$ Hz):

$$\text{Node1: } s(t) = m(t)\cos(2\pi f_c t + \theta)$$

$$\text{Node2: } m^2(t)\cos^2(2\pi f_c t + \theta) = \tfrac{1}{2} + \tfrac{1}{2}\cos[2(2\pi f_c t + \theta)]$$

$$\text{Node3: } \tfrac{1}{2}\cos[2(2\pi f_c t + \theta)]$$

$$\text{Node4: } \tfrac{1}{2}\cos(2\pi f_c t + \theta)$$

$$\text{Node5: } \tfrac{1}{2}\cos(2\pi f_c t + \theta)m(t)\cos(2\pi f_c t + \theta) = \frac{m(t)}{4}\{\cos[2(2\pi f_c t + \theta)] + 1\}$$

$$\text{Node6: } \frac{m(t)}{4}$$

For the demodulator, the sampling frequency is set at 48 kHz (in lieu of 32 kHz) to prevent aliasing and allow for the use of a bandpass filter at node 2, since the output of the first mixer is at 16 kHz.

The signal at node 1 is the output of the modulator: a cosine wave (with an M or W) due to any phase shift. At node 2, it is a 16-kHz signal with a DC component. At node 3, the signal is filtered by a 30th-order least squares FIR bandpass filter centered at 16 kHz. The FIR filter uses a least squares design with MATLAB's SPTool. The 16-kHz filtered signal is down-sampled (decimated) to obtain an 8-kHz signal at node 4. The down-sampling is achieved by setting every other input value to zero. The last stage of demodulation uses a product detector—a combination of a mixer and a lowpass filter—to recover the original binary input. The mixer multiplies the 8-kHz signal with the original input signal. This yields two signals: one at twice the carrier frequency and the other as a DC component with the original $m(t)$ input signal. This signal is then lowpass filtered to yield the original binary signal, regardless of the input phase. The lowpass filter is a 30th-order Kaiser FIR filter, also designed with MATLAB's SPTool. The output at node 6 is then a 100-Hz square wave, the same as the modulator input signal. Figure 10.34 shows the core of the C source program *bpsk_demodulate.c* for the demodulator.

```
//BPSK_demodulate.c  Core C program for BPSK demodulation
...
double mixer_out, pd;
interrupt void c_int11()
{
 input_signal=((short)input_sample()/10);
 mixer_out = input_signal*input_signal;
 dly[0] = mixer_out;
 ..
 filter_output = (yn >> 15);        //output of 16 kHz BP filter
 x = 0;                             //init downsampled value
 if (flag == 0)                     //discard input sample value
   flag = 1;                        //don't discard at next sampling
 else {
   x = filter_output;              //downsampled value is input value
   flag = 0;
 }
 pd = x * input_signal;             //product detector
 dly2[0] = ((short)pd);             //for 4 kHz LP filter
 ..
 m = (yn2 >> 15);                   //output of LP filter
 output_sample(m);
 return;
}
void main()
{ comm_intr(); while(1); }
```

FIGURE 10.34. Core C program for BPSK demodulation (bpsk_demodulate.c).

Verify that the original input signal to the modulator is recovered as the output from the demodulator. Experiment with different sampling rates, filter characteristics, and carrier frequencies to reduce the occasional output noise.

10.13 MODULATION SCHEMES—PAM AND PSK

This project implements both pulse amplitude modulation and phase shift keying schemes. See also the projects in Sections 10.11 and 10.12. The files for this project are included in the folder **modulation_schemes**.

10.13.1 Pulse Amplitude Modulation

In pulse amplitude modulation (PAM), the amplitude of the pulse conveys the information. The information symbols are transmitted at discrete and uniformly spaced time intervals. They are mapped to a train of pulses in the form of a carrier signal. The amplitude of these pulses represents a one-to-one mapping of the information symbols to the respective levels. For example, in binary PAM, bit 1 is represented by a pulse with amplitude A and bit 0 by −A.

At the receiver, the information is recovered by obtaining the amplitude of each pulse. The pulse amplitudes are then mapped back to the information symbol. Figure 10.35 shows the block diagram of a typical PAM system. This is a simplified version without the introduction of adaptive equalizers or symbol clock recovery, which takes into account the effects of the channel. The incoming bit stream (output of the DSK) is parsed into J-bit words, with different lengths of parsing, resulting in different numbers of levels. For example, there are eight levels when $J = 3$. These levels are equidistant from each other on a constellation diagram and symmetric around the zero level, as shown in Figure 10.36. The eight constellation points represent the levels, with each level coded by a sequence of 3 bits. Tables 10.5–10.7 show the mapping levels.

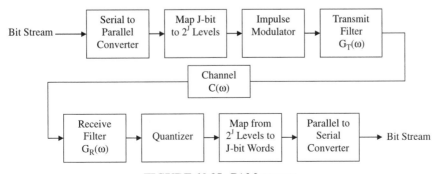

FIGURE 10.35. PAM system.

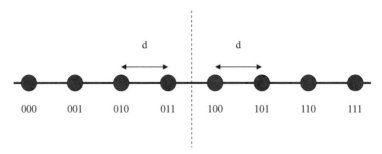

FIGURE 10.36. Constellation diagram of an eight-level PAM.

TABLE 10.5   Four-Level PAM Lookup Table for Mapping

Symbol Block	Level (in hex)
0000	0x7FFF
0101	0x2AAA
1010	−0x2AAB
1111	−0x8000

TABLE 10.6   Eight-Level PAM Lookup Table for Mapping

Symbol Block	Level (in hex)
000	0x7FFF
001	0x5B6D
010	0x36DB
011	0x1249
100	−0x1249
101	−0x36DB
110	−0x5B6D
111	−0x7FFF

Transmitter/Receiver Algorithm

An input sample is composed of 16 bits. Depending on the type of PAM, an appropriate masking is used. The same transmitter and receiver implementations apply to four-level and eight-level PAM with differences in masking, shifting, and lookup tables (see Tables 10.5–10.7). For the 8-PAM, the LSB of the input sample is discarded so that the remaining number of bits (15) is an integer multiple of 3, which does not have a noticeable effect on the modulated waveform and on the recovered voice.

Consider the specific case of a 16-PAM. In order to achieve the desired symbol rate, the input sample is decomposed into segments 4 bits long. Each input sample is composed of four segments. Parsing the input sample is achieved through the use

**TABLE 10.7   Sixteen-Level PAM Lookup
Table for Mapping**

Symbol Block	Level (in hex)
0000	0x7FFF
0001	0x6EEE
0010	0x5DDD
0011	0x4CCC
0100	0x3BBB
0101	0x2AAA
0110	0x1999
0111	0x0888
1000	–0x0889
1001	–0x199A
1010	–0x2AAB
1011	–0x3BBC
1100	–0x4CCD
1101	–0x5DDE
1110	–0x6EEF
1111	–0x8000

of masking and shifting. The first symbol block is obtained with masking of the four least significant bits by *anding* the input sample with 0x000F. The second symbol block is obtained through shifting the original input sample by four to the right and masking the four LSBs. These steps are repeated until the end of the input sample length and produce four symbol blocks. Assume that the input sample is 0xA52E. In this case, 1110 (after masking the four LSBs) is mapped to –0x6EEF, as shown in Table 10.7. Each symbol block is composed of 4 bits mapped into the 16 uniformly spaced levels between –0x8000 and 0x7FFF. The spacing between each level is 0x1111, selected for uniform spacing. The selected level is then transmitted as a square wave. The period of the square wave is achieved by outputting the same level many times to ensure a smooth-looking square wave at the output of the transmitter.

The receiver is implemented with the assumption that the effects of the channel and noise are neglected. As a result, the received sample is composed of individual transmitted symbols or levels. Each transmitted symbol is a 4-bit segment, demodulated by mapping it back to the original sequence of bits. The demodulated symbols are then arranged in a buffer in order to reproduce the original transmitted sequence. The least significant transmitted segment is placed in the least significant received sequence (by adding and shifting). The first segment is shifted by 12 to the left in order to place it at the most significant segment, and subsequently shifted by 4 to the right. The process is repeated until the four segments are in the right order the way they were transmitted. The sample is then sent to the codec, and the original waveform is reconstructed.

10.13.2 Phase Shift Keying

Phase shift keying (PSK) is a method of transmitting and receiving digital signals in which the phase of a transmitted signal is varied to convey information. Several schemes can be used to accomplish PSK, the simplest one being binary PSK (BPSK), using only two signal phases: $0°$ and $180°$. If the phase of the wave is $0°$, then the signal state is low, and if the phase of the wave is $180°$ (if phase reverses), the signal state is high (*biphase modulation*). More complex forms of PSK employ four- or eight-wave phases, allowing binary data to be transmitted at a faster rate per phase change. In four-phase modulation, the possible phase angles are $0°$, $+90°$, $-90°$, and $180°$; each phase shift can represent two bits per symbol. In eight-phase modulation, the possible phase angles are $0°$, $+45°$, $-45°$, $+90°$, $-90°$, $+135°$, $-135°$, and $180°$; each phase shift can represent 4 bits per symbol.

Binary Phase Shift Keying
A single data channel modulates the carrier. A single bit transition, 1 to 0 or 0 to 1, causes a $180°$ phase shift in the carrier. Thus, the carrier is modulated by the data. Detection of a BPSK signal uses the following: (1) a squarer that yields a DC component and a component at $2f_c$; (2) a bandpass filter to extract the f_c component; (3) a frequency divider, the output of which is multiplied by the input. The result is lowpass filtered to yield a PCM signal.

Quadrature Phase Shift Keying
Quadrature phase shift keying (QPSK) is a modulation scheme in which the phase is modulated while the frequency and the amplitude are kept fixed. There are four phases, each of which is separated by $90°$. These phases are sometimes referred to as *states* and are represented by a pair of bits. Each pair is represented by a particular waveform, called a *symbol*, to be sent across the channel after modulating the carrier. The receiver demodulates the signal and look at the recovered symbol to determine which pair of bits was sent. This requires a unique symbol for each possible combination of data bits in a pair. Because there are four possible combinations of data bits in a pair, QPSK creates four different symbols, one for each pair, by changing an in-phase (I) gain and a quadrature (Q) gain.

The QPSK transmitter system uses both sine and cosine at the carrier frequency to transmit two separate message signals, sI[n] and sQ[n], referred to as the *in-phase* and *quadrature* signals, respectively. Both the in-phase and quadrature signals can be recovered, allowing transmission with twice the amount of signal information at the same carrier frequency.

Transmitter/Receiver Algorithm
An input sample is obtained and stored in a memory location, which contains 16 bits. Depending on the type of PSK (two-level or four-level), appropriate masking

is used. For BPSK, an input value is segmented into sixteen 1-bit components; for QPSK, it is fractioned into 8 dibits. This is achieved by masking the input with the appropriate values, 0x0001, and 0x0003, respectively. In order to obtain the next segment to be processed, the previous input data is shifted once for BPSK or twice for QPSK.

Following the extraction of segments, values are assigned to sinusoids with corresponding phases. In BPSK, there are only two phases: 0° and 180° for bits 0 and 1, respectively. However, for QPSK, we need four phases (0°, 90°, 180°, and 270°) corresponding to 00, 01, 11, and 10. This mapping is used in accordance with *gray encoding*. This minimizes the error caused by interference during the transmission of the signal by maximizing the distance between symbols with the most different bits on the constellation diagram. Each input sample is represented with 16 bits. Every sampled data contains 16 segments for BPSK, and 8 segments for QPSK. Since each symbol is transmitted by a sinusoid generated digitally by four points, an input sample is acquired every 64 and 32 output samples for BPSK and QPSK, respectively.

At the PSK receiver, each sinusoid is mapped into the corresponding symbols composed of 1 bit for BPSK or 2 bits for QPSK. The extracted symbols are then aligned in the newly constructed 16-bit value by appropriate left shifts. The sample is then sent to the codec, and the original waveform is regenerated.

Implementation Results

The necessary files are in the folder **modulation_schemes**. The C source file `modulation_scheme.c` contains all the schemes for both modulation and demodulation, and a *gel* file to select the specific case. The 10 cases implement the 4-, 8-, and 16-PAM, BPSK, and QPSK for both modulation and demodulation. For example, the slider in positions 1 and 2 implements the four-PAM scheme for modulation and demodulation, respectively.

PAM

Three PAM modulation and demodulation schemes are implemented, based on a lookup table and level assignment. The demodulation process is designed on the same DSK, with the output of the modulator fed into the input of the demodulator. The modulation output for each PAM scheme is obtained using a 1.3-kHz sinusoid as input, with the output to a scope. For the four-PAM scheme, the output is shown in Figure 10.37a. The four levels are labeled to indicate the modulation process. The 2's complement format of the codec reverses the negative and positive values. For example, -0x8000 is shown as the most positive value. Figure 10.37b shows the modulation levels for the eight-PAM output with the same sinusoidal input. Figure 10.37c shows the output of the 16-PAM modulator, where 12 of the 16 levels are present. This describes the effect of increasing the number of levels. The spacing between levels is smaller than in the other two PAM schemes. The higher the number of levels, the harder it is to distinguish and demodulate the signal.

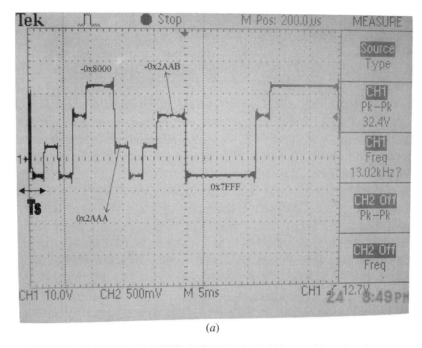

(*a*)

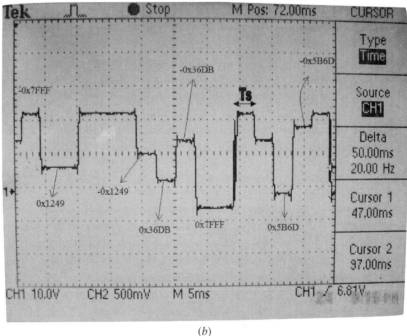

(*b*)

FIGURE 10.37. PAM output obtained with a scope: (*a*) 4-level; (*b*) 8-level; (*c*) 16-level.

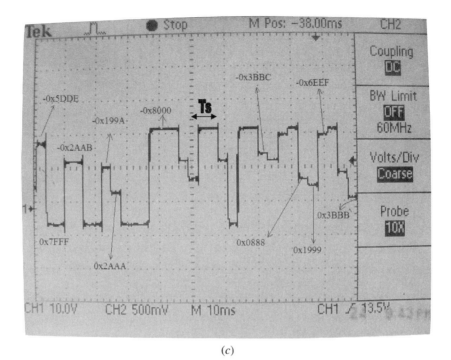

(c)

FIGURE 10.37. (*Continued*)

BPSK

The waveforms generated from the BPSK modulator are sinusoids phase-shifted by 180°. Figure 10.38 shows the BPSK modulator output. When the sinusoid has a 0° phase shift, it represents a binary 0, and when it is shifted by 180°, it represents a binary 1. Using the lookup table, the symbol is demodulated into "0" or "1." When similar symbols follow each other, the waveform is continuous; when different symbols follow each other, the waveform shows an abrupt shift at that point.

QPSK

The output of the QPSK modulator is shown in Figure 10.39. The major drawback of the QPSK implementation on the DSK concerns interpolation. Since the phases are 90° phase-shifted with respect to each other, the waveforms are not continuous. As a result, when one waveform ends with a 0 and the other starts with a 0, there is a slight perturbation (in the case of 01 followed by 00 in Figure 10.39). The narrow spacings are transitions created by the interpolation filter. Note that 01 has a 180° phase shift with respect to 10, and 00 is 90° out of phase with both of them.

Modulation and demodulation for each scheme were also tested using *TheForce.wav* as input. The quality of the output voice indicates a successful demodulator (with the output of the modulator as input to the demodulator).

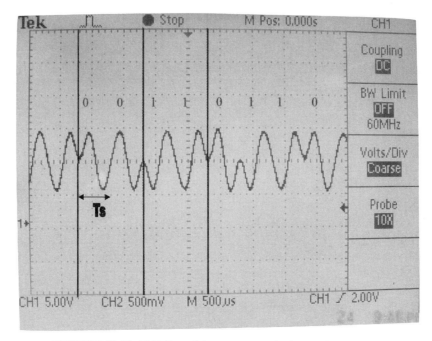

FIGURE 10.38. BPSK modulator output obtained with a scope.

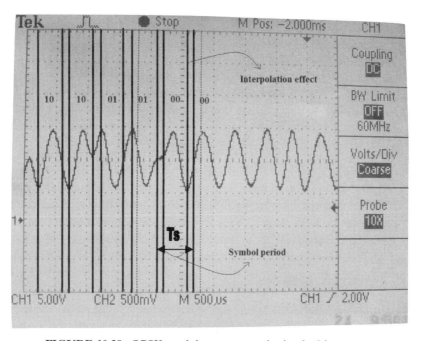

FIGURE 10.39. QPSK modulator output obtained with a scope.

Implementation Issues

Each input sample was parsed into four levels. Each level was sent to the output of the codec 12 times (for an acceptable square wave). As a result, for each input sample there are 48 output samples (4×12). The output sample rate is 48 times the input sample rate (using down-sampling). For the PSK cases, the output waveform is a four-sample sinusoid with different phases. Each input sample is parsed into symbols, and each symbol is sent to the output of the codec four times. For BPSK, the symbol is 1 bit with an output-to-input ratio of 64 (4×16), and for QPSK, the symbol consists of 2 bits with a ratio of 32 (4×8).

For the PAM cases, a square wave pulse was chosen and implemented by outputting the level 12 times. For BPSK and QPSK, the output was a sinusoid composed of four output samples with different phases (to represent the sinusoid appropriately). It is more efficient than the PAM case.

Transmitting from one DSK and receiving from another DSK involves synchronization issues that requires symbol clock recovery and an adaptive equalizer (using a PLL).

10.14 SELECTABLE IIR FILTER AND SCRAMBLING SCHEME USING ONBOARD SWITCHES

This mini-project implements one of several IIR filters using the onboard DIP switches to select a specific filter type. Furthermore, one of the switch options implements a scrambling scheme with voice as input. With the DSK output of the voice scrambler as the input to a second DSK to unscramble, the original voice signal can be recovered.

Four 10th-order IIR Butterworth filters of varying bandwidths are designed using MATLAB's SPTool described in Appendix D (utilized for FIR and IIR filter designs in Chapters 4 and 5). Table 10.8 shows the assignments of the DIP switches and the corresponding implementations. A "1" represents a switch in the up position, while a "0" represents a switch in the down or pressed position. For example, the switch combinations of "0011" (binary 3) and 0101 (binary 5) select a 3-kHz lowpass IIR filter and a voice scrambling scheme, respectively, for implementation.

TABLE 10.8   Dip Switch Assignments and Corresponding Implementations

Dip Switch Combination	Type	f_c or Bandwidth
0000	Original signal	N/A
0001	Lowpass	2 kHz
0010	Highpass	2 kHz
0011	Lowpass	3 kHz
0100	Bandpass	1.5–3 kHz
0101	Voice scrambler	N/A
0110–1111	No output	N/A

Figure 10.40 shows the core of the C source program *IIR_ctrl.c* that implements the four IIR filters as well as the scrambling scheme. The code section of the program that implements the four IIR filters can be found in the program example *IIR.c* in Chapter 5. The complete code section for the scrambling scheme is included in *IIR_ctrl.c*. From Figure 10.40, if *DIP_Mask* is 3 or 5, a 3-kHz IIR lowpass filter or a voice-scrambling scheme is selected and implemented.

Scrambling/Unscrambling

By setting the sample rate to 16 kHz and taking every other input sample in the voice scrambler scheme, input samples are effectively acquired at 8 kHz and output samples intermittently at 16 kHz. The input samples are stored in a buffer. The samples from the buffer are output in quick bursts, independently of the input. When it is nearly full, the buffer is emptied by outputting a sample every sampling period. The buffer is then refilled and the process is repeated. This results in an output that sounds as if the signal frequency had doubled. Table 10.9 illustrates the input and output scheme for a buffer size of 4. This is neither an up-sampling (interpolating) nor a down-sampling (decimating) scheme, since no data are added or ignored by the program. After period 8, the buffer is emptied and the cycle restarts at period 1. For a buffer size of 4, there is no pronounced difference between the input and output voice signals. However, for a buffer size of 512 or greater, the output voice signal is quite unrecognizable.

The scrambled output signal can be recovered. The complete unscrambling C source program *IIR_recov.c* is on the CD. The output of the voice scrambler becomes the input to the second DSK running the program *IIR_recov.c.* (Chapter 4 includes an example using modulation and FIR filtering to scramble and unscramble a voice signal.) The unscrambling program assumes that *DIP_Mask* is equal to 5 in the scrambler program. The buffer size of 512 used by the scrambler must be known in order to recover the original input voice signal. The samples are lowpass filtered by 4 kHz in order to reduce some high-frequency noise incurred with the scrambling process before being outputted. There is still a small amount of high-frequency noise in the output. Note that the scrambling scheme uses bit manipulation that requires no external synchronization between the scrambling transmitter and the unscrambling receiver.

The (complete) executable file for the IIR and scrambling implementations is on the CD as `minimicro.out`, and the unscrambling executable file is on the CD as `minimicrob.out`. These executable files can be used first to test the different implementations for IIR filtering and the scrambling/unscrambling scheme. The appropriate support files are included in the folder **IIR_ctrl**.

DIP switch values 6 to 15 yield no output, and can be used for expanding this project to implement additional IIR or FIR filters and/or another scrambling scheme. RTDX can be used to pass the designed coefficients (see the FIR project incorporating RTDX and Chapter 9).

```
//IIR_ctrl.c Selectable IIR filter with scrambling option using DIP SW
. . .
short DIP_Mask = 20;              //any DIP SW value except 0-15
short BUFFER_SIZE = 512;          //size of buffer
short buffer[512];               //buffer for voice scrambler
short index=0,input_index=0,output_index=0;//index for sample #,buffer
interrupt void c_int11()
{
 short i, input;
 int un, yn;
 input = (short)input_sample();        //external input
 if (DIP_Mask == 0) {             //output = input (no filtering)
  {. . .    yn=input; }           //like a loop program
 }else if (DIP_Mask == 1) {       //2kHz filter if DIP=1
   for(i=0;i<stages;i++) {un=input-... yn=...update delays- See IIR.c}
   ...
   }else if (DIP_Mask == 2) {...            //...for other filters
   }
   else if (DIP_Mask == 5){                 //for voice scrambler
   if((index % 2) == 0) {                   //every other sample
    buffer[input_index++] = input;          //input sample->buffer
    if(input_index==BUFFER_SIZE) {input_index=0;} //reset when full
   }
   if (index >= BUFFER_SIZE) {      //if buffer is at least half full
     yn = buffer[output_index++];                 //output next value
     if(output_index==BUFFER_SIZE) {output_index=0;} //reset if at end
   }
   index++;                               //incr overall sample index
   if(index>=(BUFFER_SIZE*2)) {index=0; } //reinit sample index if end
 }else { yn = 0;   }                       //no output if other DIP #
 output_sample((short)(yn));               // output
 return;
}
void main()
{
  comm_intr();
  while(1) {
  short newMask = 0;
  newMask += DSK6713_DIP_get(3) * 1;
  newMask += DSK6713_DIP_get(2) * 2;
  newMask += DSK6713_DIP_get(1) * 4;
  newMask += DSK6713_DIP_get(0) * 8;     //hex value of DIP switch
  if (DIP_Mask != newMask) {             //wait for change
   DIP_Mask = newMask;                   //load DIP switch value
   if (DIP_Mask == 5) {
      DSK6713_LED_on(3);
      DSK6713_LED_off(2);
      DSK6713_LED_on(1);
      DSK6713_LED_off(0);
   } else if (DIP_Mask == 4) { ...       //for other SWs
   }                                     //and all LEDs off
  }                                      //end of 1st if
 }                                       //end of while(1)
}                                        //end of main
```

FIGURE 10.40. Core C program to select an implement IIR filters using the onboard switches with an optional scrambling scheme.

TABLE 10.9 Input and Output Scheme for Voice Scrambler

	Period 1	Period 2	Period 3	Period 4	Period 5	Period 6	Period 7	Period 8
Input	Sample 1	X	Sample 2	X	Sample 3	X	Sample 4	X
Output	X	X	X	X	Sample 1	Sample 2	Sample 3	Sample 4

FIGURE 10.41. Hard-decision decoding setup.

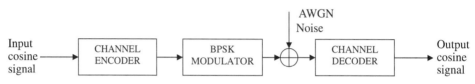

FIGURE 10.42. Soft-decision decoding setup.

10.15 CONVOLUTIONAL ENCODING AND VITERBI DECODING

Channel coding schemes widely used in communication systems mostly consist of the convolutional encoding and Viterbi decoding algorithms to reduce the bit errors on noisy channels. This project implements a 3-output, 1-input, 2-shift register (3,1,2) convolutional encoder used for channel encoding and a channel decoder employing soft-decision and basic Viterbi decoding techniques.

Soft Decision and Basic Viterbi Decoding

The system setups are used for soft decision and Viterbi decoding techniques. In Figures 10.41 and 10.42, the channel encoder represents a (3,1,2) convolutional encoding algorithm, and the channel decoder represents the Viterbi decoding algorithm.

In the Viterbi decoding setup shown in Figure 10.41, a cosine signal is the input to the channel encoder algorithm. The encoded output is stored in a buffer. The elements of this buffer provide the input to the channel decoder algorithm that decodes it and returns the original cosine signal. Both the encoder and decoder outputs are displayed within CCS.

In the soft decision decoding setup shown in Figure 10.42, a cosine signal is given as input to the channel encoder algorithm. The binary output of the channel encoder is modulated using the BPSK technique, whereby the 0 output of the channel encoder is translated into –1 and the 1 output is translated into +1. Additive white

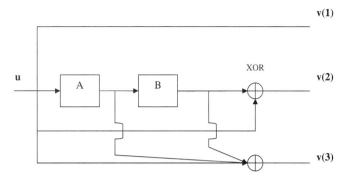

FIGURE 10.43. A (3,1,2) convolutional encoder.

Gaussian noise (AWGN) is generated and added to the modulated output. The signal that is corrupted by the additive noise is fed to the channel decoder. Both the encoder and decoder outputs are displayed within CCS. The variance of AWGN is varied, and the decoder's performance is observed.

(3,1,2) Convolutional Encoder

Convolutional coding provides error correction capability by adding redundancy bits to the information bits. The convolutional encoding is usually implemented by the shift register method and associated combinatorial logic that performs modulo-two addition, an XOR operation. A block diagram of the implemented (3,1,2) convolutional encoder is shown in Figure 10.43, where u is the input, $v(1), v(2), v(3)$ are the outputs, and A, B are the shift registers. The outputs are,

$$v(1) = u$$
$$v(2) = u \oplus b$$
$$v(3) = u \oplus a \oplus b$$

where, a and b are the contents of the shift registers A and B, respectively. Initially the contents of the shift registers are 0s. The shift registers go through four different states, depending upon the input (0 or 1) received. Once all the input bits are processed, the contents of the shift registers are again reset to zero by feeding two 0s (since we have two shift registers) at the input.

State Diagram

The basic state diagram of the encoder is shown in Figure 10.44, where $S_0, S_1, S_2,$ and S_3 represent the different states of the shift registers. Furthermore, m/xyz indicates that on receiving an input bit m, the output of the encoder is xyz; that is, if $u = m \Rightarrow v(1) = x, v(2) = y, v(3) = z$ for that particular state of shift registers A and B. The arrows indicate the state changes on receiving the inputs.

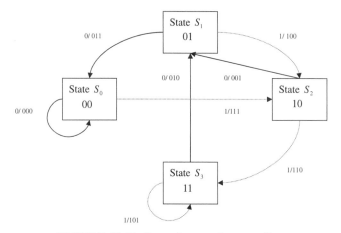

FIGURE 10.44. State diagram for encoding.

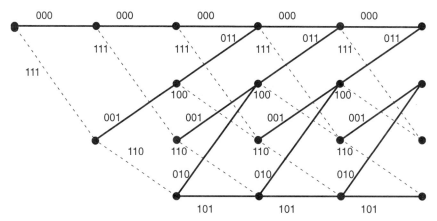

FIGURE 10.45. Trellis diagram for encoding.

Trellis Diagram

The corresponding trellis diagram for the state diagram is shown in Figure 10.45. The four possible states of the encoder are shown as four rows of horizontal dots. There is one column of four dots for the initial state of the encoder and one for each time instant during the message. The solid lines connecting the dots in the diagram represent state transitions when the input bit is a 0. The dotted lines represent transitions when the input bit is a 1. For this encoding scheme, each encoding state at time n is linked to two states at time $n + 1$. The Viterbi algorithm is used for decoding this trellis-coded information bits by expanding the trellis over the received symbols. The Viterbi algorithm reduces the computational load by taking advantage of the special structure of the trellis codes.

Modulation and AWGN for Soft Decision

In the soft decision decoding setup, the 1/0 output of the convolutional encoder is mapped into an antipodal baseband signaling scheme (BPSK) by translating 0s to −1s and 1s to +1s. This can be accomplished by performing the operation $y = 2x - 1$ on each convolutional encoder output symbol, where x is the encoder output symbol and y is the output of the BPSK modulator.

AWGN is added to this modulated signal to create the effect of channel noise. AWGN is a noise whose voltage distribution over time has characteristics that can be described using a Gaussian distribution, that is, a bell curve. This voltage distribution has zero mean and a standard deviation that is a function of the SNR of the received signal. The standard deviation of this noise can be varied to obtain signals with different SNRs at the decoder input.

A zero-mean Gaussian noise with standard deviation σ can be generated as follows. In order to obtain Gaussian random numbers, we take advantage of the relationships between uniform, Rayleigh, and Gaussian distributions. C only provides a uniform random number generator, `rand( )`. Given a uniform random variable U, a Rayleigh random variable R can be obtained using

$$R = \sqrt{2 \cdot \sigma^2 \cdot \ln(1/(1-U))} = \sigma \cdot \sqrt{2 \cdot \ln(1/(1-U))}$$

where σ^2 is the variance of the Rayleigh random variable. Given R and a second uniform random variable V, a Gaussian random variable G can be obtained using

$$G = R \cos V$$

Viterbi Decoding Algorithm

The Viterbi decoding algorithm uses the trellis diagram to perform the decoding. The basic cycle repeated by the algorithm at each stage into the trellis is

1. *Add*: At each cycle of decoding, the branch metrics enumerating from the nodes (states) of the previous stage are computed. These branch metrics are added to the previously accumulated and saved path metrics.
2. *Compare*: The path metrics leading to each of the encoder's states are compared.
3. *Select*: The highest-likelihood path (survivor) leading to each of the encoder's states is selected, and the lower-likelihood paths are discarded.

A metric is a measure of the "distance" between what is received and all of the possible channel symbols that could have been received. The metrics for the soft decision and the basic Viterbi decoding techniques are computed using different methods. For basic Viterbi decoding, the metric used is the Hamming distance, which specifies the number of bits by which two symbols differ. For the soft decision technique, the metric used is the Euclidean distance between the signal points in a signal constellation. More details of the decoding algorithm are presented elsewhere [40,41].

Implementation

Build this project as **viterbi**. The complete C source program and the executable (.out) files are included on the CD in the folder `Viterbi`. Several functions are included in the program to perform convolutional encoding and BPSK modulation, add white Gaussian noise, and implement the Viterbi decoding algorithm (the more extensive function).

The following time-domain graphs can be viewed within CCS—input, encoder output, and decoder output—using the addresses *input, enc_output*, and *dec_output*, respectively. For the graphs, use an acquisition buffer size of 128, a sampling frequency of 8000, a 16-bit signed integer for both input and decoder output, and a 32-bit float for the encoder output.

Three gel files are used (included on the CD):

1. *Input.gel*: to select one of the following three input signals: $cos\,666$ (default), $cos\,666 + cos\,1500$, and $cos\,666 + cos\,2200$, where 666 represents a 666-Hz cosine.

2. *Technique.gel*: to select between soft decision and basic Viterbi decoding.

3. *Noise.gel*: to select a suitable standard deviation for AWGN. One of five different values $(0, 0.3, 0.4, 2.0, 3.0)$ of the standard deviation of the AWGN can be selected.

Results

The following results are obtained:

Case 1:   input = cosine 666 Hz, using soft-decision

Case 2:   input = cosine 666 Hz, standard deviation $\sigma = 0.4$

Case 3:   input = cosine 666 Hz, standard deviation $\sigma = 3.0$

Case 4:   input = cosine (666 + 1500) Hz, using basic Viterbi decoding (noise level 0)

With the default settings, the encoded output will appear between the +1 and −1 voltage levels, as shown in Figure 10.46a. The output of the Viterbi decoder is shown in Figure 10.46b). With an increase in the noise level, slight variations will be observed around the +1 and −1 voltage levels at the encoder output. These variations will increase with an increase in noise level. It can be observed from the decoder outputs that it is able to recover the original cosine signal. With the noise level set at 0, 0.3, or 0.4 using the *noise.gel* slider, the decoder is still able to recover the original cosine signal, even though there is some degradation in the corresponding encoder output, as shown in Figure 10.47. With further increase

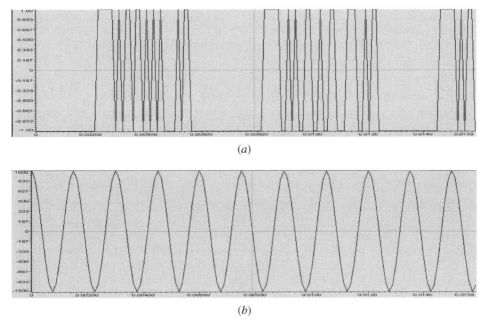

(a)

(b)

FIGURE 10.46. CCS plots of output using case 1: (a) convolutional encoder varying between +1/−1; (b) Viterbi decoder.

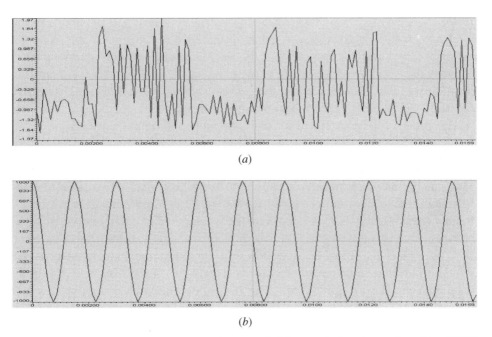

(a)

(b)

FIGURE 10.47. CCS plots of output using case 2: (a) convolutional encoder with AWGN (sigma = 0.4); (b) Viterbi decoder.

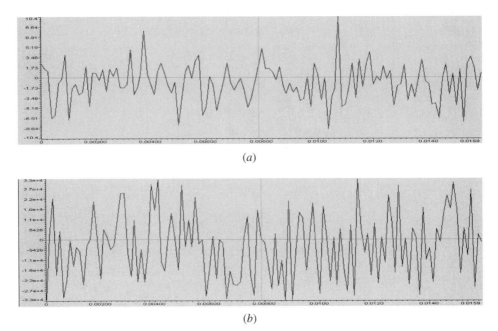

FIGURE 10.48. CCS plots of output using case 3: (*a*) convolutional encoder with AWGN (sigma = 3.0); (*b*) Viterbi decoder.

in the noise level with $\sigma = 3.0$, the decoder output is degraded, as shown in Figure 10.48.

Figure 10.49 illustrates case 4 using cosine (666 + 1500) as input. With the *technique.gel* slider selected for Viterbi decoding, the encoder output appears between the 0 and 1 voltage levels, as shown in Figure 10.49*b*, since the input is of plain binary form. The decoded output is the restored input cosine signal shown in Figure 10.49*c*. There is no additive noise added in this case.

This project can be extended for real-time input and output signals.

Illustration of the Viterbi Decoding Algorithm

Much of the material introduced here can be found in Ref. 41. To illustrate the Viterbi decoding algorithm, consider the basic Viterbi symbol inputs. Each time a triad of channel symbols is received, a metric is computed to measure the "distance" between what is received and all of the possible channel symbol triads that could have been received. Going from $t = 0$ to $t = 1$, there are only two possible channel symbol triads that could have been received: 000 and 111. This is because the convolutional encoder was initialized to the all-0s state, and given one input bit = 1 or 0, there are only two states to transition to and two possible outputs of the encoder: 000 and 111.

The metric used is the Hamming distance between the received channel symbol triad and the possible channel symbol triad. The Hamming distance is computed by

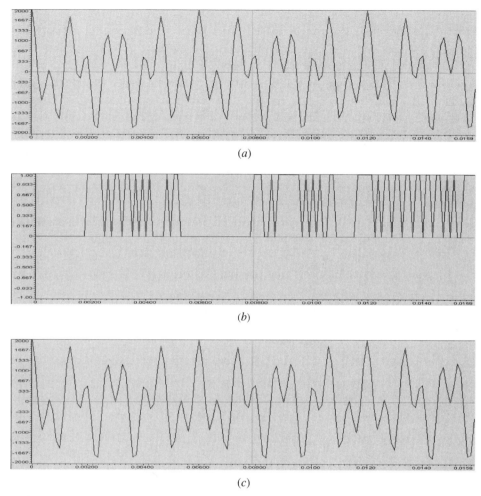

FIGURE 10.49. CCS plots using case 4: (*a*) input to convolutional encoder; (*b*) output from convolutional encoder (between 0 and 1); (*c*) output from a Viterbi decoder.

simply counting how many bits are different between the received channel symbol triad and the possible channel symbol triad. The results can only be zero, one, two, or three. The Hamming distance (or other metric) values computed at each time instant, for the paths between the states at the previous time instant and the states at the current time instant, are called *branch metrics*. For the first time instant, these results are saved as *accumulated error metric* values associated with states. From the second time instant on, the accumulated error metrics are computed by adding the previous accumulated error metrics to the current branch metrics.

Consider that at $t = 1$, 000 is received at the input of the decoder. The only possible channel symbol triads that could have been received are 000 and 111. The Hamming distance between 000 and 000 is zero. The Hamming distance between

000 and 111 is three. Therefore, the branch metric value for the branch from State 00 to State 00 is zero, and for the branch from State 00 to State 10 it is two. Since the previous accumulated error metric values are equal to zero, the accumulated metric values for State 00 and for State 10 are equal to the branch metric values. The accumulated error metric values for the other two states are undefined (in the program, this undefined value is initialized to be the maximum value for integer). The path history table is updated for every time instant. This table, which has an entry for each state, stores the surviving path for that state at each time instant. These results at $t = 1$ are shown in Figure 10.50a.

Consider that at $t = 2$, 110 is received at the input of the decoder. The possible channel symbol triads that could have been received in going from $t = 1$ to $t = 2$ are 000 going from State 00 to State 00, 111 going from State 00 to State 10, 001 going from State 10 to State 01, and 110 going from State 10 to State 11. The Hamming distance is two between 000 and 110, one between 111 and 110, three between 001 and 110, and zero between 110 and 110. These branch metric values are added to the previous accumulated error metric values associated with each state that we came from to get to the current states. At $t = 1$, we can only be at State 00 or State 10. The accumulated error metric values associated with those states were 0 and 2, respectively. The calculation of the accumulated error metric associated with each state at $t = 2$ is shown in Figure 10.50b.

Consider that at $t = 3$, 010 is received. There are now two different ways that we can get from each of the four states that were valid at $t = 2$ to the four states that are valid at $t = 3$. To handle that, we compare the accumulated error metrics associated with each branch and discard the larger one of each pair of branches leading into a given state. If the members of a pair of accumulated error metrics going into a particular state are equal, that value is saved. The operation of adding the previously accumulated error metrics to the new branch metrics, comparing the results, and selecting the smaller accumulated error metric to be retained for the next time instant is called the *add-compare-select* operation. The path history for a state is also updated by selecting the path corresponding to the smallest path metric for that state. This can be found by adding the current selected path transition to the path history of its previous state. The result for $t = 3$ follows.

At $t = 3$, the decoder has reached its steady state; that is, it is possible to have eight possible state transitions. For every other time instant from now on, the same process gets repeated until the end of input is reached. The last two inputs that are received in a Viterbi decoder are also considered special cases. At the convolutional encoder, when the end of input is reached, we input two trailing zeros in order to reset the shift register states to zero. As a consequence of this, in a Viterbi decoder, in the last but one time instant, the only possible states in the Viterbi decoder are State 00 and State 01. Therefore, the expected inputs are 000, 011, 001, and 010. And for the last time instant, the only possible state is 00. Therefore, the expected inputs are only 000 and 011. This case is illustrated in Figure 10.50c.

In the program, it is assumed that the decoder has a memory of only 16, meaning that at any one time, the path history can store only 16 paths. As soon as the first

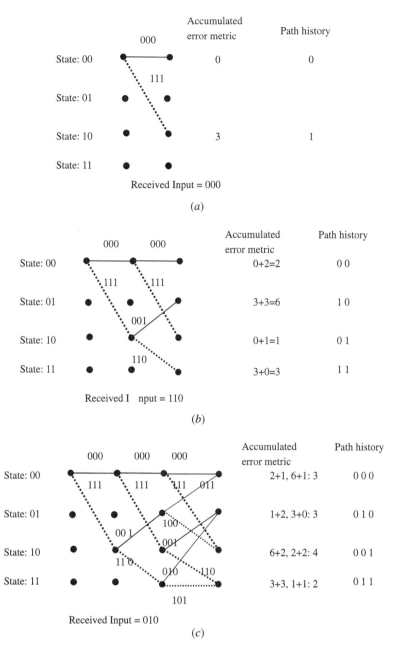

FIGURE 10.50. Trellis diagrams to illustrate Viterbi decoding:(*a*) *t* = 1; (*b*) *t* = 2; (*c*) *t* = 3.

16 channel symbol triads are read, the path history becomes full. The path history in this source code is an array named *path_history*. Each variable of this array maintains the path history for a particular state, with each bit in the variable storing a selected path with the rightmost bit storing the most recent path. Therefore, before processing the 17th channel symbol triad, the minimum branch metric state is found,

and the leftmost bit in the path history of this state is output into a variable *dec_output*. For every other time instant afterward, this process is repeated and the leftmost bit of the selected *path_history* variable is output to *dec_output*. On completing the decoding algorithm, *dec_output* contains the desired decoder output.

A variable named *output_table*, lists the output symbols for every input at a particular state, as shown in the following table:

Current State	Output Symbols If:	
	Input = 0	Input = 1
00	000	111
01	011	100
10	001	110
11	010	101

The soft decision Viterbi algorithm functions in a similar fashion, except that the metric is computed in a different way. The metric is specified using the Euclidean distance between the signal points in a signal constellation. In the soft decision algorithm, the output of the encoder is sent in the form of BPSK-modulated symbols, that is, 0 is sent as −1 and 1 is sent as +1. Before this distance is found, BPSK modulation is performed on the possible channel symbol triad. Assume that a channel symbol triad containing $\{a1, a2, a3\}$ is received, and the expected input channel symbol triad is 001. After BPSK modulation, it can be written as $\{b1, b2, b3\}$, where $b1 = −1$, $b2 = −1$, and $b3 = +1$. Then, the distance between these two channel symbols is found using

$$\text{distance} = abs(b1 - a1) + abs(b2 - a2) + abs(b3 - a3)$$

10.16 SPEECH SYNTHESIS USING LINEAR PREDICTION OF SPEECH SIGNALS

Speech synthesis is based on the reproduction of human intelligible speech through artificial means [42–45]. Examples of speech synthesis technology include *text-to-speech* systems. The creation of synthetic speech covers a range of processes; and even though they are often lumped under the general term *text-to-speech*, a lot of work has been done to generate speech from sequences of speech sounds. This would be a speech-sound (phoneme) to audio waveform synthesis, rather than going from text to phonemes (speech sounds) and then to sound. One of the first practical applications of speech synthesis was a speaking clock. It used optical storage for phrases and words (noun, verb, etc.), concatenated to form complete sentences. This led to a series of innovative products such as vocoders, speech toys, and so on.

Advances in the understanding of speech production mechanism in humans, coupled with similar advances in DSP, have had an impact on speech synthesis techniques. Perhaps the most singular factors that started a new era in this field were the computer processing and storage technologies. While speech and language were already important parts of daily life before the invention of the computer, the equipment and technology that developed over the last several years have made it possible to produce machines that speak, read, and even carry out dialogs. A number of vendors provide both recognition and speech technology. Some of the latest applications of speech synthesis are in cellular phones, security networks, and robotics.

There are different methods of speech synthesis based on the source. In a text-to-speech system, the source is a text string of characters read by the program to generate voice. Another approach is to associate intelligence in the program so that it can generate voice without external excitation. One of the earliest techniques was *Formant synthesis*. This method was limited in its ability to represent voice with high fidelity due to its inherent drawback of representing phonemes by three frequencies. This method, and several analog technologies that followed, were replaced by digital methods. Some early digital technologies were RELP (residue excited) and VELP (voice excited). These were replaced by new technologies, such as LPC (linear predictive coding), CELP (code excited), and PSOLA (pitch synchronous overlap-add). These technologies have been extensively used to generate artificial voice.

Linear Predictive Coding

Most methods that are used for analyzing speech start by transforming acoustic data into spectral form by performing short time Fourier analysis of the speech wave. Although this type of spectral analysis is a well-known technique for studying signals, its application to speech signal suffers from limitations due to the nonstationary and quasi-periodic properties of the speech wave. As a result, methods based on spectral analysis often do not provide a sufficiently accurate description of speech articulation. Linear predictive coding (LPC) represents the speech waveform directly in terms of time-varying parameters related to the transfer function of the vocal tract and the characteristics of the source function. It uses the knowledge that any speech can be represented by certain types of parametric information, including the filter coefficients (that model the vocal tract) and the excitation signal (that maps the source signals). The implementation of LPC reduces to the calculation of the filter coefficients and excitation signals, making it suitable for digital implementation.

Speech sounds are produced as a result of acoustical excitation of the human vocal tract. During production of the voiced sounds, the vocal chord is excited by a series of nearly periodic pulses generated by the vocal cords. In unvoiced sounds, excitation is provided by the air passing turbulently through constrictions in the tract. A simple model of the vocal tract is a discrete time-varying linear filter.

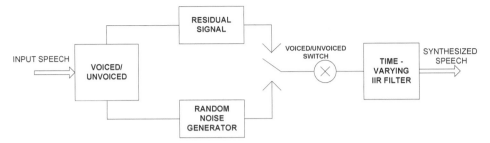

FIGURE 10.51. Diagram of the speech synthesis process.

Figure 10.51 is a diagram of the LPC speech synthesis. To reproduce the voice signal, the following are required:

1. An excitation signal
2. The LPC filter coefficients

The excitation mechanism can be approximated using a residual signal generator (for voiced signals) or a white Gaussian noise generator (for unvoiced signals) with adjustable amplitudes and periods. The linear predictor P, a transversal filter with p delays of one sample interval each, forms a weighed sum of past samples as the input of the predictor. The output of the predictor at the nth sampling instant is given by

$$s_n = \sum_{k=1}^{p} a_k \cdot (s_m) + \delta_n$$

where $m = n - k$ and δ_n represents the nth excitation sample.

Implementation

The input to the program is a sampled array of input speech using an 8-kHz sampling rate. The samples are stored in a header file. The length of the input speech array is 10,000 samples, translating into approximately 1.25 seconds of speech. The input array is segmented into a large number of frames, each 80 B long with an overlap of 40 B for each frame. Each frame is then passed to the following modules: windowing, autocorrelation, LPC, residual, IIR, and accumulate. External memory is utilized. A block diagram of the LPC speech synthesis algorithm with the various modules is shown in Figure 10.52.

1. *Segmentation.* This module separates the input voice into overlapping segments. The length of the segment is such that the speech segment appears stationary as well as quasi-periodic. The overlap provides a smooth transition between consecutive speech frames.

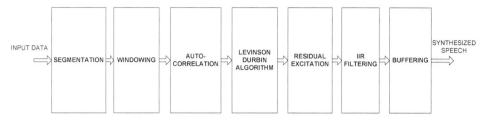

FIGURE 10.52 Speech synthesis algorithm with various modules.

2. *Windowing.* The speech waveform is decomposed into smaller frames using the Hamming window. This suppresses the side lobes in the frequency domain.

3. *Levinson–Durbin algorithm.* To calculate the LPC coefficients, the autocorrelation matrix of the speech frame is required. From this matrix, the LPC coefficients can be obtained using

$$r(i) = \sum_{k=1}^{p} a_k \cdot r(|i - k|)$$

where $r(i)$ and a_k represent the autocorrelation array and the coefficients, respectively.

4. *Residual signal.* For synthesis of the artificial voice, the excitation is given by the residual signal, which is obtained by passing the input speech frame through an FIR filter. It serves as an excitation signal for both voiced and unvoiced signals. This limits the algorithm due to the energy and frequency calculations required for making decisions about voiced/unvoiced excitation since, even for an unvoiced excitation that has a random signal as its source, the same principle of residue signal can still be used. This is because, in the case of unvoiced excitation, even the residue signal obtained will be random.

5. *Speech synthesis.* With the representation of the speech frame in the form of the LPC filter coefficients and the excitation signal, speech can be synthesized. This is done by passing the excitation signal (the residual signal) through an IIR filter. The residual signal generation and the speech synthesis modules imitate the vocal chord and the vocal tract of the speech production system in humans.

6. *Accumulation and buffering.* Since speech is segmented at the beginning, the synthesized voice needs to be concatenated. This is performed by the accumulation and buffering module.

7. *Output.* When the entire synthesized speech segment is obtained, it is played. During playback, the data are down-sampled to 4 kHz to restore the intelligibility of the speech.

Implementation

The complete support files are on the CD in the folder speech_syn. Generate a *.wav* file of the speech sample to be synthesized. For example, include *goaway.wav* in the MATLAB file input_read.m. The MATLAB file samples it for 8 kHz and stores the input samples array in the header file *input.h*. Include this generated header file in the main C source program *speech.c*. Build this project as **speech_syn**. Run the MATLAB program *input_read.m* to generate the two header files *input.h* (containing the input samples) and *hamming.h* (for the Hamming coefficients). Load/run *speech_syn*.out and verify the synthesized speech "go away" from a speaker connected to the DSK output. Three other *.wav* files are included in the folder and can be tested readily.

Results

Speech is synthesized for the following: "Go away," "Hello, professor," "Good evening," and "Vacation." The synthesized output voice is found to have considerable fidelity to the original speech. The voice/unvoiced speech phonemes are reproduced with considerable accuracy. This project can be improved with a larger buffer size for the samples and noise suppression filters. There is noise after each time the sentence is played. A speech recognition algorithm can be implemented in conjunction with the speech synthesis to facilitate a dialogue.

10.17 AUTOMATIC SPEAKER RECOGNITION

This project implements an automatic speaker recognition system [46–50]. *Speaker recognition* refers to the concept of recognizing a speaker by his/her voice or speech samples. This is different from speech recognition. In automatic speaker recognition, an algorithm generates a hypothesis concerning the speaker's identity or authenticity. The speaker's voice can be used for ID and to gain access to services such as banking, voice mail, and so on.

Speaker recognition systems contain two main modules: *feature extraction* and *classification*.

1. Feature extraction is a process that extracts a small amount of data from the voice signal that can be used to represent each speaker. This module converts a speech waveform to some type of parametric representation for further analysis and processing. Short-time spectral analysis is the most common way to characterize a speech signal. The Mel-frequency cepstrum coefficients (MFCC) are used to parametrically represent the speech signal for the speaker recognition task. The steps in this process are shown in Figure 10.53:

   (a) Block the speech signal into frames, each consisting of a fixed number of samples.

   (b) Window each frame to minimize the signal discontinuities at the beginning and end of the frame.

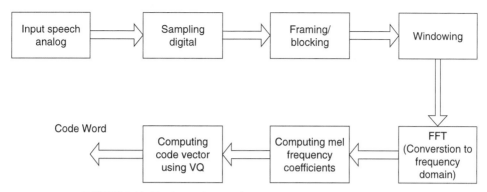

FIGURE 10.53. Steps for speaker recognition implementation.

(c) Use FFT to convert each frame from time to frequency domain.

(d) Convert the resulting spectrum into a Mel-frequency scale.

(e) Convert the Mel spectrum back to the time domain.

2. Classification consists of models for each speaker and a decision logic necessary to render a decision. This module classifies extracted features according to the individual speakers whose voices have been stored. The recorded voice patterns of the speakers are used to derive a classification algorithm. Vector quantization (VQ) is used. This is a process of mapping vectors from a large vector space to a finite number of regions in that space. Each region is called a *cluster* and can be represented by its center, called a *codeword*. The collection of all clusters is a *codebook*. In the training phase, a speaker-specific VQ codebook is generated for each known speaker by clustering his/her training acoustic vectors. The distance from a vector to the closest codeword of a codebook is called a *VQ distortion*. In the recognition phase, an input utterance of an unknown voice is vector-quantized using each trained codebook, and the total VQ distortion is computed. The speaker corresponding to the VQ codebook with the smallest total distortion is identified.

Speaker recognition can be classified with identification and verification. *Speaker identification* is the process of determining which registered speaker provides a given utterance. *Speaker verification* is the process of accepting or rejecting the identity claim of a speaker. This project implements only the speaker identification (ID) process. The speaker ID process can be further subdivided into *closed set* and *open set*. The *closed set* speaker ID problem refers to a case where the speaker is known a priori to belong to a set of M speakers. In the *open set* case, the speaker may be out of the set and, hence, a "none of the above" category is necessary. In this project, only the simpler closed set speaker ID is used.

Speaker ID systems can be either *text-independent* or *text-dependent*. In the *text-independent* case, there is no restriction on the sentence or phrase to be spoken, whereas in the *text-dependent* case, the input sentence or phrase is indexed for each speaker. The text-dependent system, implemented in this project, is commonly found in speaker verification systems in which a person's password is critical for verifying his/her identity.

In the *training phase*, the feature vectors are used to create a model for each speaker. During the *testing phase*, when the test feature vector is used, a number will be associated with each speaker model indicating the degree of match with that speaker's model. This is done for a set of feature vectors, and the derived numbers can be used to find a likelihood score for each speaker's model. For the speaker ID problem, the feature vectors of the test utterance are passed through all the speakers' models and the scores are calculated. The model having the best score gives the speaker's identity (which is the decision component).

This project uses MFCC for feature extraction, VQ for classification/training, and the Euclidean distance between MFCC and the trained vectors (from VQ) for speaker ID. Much of this project was implemented with MATLAB [47].

Mel-Frequency Cepstrum Coefficients

MFCCs are based on the known variation of the human ear's critical bandwidths. A Mel-frequency scale is used with a linear frequency spacing below 1000 Hz and a logarithmic spacing above that level. The steps used to obtain the MFCCs follow.

1. *Level detection.* The start of an input speech signal is identified based on a pre-stored threshold value. It is captured after it starts and is passed on to the framing stage.

2. *Frame blocking.* The continuous speech signal is blocked into frames of N samples, with adjacent frames being separated by M ($M < N$). The first frame consists of the first N samples. The second frame begins M samples after the first frame and overlaps it by $N - M$ samples. Each frame consists of 256 samples of speech signal, and the subsequent frame starts from the 100th sample of the previous frame. Thus, each frame overlaps with two other subsequent frames. This technique is called *framing*. The speech sample in one frame is considered to be stationary.

3. *Windowing.* After framing, windowing is applied to prevent spectral leakage. A Hamming window with 256 coefficients is used.

4. *Fast Fourier transform.* The FFT converts the time-domain speech signal into a frequency domain to yield a complex signal. Speech is a real signal, but its FFT has both real and imaginary components.

5. *Power spectrum calculation.* The power of the frequency domain is calculated by summing the square of the real and imaginary components of the signal to yield a real signal. The second half of the samples in the frame are ignored since they are symmetric to the first half (the speech signal being real).

6. *Mel-frequency wrapping.* Triangular filters are designed using the Mel-frequency scale with a bank of filters to approximate the human ear. The

power signal is then applied to this bank of filters to determine the frequency content across each filter. Twenty filters are chosen, uniformly spaced in the Mel-frequency scale between 0 and 4 kHz. The Mel-frequency spectrum is computed by multiplying the signal spectrum with a set of triangular filters designed using the Mel scale. For a given frequency f, the mel of the frequency is given by

$$B(f) = [1125 \ln(1 + f/700)] \text{ mels}$$

If m is the mel, then the corresponding frequency is

$$B^{-1}(m) = [700 \exp(m/1125) - 700] \text{ Hz}$$

The frequency edge of each filter is computed by substituting the corresponding mel. Once the edge frequencies and the center frequencies of the filter are found, boundary points are computed to determine the transfer function of the filter.

7. *Mel-frequency cepstral coefficients.* The log mel spectrum is converted back to time. The discrete cosine transform (DCT) of the log of the signal yields the MFCC.

Speaker Training—VQ

VQ is a process of mapping vectors from a large vector space to a finite number of regions in that space. Each region is called a *cluster* and can be represented by its center, the codeword. As noted earlier, a codebook is the collection of all the clusters. An example of a one-dimensional VQ has every number less than −2 approximated by −3; every number between −2 and 0 approximated by −1; every number between 0 and 2 approximated by +1; and every number greater than 2 approximated by +3. These approximate values are uniquely represented by 2 bits, yielding a one-dimensional, 2-bit VQ. An example of a two-dimensional VQ consists of 16 regions and 16 stars, each of which can be uniquely represented by 4 bits (a two-dimensional 4-bit VQ). Each pair of numbers that fall into a region are approximated by a star associated with that region. The stars are called *codevectors*, and the regions are called *encoding regions*. The set of all the codevectors is called the *codebook*, and the set of all encoding regions is called the *partition* of the space.

Speaker Identification (Using Euclidean Distances)

After computing the MFCCs, the speaker is identified using a set of trained vectors (samples of registered speakers) in an array. To identify the speaker, the Euclidean distance between the trained vectors and the MFCCs is computed for each trained vector. The trained vector that produces the smallest Euclidean distance is identified as the speaker.

Implementation

The design is first tested with MATLAB. A total of eight speech samples from eight different people (eight speakers, labeled S1 to S8) are used to test this project. Each speaker utters the same single digit, *zero*, once in a training session (then also in a testing session). A digit is often used for testing in speaker recognition systems because of its applicability to many security applications. This project was implemented on the C6711 DSK and can be transported to the C6713 DSK. Of the eight speakers, the system identified six correctly (a 75% identification rate). The identification rate can be improved by adding more vectors to the training codewords. The performance of the system may be improved by using two-dimensional or four-dimensional VQ (training header file would be $8 \times 20 \times 4$) or by changing the quantization method to dynamic time wrapping or hidden Markov modeling. A `readme` file to test this project is on the CD in the folder **speaker_recognition**, along with all the appropriate support files. These support files include several modules for framing and windowing, power spectrum, threshold detection, VQ, and the Mel-frequency spectrum.

10.18 μ-LAW FOR SPEECH COMPANDING

An analog input such as speech is converted into digital form and compressed into 8-bit data. μ-Law *encoding* is a nonuniform quantizing logarithmic compression scheme for audio signals. It is used in the United States to compress a signal into a logarithmic scale when coding for transmission. It is widely used in the telecommunications field because it improves the SNR without increasing the amount of data.

The dynamic range increases, while the number of bits for quantization remains the same. Typically, μ-law compressed speech is carried in 8-bit samples. It carries more information about smaller signals than about larger signals. It is based on the observation that many signals are statistically more likely to be near a low-signal level than a high-signal level. As a result, there are more quantization points closer to the low level.

A lookup table with 256 values is used to obtain the quantization levels from 0 to 7. The table consists of a 16×16 set of numbers: Two 0's, two 1's, four 2's, eight 3's, sixteen 4's, thirty-two 5's, sixty-four 6's, and one hundred twenty-eight 7's. More higher-level signals are represented by 7 (from the lookup table). Three exponent bits are used to represent the levels from 0 to 7, 4 mantissa bits are used to represent the next four significant bits, and 1 bit is used for the sign bit.

The 16-bit input data are converted from linear to 8-bit μ-law (simulated for transmission), then converted back from μ-law to 16-bit linear (simulated as receiving), and then output to the codec.

From the 16-bit sample signal, the eight MSBs are used to choose a quantization level from the lookup table of 256 values. The quantization is from 0 to 7 so that 0 and 1 range across 2 values, . . . , 2 ranges across 4 values, 3 ranges across 8 values, . . . , and 7 ranges across 128 values. This is a logarithmic companding scheme.

Build this project as **Mulaw**. The C source file for this project, `Mulaw.c`, is included on the CD.

10.19 VOICE SCRAMBLER USING DMA AND USER SWITCHES

The project **scram16k_sw** (on the CD) is an extension of the voice scrambler example in Chapter 4. It was implemented on the C6711 DSK and can be transported to the C6713 DSK. It uses the three dip switches, USER_SW1 through USER_SW3 (the fourth switch is not used), available on board the C6711-based DSK. Using the BSL utilities for testing whether a switch is pressed on the C6713 DSK, one can implement this project on the C6713 DSK. With voice as input, the output can be unscrambled voice (based on the user switch settings).

The user dip switches are used to determine whether or not to up-sample. The program can also be used as a loop or filter program, depending on the position of the switches. USER_SW1 corresponds to the LSB. A setting such as "down/down/up" represents $(001)_b$ and is the first one tested in the program. If it is true, the output is scrambled with up-sampling at 16 kHz (the sampling rate is set at 8 kHz). The following switch positions are used:

USER_SW1	USER_SW2	USER_SW3	
0	0	1	Output scrambled with $F_s = 16$ kHz
1	0	1	Output unscrambled with $F_s = 16$ kHz
1	1	1	Lowpass filtering with $F_s = 16$ kHz
0	1	0	Output scrambled with $F_s = 8$ kHz
1	1	0	Output unscrambled with $F_s = 8$ kHz
0	0	0	Lowpass filtering with $F_s = 8$ kHz
1	0	0	Loop program

scram8k_DMA

The alternative project **scram8k_DMA** (on the CD) implements the voice scrambling scheme using DMA and sampling at 8 kHz. It illustrates the use of DMA with options within the program to implement a loop program, a filter, or the voice scrambling scheme (without up-sampling).

10.20 SB-ADPCM ENCODER/DECODER: IMPLEMENTATION OF G.722 AUDIO CODING

An audio signal is sampled at 16 kHz, transmitted at a rate of 64 kbits/s, and reconstructed at the receiving end [51,52].

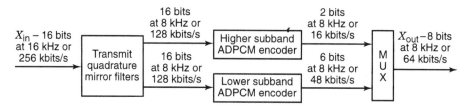

FIGURE 10.54. Block diagram of the ADPCM encoder.

Encoder

The subband adaptive differential pulse code-modulated (SB-ADPCM) encoder consists of a transmit quadrature mirror filter that splits the input signal into a low-frequency band, 0 to 4 kHz, and a high-frequency band, 4 to 8 kHz. The low- and high-frequency signals are encoded separately by dynamically quantizing an adaptive predictor's output error. The low and high encoder error signals are encoded with 6 and 2 bits, respectively. As long as the error signal is small, a negligible amount of overall quantization noise and good performance can be obtained. The low- and high-band bits are multiplexed, and the result is 8 bits sampled at 8 kHz for a bit rate of 64 kbits/s. Figure 10.54 shows a block diagram of an SB-ADPCM encoder.

Transmit Quadrature Mirror Filter

The transmit quadrature mirror filter (QMF) takes a 16-bit audio signal sampled at 16 kHz and separates it into a low band and a high band. The filter coefficients represent a 4-kHz lowpass filter. The sampled signal is separated into odd and even samples, with the effect of aliasing the signals from 4 to 8 kHz. This aliasing causes the high-frequency odd samples to be 180° out of phase with the high-frequency even samples. The low-frequency even and odd samples are in phase. When the odd and even samples are added after being filtered, the low-frequency signals constructively add, while the high-frequency signals cancel each other, producing a low-band signal sampled at 8 kHz.

The low subband encoder converts the low frequencies from the QMF into an error signal that is quantized to 6 bits.

Decoder

The decoder decomposes a 64-kbits/s signal into two signals to form the inputs to the lower and higher SB-ADPCM decoder, as shown in Figure 10.55. The receive QMF consists of two digital filters to interpolate the lower- and higher-subband ADPCM decoders from 8 to 16 kHz and produce output at a rate of 16 kHz. In the higher SB-ADPCM decoder, adding the quantized difference signal to the signal estimate produces the reconstructed signal.

Components of the ADPCM decoder include an inverse adaptive quantizer, quantizer adaptation, adaptive prediction, predicted value computation, and recon-

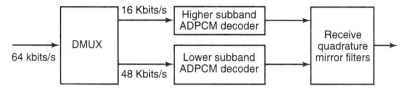

FIGURE 10.55. Block diagram of the ADPCM decoder.

structed signal computation. With input from a CD player, the DSK reconstructed output signal sound quality was good. Buffered input and reconstructed output data also confirmed successful results from the decoder.

Build this project as **G722**. The support files (encoder and decoder functions, etc.) to implement this project are included on the CD in the folder G722.

10.21 ENCRYPTION USING THE DATA ENCRYPTION STANDARD ALGORITHM

Cryptography is the art of communicating with secret data. In voice communication, cryptography refers to the encrypting and decrypting of voice data through a possibly insecure data line. The goal is to prevent anyone who does not have a "key" from receiving and understanding a transmitted message.

The data encryption standard (DES) is an algorithm that was formerly considered to be the most popular method for private key encryption. DES is still appropriate for moderately secured communication. However, with current computational power, one would be able to break (decrypt) the 56-bit key in a relatively short period of time. As a result, for very secure communication, the DES algorithm has been modified into the triple-DES or (AES) standards. DES is a very popular private-key encryption algorithm and was an industry-standard until 1998, after which it was replaced by triple-DES and AES, two slightly more complex algorithms derived from DES [53–56]. Triple-DES increases the size of the key and the data blocks used in this project, essentially performing the same algorithm three times before sending the ciphered data. AES encryption, known as the *Rijndael algorithm*, is the new standard formally implemented by the National Institute of Standards and Technology (NIST) for data encryption in high-level security communications.

DES is a bit-manipulation technique with a 64-bit block cipher that uses an effective key of 56 bits. It is an iterated Feistel-type cipher with 16 rounds. The general model of DES has three main components for (see Figure 10.56): (1) initial permutation; (2) encryption—the core iteration/f-function (16 rounds); and (3) final permutation. X and Y are the input and output data streams in 64-bit block segments, respectively, and $K1$ through $K16$ are distinct keys used in the encryption algorithm. The initial permutation is based on the predefined Table 10.10. The value at each position is used to scramble the input before the encryption routine. For example,

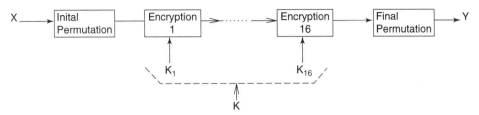

FIGURE 10.56. DES model.

TABLE 10.10 Initial Permutation

				IP			
58	50	42	34	26	18	10	2
60	52	44	36	28	20	12	4
62	54	46	38	30	22	14	6
64	56	48	40	32	24	16	8
57	49	41	33	25	17	9	1
59	51	43	35	27	19	11	3
61	53	45	37	29	21	13	5
63	55	47	39	31	23	15	7

the 58th bit of data is moved into the first position of a 64-bit array, the 50th bit into position 2, and so on. The input stream is permutated using a nonrepetitive random table of 64 integers (1–64) that corresponds to a new position of each bit in the 64-bit data block. The final permutation is the reverse of the initial permutation to reorder the samples into the correct original formation. The initial permutation is followed by the actual encryption. The permutated 64-bit block is divided into a left and a right block of 32 bits each. Sixteen rounds take place, each undergoing a similar procedure, as illustrated in Figure 10.57. The right block is placed into the left block of the next round, and the left block is combined with an encoded version of the right block and placed into the right block of the next round, or

$$L_i = R_{i-1}$$
$$R_i = L_{i-1} \oplus f(R_{i-1}, k_i)$$

where L_{i-1} and R_{i-1} are the left and right blocks, respectively, each with 32 bits, and k_i is the distinct key for the particular round of encryption. The original key is sent through a key scheduler that alters the key for each round of encryption. The left block is not utilized until the very end, when it is XORed with the encrypted right block.

The f-function operating on a 32-bit quantity expands these 32 bits into 48 bits using the expansion table (see Table 10.11). This expansion table performs a permutation while duplicating 16 of the bits (the rightmost two columns). For example,

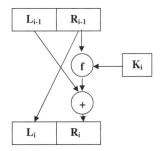

FIGURE 10.57. Encryption process—one round.

TABLE 10.11   Expansion of 32 Bits to 48

32	1	2	3	4	5
4	5	6	7	8	9
8	9	10	11	12	13
12	13	14	15	16	17
16	17	18	19	20	21
20	21	22	23	24	25
24	25	26	27	28	29
28	29	30	31	32	1

TABLE 10.12 S-Box Example, S_1

14	4	13	1	2	15	11	8	3	10	6	12	5	9	0	7
0	15	7	4	14	2	13	1	10	6	12	11	9	5	3	8
4	1	14	8	13	6	2	11	15	12	9	7	3	10	5	0
15	12	8	2	4	9	1	7	5	11	3	14	10	0	6	14

the first integer is 32, so that the first bit in the output block will be bit 32; the second integer is 1, so that the second bit in the output block will be bit 1; and so on.

The 48-bit key transformations are XORed with these expanded data, and the results are used as the input to eight different S-boxes. Each S-box takes 6 consecutive bits and outputs only 4 bits. The 4 output bits are taken directly from the numbers found in a corresponding S-box table. This process is similar to that of a decoder where the 6 bits act as a table address and the output is a binary representation of the value at that address. The zeroth and fifth bits determine the row of the S-box, and the first through fourth bits determine which column the number is located in. For example, 110100 points to the third row (10) and 10th column (1010). The first 6 bits of data correspond to the first of eight S-box tables, shown in Table 10.12. The 32 bits of output from the S-boxes are permutated according to the P-box shown in Table 10.13, and then output from the f-function shown in Figure 10.58. For example, from Table 10.13, bits 1 and 2 from the input block will be moved to bits 16 and 7 in the output, respectively. After the 16 rounds of encryption, a final

TABLE 10.13 *P*-Box

16	7	20	21	29	12	28	17	1	15	23	26	5	18	31	10
2	8	24	14	32	27	3	9	19	13	30	6	22	11	4	25

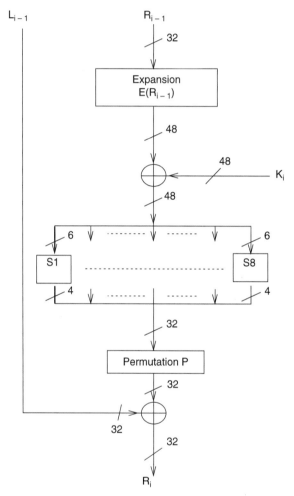

FIGURE 10.58. Core *f*-function of DES.

permutation occurs, which reverses the initial permutation, yielding an encrypted data signal.

The signal output from the encryption algorithm is not decipherable by the human ear even if the signal is filtered in any way. For testing purposes, the first three onboard switches were utilized: *sw*0 for selecting different keys; *sw*1 to enable encryption only, or both encryption and decryption; and *sw*2 as an on/off switch (a loop program).

This project was successfully implemented on the C6711 DSK with a different onboard codec and can be transported to a C6713 DSK. All the necessary files are in the folder **encryption**. The sections of code associated with the onboard switches need to be modified so that the corresponding available library support functions are utilized. The highest level of compiler optimization (-o3) was utilized in building this project.

10.22 PHASE-LOCKED LOOP

The PLL project implements a software-based linear PLL. The basic PLL causes a particular system to track another PLL. It consists of a phase detector, a loop filter, and a voltage-controlled oscillator. The software PLL is more versatile. However, it is limited by the range in frequency that can be covered, since the PLL function must be executed at least once every period of the input signal [57–59].

Initially, the PLL was tested using MATLAB, then ported to the C6x using C. The PLL locks to a sine wave, generated either internally within the program or from an external source. Output signals are viewed on a scope or on a PC using RTDX.

Figure 10.59 shows a block diagram of the linear PLL implemented in two versions:

1. Using an external input source, with the output of the digitally controlled oscillator (DCO) to an oscilloscope
2. Using RTDX with an input sine wave generated from a lookup table and various signals viewed using Excel

The phase detector, from Figure 10.59, multiplies the input sine wave by the square wave output of the DCO. The sum and difference frequencies of the two inputs to the phase detector produce an output with a high- and a low-frequency component, respectively. The low-frequency component is used to control the loop, while the high-frequency component is filtered out. When the PLL is locked, the two inputs to the phase detector are at the same frequency but with a quadrature (90°) relationship.

The loop filter is a lowpass filter that passes the low-frequency output component of the phase detector while it attenuates the undesired high-frequency component. The loop filter is implemented as a single-pole IIR filter with a zero to improve the loop's dynamics and stability. The scaled output of the loop filter represents the instantaneous incremental phase step the DCO is to take. The DCO outputs a square wave as a Walsh function: +1 for phase between 0 and π and -1 for phase between $-\pi$ and 0, with an incremental phase proportional to the number at its input.

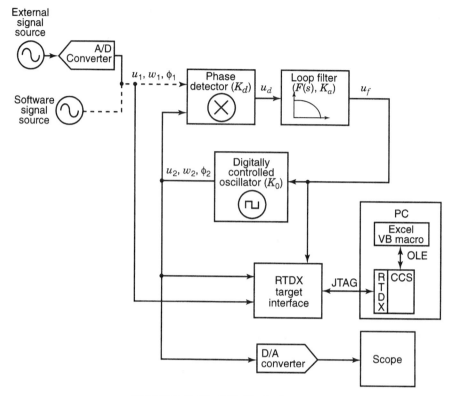

FIGURE 10.59. PLL block diagram.

RTDX for Real-Time Data Transfer

The RTDX feature was used to transfer data to the PC host using a sine wave from a lookup table as input. A single output channel was created to pass to CCS the input signal, the output of both the loop filter and the DCO, and time stamps. CCS buffers these data so that they can be accessed by other applications on the PC host. CCS has an interface that allows PC applications to access buffered RTDX data. Visual Basic Excel was used to display the results on the PC monitor. Chapter 9 introduced RTDX with several examples using different schemes.

This project was implemented on the C6211 DSK and can be transported to the C6713 DSK. All the necessary files, including the MATLAB file to test the project, are on the CD in the folder **PLL**.

10.23 MISCELLANEOUS PROJECTS

The following projects can also be used as a source of ideas to implement other projects.

10.23.1 Multirate Filter

With multirate processing, a filter can be realized with fewer coefficients than with an equivalent single-rate approach. Possible applications include a controlled noise source and background noise synthesis.

Introduction

Multirate processing uses more than one sampling frequency to perform a desired processing operation. The two basic operations are *decimation*, which is a sampling-rate reduction, and *interpolation*, which is a sampling-rate increase. Decimation techniques have been used in filtering. Multirate decimators can reduce the computational requirements of the filter. Interpolation can be used to obtain a sampling-rate increase. For example, a sampling-rate increase by a factor of K can be achieved by padding $K - 1$ zeros between pairs of consecutive input samples x_i and x_{i+1}. We can also obtain a noninteger sampling-rate increase or decrease by cascading the decimation process with the interpolation process. For example, if a net sampling-rate increase of 1.5 is desired, we would interpolate by a factor of 3, padding (adding) two zeros between each input sample, and then decimate with the interpolated input samples shifted by 2 before each calculation. Decimating or interpolating over several stages generally results in better efficiency [60–67].

Design Considerations

A binary random signal is fed into a bank of filters that are used to shape the output spectrum. The functional block diagram of the multirate filter is shown in Figure 10.60. The frequency range is divided into 10 octave bands, with each band $\frac{1}{3}$-octave controllable. The control of each octave band is achieved with three filters. The coefficients of these filters are combined to yield a composite filter with one set of coefficients for each octave. Only three unique sets of filter coefficients (low, middle, and high) are required, because the center frequency and the bandwidth are proportional to the sampling frequency. Each of the $\frac{1}{3}$-octave filters has a bandwidth of approximately 23% of its center frequency, a stopband rejection of greater than 45 dB, with an amplitude that can be controlled individually. This control provides the capability of shaping an output pseudorandom noise spectrum. The sampling rate of the output is chosen to be 16,384 Hz. Forty-one coefficients are used for the highest $\frac{1}{3}$-octave filter to achieve these requirements. The middle $\frac{1}{3}$-octave filter coefficients were used as `BP41.cof` in Chapter 4.

In order to meet the filter specifications in each region with a *constant* sampling rate, the number of filter coefficients must be doubled from one octave filter to the next lower one. As a result, the lowest-octave filter would require 41×2^9 coefficients. With 10 filters ranging from 41 to 41×2^9 coefficients, the computational requirements would be considerable. To reduce these computational requirements, a multirate approach is used, as shown in Figure 10.60.

432

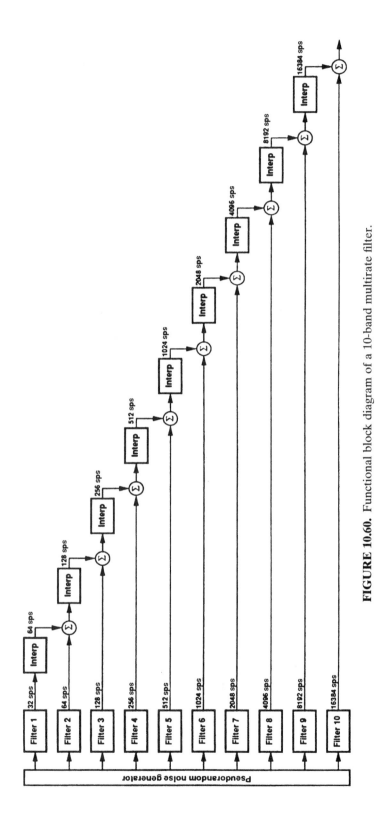

FIGURE 10.60. Functional block diagram of a 10-band multirate filter.

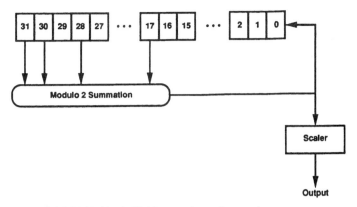

FIGURE 10.61. A 32-bit pseudorandom noise generator.

The noise generator is a software-based implementation of a maximal length sequence technique used for generating pseudorandom numbers. This pseudo-random noise generator was implemented in Example 3.4. The output of the noise generator provides uncorrelated noise input to each of the 10 sets of bandpass filters. The noise generation example in Chapter 3 uses the process shown in Figure 10.61.

Because each $\frac{1}{3}$-octave filer can be scaled individually, a total of 30 levels can be controlled. The output of each octave bandpass filter (except the last one) becomes the input to an interpolation lowpass filter, using a 2:1 interpolation factor. The ripple in the output spectrum is minimized by having each adjacent $\frac{1}{3}$-octave filter with crossover frequencies at the 3-dB points.

The center frequency and bandwidth of each filter are determined by the sampling rate. The sampling rate of the highest-octave filter is processed at 16,384 samples per second (you can use a sampling rate of 16 kHz, 48 kHz, etc.), and each successively lower-octave band is processed at half the rate of the next higher band.

Only three separate sets of 41 coefficients are used for the lower, middle, and higher $\frac{1}{3}$-octave bands. For each octave band, the coefficients are combined as follows:

$$H_{ij} = (H_{lj})(L_{3i-2}) + (H_{mj})(L_{3i-1}) + (H_{hj})(L_{3i})$$

where $i = 1, 2, \ldots, 10$ bands and $j = 0, 1, \ldots, 40$ coefficients; $L_1, L_2, \ldots, L_{30}$ represent the level of each $\frac{1}{3}$-octave band filter; and H_{lj}, H_{mj}, H_{hj} represent the jth coefficient of the lower, middle, and higher $\frac{1}{3}$-octave band FIR filter. For example, for the first band ($i = 1$),

$$H_0 = (H_{l0})(L_1) + (H_{m0})(L_2) + (H_{h0})(L_3)$$
$$H_1 = (H_{l1})(L_1) + (H_{m1})(L_2) + (H_{h1})(L_3)$$
$$\vdots$$
$$H_{40} = (H_{l40})(L_1) + (H_{m40})(L_2) + (H_{h40})(L_3)$$

and, for band 10 ($i = 10$),

$$H_0 = (H_{l0})(L_{28}) + (H_{m0})(L_{29}) + (H_{h0})(L_{30})$$
$$H_1 = (H_{l1})(L_{28}) + (H_{m1})(L_{29}) + (H_{h1})(L_{30})$$
$$\vdots$$
$$H_{40} = (H_{l40})(L_{28}) + (H_{m40})(L_{29}) + (H_{h40})(L_{30})$$

For an efficient design with the multirate technique, lower-octave bands are processed at a lower sampling rate, then interpolated up to a higher sampling rate, by a factor of 2, to be summed with the next higher octave band filter output, as shown in Figure 10.60. Each interpolation filter is a 21-coefficient FIR lowpass filter, with a cutoff frequency of approximately one-fourth of the sampling rate. For each input, the interpolation filter provides two outputs, or

$$y_1 = x_0 I_0 + 0 I_1 + x_1 I_2 + 0 I_3 + \cdots + x_{10} I_{20}$$
$$y_2 = 0 I_0 + x_0 I_1 + 0 I_2 + x_1 I_3 + \cdots + x_9 I_{19}$$

where y_1 and y_2 are the first and second interpolated outputs, respectively, x_n are the filter inputs, and I_n are the interpolation filter coefficients. The interpolator is processed in two sections to provide the data-rate increase by a factor of 2.

For the multirate filter, the approximate number of multiplication operations (with accumulation) per second is

$$\begin{aligned} \text{MAC/S} = &(41+21)(32+64+128+256+512+1{,}024+2{,}048+4{,}096+8{,}192) \\ &+ (41)(16{,}384) \\ \approx &\, 1.686 \times 10^6 \end{aligned}$$

The approximate number of multiplications/accumulation per second for an equivalent single-rate filter is then

$$\text{MAC/S} = F_s \times 41(1 + 2 + 2^2 + 2^3 + \ldots + 2^9) = 687 \times 10^6$$

which would considerably increase the processing time requirements.

A brief description (recipe) of the main processing follows, for the first time through (using three buffers B_1, B_2, B_3).

Band 1

1. Run the bandpass filter and obtain one output sample.
2. Run the lowpass interpolation filter twice and obtain two outputs. The interpolator provides two sample outputs for each input sample.
3. Store in buffer B_2, size 512, at locations 1 and 2 (in memory).

Band 2

1. Run the bandpass filter two times and sum with the two previous outputs stored in B_2 from band 1.

2. Store the summed values in B_2 at the same locations 1 and 2 (again).

3. Pass the sample in B_2 at location 1 to the interpolation filter twice and obtain two outputs.

4. Store these two outputs in buffer B_3, size 256, at locations 1 and 2.

5. Pass the sample in B_2 at location 2 to the interpolation filter twice and obtain two outputs.

6. Store these two outputs in buffer B_3 at locations 3 and 4.

Band 3

1. Run the bandpass filter four times and sum with the previous four outputs stored in B_3 from band 2.

2. Store the summed values in B_3 at locations 1 through 4.

3. Pass the sample in B_3 at location 1 to the interpolation filter twice and obtain two outputs.

4. Store these two outputs in buffer B_2 at locations 1 and 2.

5. Pass the sample in B_3 at location 2 to the interpolation filter twice and obtain two outputs.

6. Store these two outputs in buffer B_2 at locations 3 and 4.

7. Repeat steps 3 and 4 for the other two samples at locations 3 and 4 in B_3. For each of these samples, obtain two outputs, and store each set of two outputs in buffer B_2 at locations 5 through 8.

Bank 10

1. Run the bandpass filter 512 times and sum with the previous 512 outputs stored in B_2 from band 9.

2. Store the summed values in B_2 at locations 1 through 512.

No interpolation is required for band 10. After all the bands are processed, wait for the output buffer B_1, size 512, to be empty. Then switch the buffers B_1 and B_2—the last working buffer with the last output buffer. The main processing is then repeated.

The multirate filter was implemented on the C25 processor using 9 bands and on the C30 processor using 10 bands [8], and can be transported to the C6x. Using a total of 30 different levels, any specific $\frac{1}{3}$-octave filter can be turned on or off. For example, all the filter bands can be turned on except bands 2 and 5. Figure 10.62 shows the frequency response of the three $\frac{1}{3}$-octave filters of band 9 implemented on the C30. Note that if a sampling rate of 8 kHz is set (for the highest band), the middle $\frac{1}{3}$-octave band 1 filter would have a center frequency of 4 Hz (one-fourth of the equivalent sampling rate for band 1).

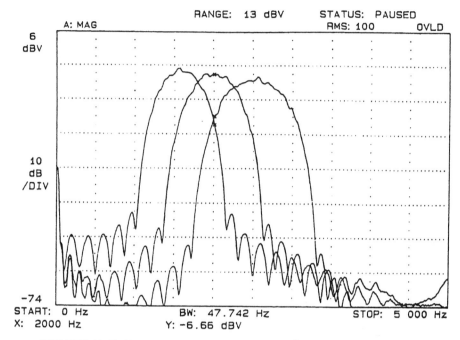

FIGURE 10.62. Frequency response of the three $\frac{1}{3}$-octave filters of band 9.

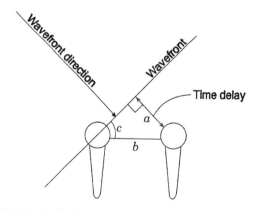

FIGURE 10.63. Signal reception with two microphones.

10.23.2 Acoustic Direction Tracker

This project uses two microphones to capture an audio signal. From the delay associated with the signal reaching one of the microphones before the other, a relative angle where the source is located can be determined. A signal radiated at a distance from its source can be considered to have a plane wavefront, as shown in Figure 10.63. This allows the use of equally spaced sensors (many microphones can be used as acoustical sensors) in a line to ascertain the angle at which the signal is radiat-

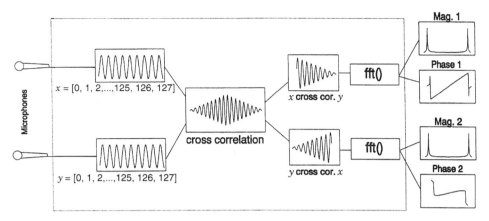

FIGURE 10.64. Block diagram of an acoustic signal tracker.

ing. Since one microphone is closer to the source than the other, the signal received by the more distant microphone is delayed in time. This time shift corresponds to the angle where the source is located and the relative distance between the microphones and the source. The angle $c = \arcsin(a/b)$, where the distance a is the product of the speed of sound and the time delay (phase/frequency).

Figure 10.64 shows a block diagram of the acoustic signal tracker. Two 128-point arrays of data are obtained, cross-correlating the first signal with the second and then the second signal with the first. The resulting cross-correlation data are decomposed into two halves, each transformed using a 128-point FFT. The resulting phase is the phase difference of the two signals.

This project was implemented on the C30 [17] and can be transported to the C6713 processor. To test this project, a speaker was positioned a few feet from the two microphones, which are separated by 1 foot. The speaker receives a 1-kHz signal from a function generator. A track of the source speaker is plotted over time on the PC monitor. Plots of the cross-correlation and the magnitude of the cross-correlation of the two microphone signals were also displayed on the PC monitor.

10.23.3 Neural Network for Signal Recognition

The goal of this project is to recognize a signal. The FFT of a signal becomes the input to a neural network that is trained to recognize the signal using the back-propagation learning rule.

Design and Implementation

The neural network consists of three layers with a total of 90 nodes: 64 input nodes in the first layer, 24 nodes in the middle or hidden layer, and 2 output nodes in the third layer. The 64 points as input to the neural network are obtained by retaining

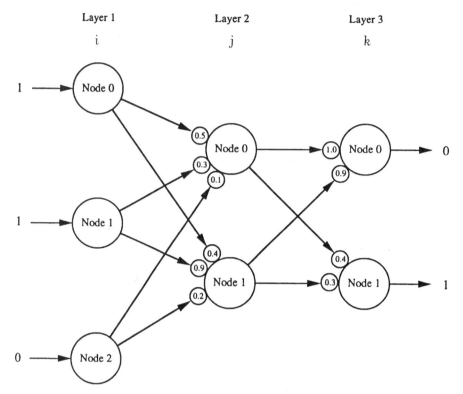

FIGURE 10.65. Three-layer neural network with seven nodes.

half of the 128 points resulting from a 128-point FFT of the signal to be recognized. In recent years, many books and articles on neural networks have been published [68,69]. Neural network products are now available from many vendors.

Many different rules have been described in the literature for training a neural network. The back-error propagation is one of the most widely used for a wide range of applications. Given a set of input, the network is trained to give a desired response. If the network gives the wrong answer, then it is corrected by adjusting its parameters so that the error is reduced. During this correction process, one starts with the output nodes and propagation is backward to the input nodes (back propagation). Then the propagation process is repeated.

To illustrate the procedure for training a neural network using the back-propagation rule, consider a simple three-layer network with seven nodes, as shown in Figure 10.65. The input layer consists of three nodes, and the hidden layer and output layer, each consists of two nodes. Given the following set of inputs: input No. 1 = 1 into node 0, input No. 2 = 1 into node 1, and input No. 3 = 0 into node 2, the network is to be trained to yield the desired output 0 at node 0 and 1 at node 1. Let the subscripts i, j, k be associated with the first, second, and third layers, respectively. A set of random weights are initially chosen, as shown in Figure 10.65. For example, the weight $w_{11} = 0.9$ represents the weight value associated with node 1 in layer 1 and node 1 in the middle or hidden layer 2. The weighted sum of the input value is

$$s_j = \sum_{i=0}^{2} w_{ji} x_i$$

where $j = 0, 1$ and $i = 0, 1, 2$. Then.

$$s_0 = w_{00} x_0 + w_{01} x_1 + w_{02} x_2 = (0.5)(1) + (0.3)(1) + (0.1)(0) = 0.8$$

Similarly, $s_1 = 1.3$. A function of the resulting weighted sum $f(s_j)$ is next computed. This transfer function f of a processing element must be differentiable. For this project, f is chosen as the hyperbolic tangent function tanh. Other functions, such as the unit step function or the smoother sigmoid function, also can be used. The output of the transfer function associated with the nodes in the middle layer is

$$x_j = f(s_j) = \tanh(s_j), \quad j = 0, 1$$

The output of node 0 in the hidden layer then becomes

$$x_0 = \tanh(0.8) = 0.664$$

Similarly, $x_1 = 0.862$. The weighted sum at each node in layer 3 is

$$s_k = \sum_{j=0}^{1} w_{kj} x_j, \quad k = 0, 1$$

to yield

$$s_0 = w_{00} x_0 + w_{01} x_1 = (1.0)(0.664) + (0.9)(0.862) = 1.44$$

Similarly, $s_1 = 0.524$. The output of the transfer function is associated with the output layer, and replacing j by k,

$$x_k = f(s_k), \quad k = 0, 1$$

Then $x_0 = \tanh(1.44) = 0.894$, and $x_1 = \tanh(0.524) = 0.481$. The error in the output layer can now be found using

$$e_k = (d_k - x_k) f'(s_k)$$

where $d_k - x_k$ reflects the amount of error, and $f'(s)$ represents the derivative of $\tanh(s)$, or

$$f'(x) = (1 + f(s))(1 - f(s))$$

Then

$$e_0 = (0 - 0.894)(1 + \tanh(1.44))(1 - \tanh(1.44)) = -0.18$$

Similarly, $e_1 = 0.399$. Based on this output error, the contribution to the error by each hidden layer node is to be found. The weights are then adjusted based on this error using

$$\Delta w_{kj} = \eta e_k x_j$$

where η is the network learning rate constant, chosen as 0.3. A large value of η can cause instability, and a very small one can make the learning process much too slow. Then

$$\Delta w_{00} = (0.3)(-0.18)(0.664) = -0.036$$

Similarly, $\Delta w_{01} = -0.046$, $\Delta w_{10} = 0.08$, and $\Delta w_{11} = 0.103$. The error associated with the hidden layer is

$$e_j = f'(s_j) \sum_{k=0}^{1} e_k w_{kj}$$

Then

$$e_0 = (1 + \tanh(0.8))(1 - \tanh(0.8))\{(-0.18)(1.0) + (0.399)(0.4)\} = -0.011$$

Similarly, $e_1 = -0.011$. Changing the weights between layers i and j,

$$\Delta w_{ji} = \eta e_j x_i$$

Then

$$\Delta w_{00} = (0.3)(-0.011)(1) = -0.0033$$

Similarly, $\Delta w_{01} = -0.0033$, $\Delta w_{02} = 0$, $\Delta w_{10} = -0.0033$, $\Delta w_{11} = -0.0033$, and $\Delta w_{12} = 0$. This gives an indication of by how much to change the original set of weights chosen. For example, the new set of coefficients becomes

$$w_{00} = w_{00} + \Delta w_{00} = 0.5 - 0.0033 = 0.4967$$

and $w_{01} = 0.2967$, $w_{02} = 0.1$, and so on.

This new set of weights represents only the values after one complete cycle. These weight values can be verified using a training program for this project. For this procedure of training the network, readjusting the weights is continuously repeated until the output values converge to the set of desired output values. For this project, the training program is such that the training process can be halted by the user, who can still use the resulting weights.

This project was implemented on the C30 and can be transported to the C6x. Two sets of inputs were chosen: a sinusoidal and a square wave input. The FFT (128-point) of each input signal is captured and stored in a file, with a total of 4800 points: 200 vectors, each with 64 features (retaining one-half of the 128 points). Another program scales each set of data (sine and square wave) so that the values are between 0 and 1.

To demonstrate this project, two output values for each node are displayed on the PC screen. Values of +1 for node 0 and −1 for node 1 indicate that a sinusoidal input is recognized, and values of −1 for node 0 and +1 for node 1 indicate that a square wave input is recognized.

This project was successful but was implemented for only the two sets of chosen data. Much work remains to be done, such as training more complex sets of data and examining the effects of different training rules based on the different signals to be recognized.

10.23.4 Adaptive Temporal Attenuator

An adaptive temporal attenuator (ATA) suppresses undesired narrowband signals to achieve a maximum signal-to-interference ratio. Figure 10.66 shows a block diagram of the ATA. The input is passed through delay elements, and the outputs from selected delay elements are scaled by weights. The output is

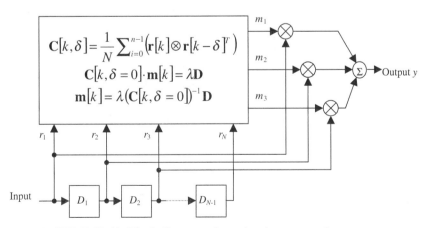

$$\mathbf{C}[k,\delta] = \frac{1}{N}\sum_{i=0}^{n-1}\left(\mathbf{r}[k]\otimes\mathbf{r}[k-\delta]^{T}\right)$$
$$\mathbf{C}[k,\delta=0]\cdot\mathbf{m}[k] = \lambda\mathbf{D}$$
$$\mathbf{m}[k] = \lambda\left(\mathbf{C}[k,\delta=0]\right)^{-1}\mathbf{D}$$

FIGURE 10.66. Block diagram of an adaptive temporal attenuator.

$$y[k] = \mathbf{m}^T \cdot \mathbf{r}[k] = \sum_{i=0}^{N-1} \left(\mathbf{m}_i \cdot \mathbf{r}[k-i] \right)$$

where $\mathbf{m}$ is a weight vector, $\mathbf{r}$ a vector of delayed samples selected from the input signal, and N the number of samples in $\mathbf{m}$ and $\mathbf{r}$. The adaptive algorithm computes the weights based on the correlation matrix and a direction vector:

$$\mathbf{C}[k, \delta = 0] \cdot \mathbf{m}[k] = \lambda \mathbf{D}$$

where $\mathbf{C}$ is a correlation matrix, $\mathbf{D}$ a direction vector, and λ a scale factor. The correlation matrix $\mathbf{C}$ is computed as an average of the signal correlation over several samples:

$$\mathbf{C}[k, \delta] = \frac{1}{N_{AV}} \sum_{i=0}^{n-1} \left(\mathbf{r}[k] \otimes \mathbf{r}[k - \delta]^T \right)$$

where N_{AV} is the number of samples included in the average. The direction vector $\mathbf{D}$ indicates the signal desired:

$$\mathbf{D} = \begin{bmatrix} 1 & \exp(j\omega_T \tau) & \cdots & \exp[j\omega_T (N-1)\tau] \end{bmatrix}^T$$

where ω_T is the angular frequency of the signal desired, τ the delay between samples that create the output, and N the order of the correlation matrix.

This procedure minimizes the undesired-to-desired ratio (UDR) [70]. UDR is defined as the ratio of the total signal power to the power of the signal desired, or

$$\text{UDR} = \frac{P_{\text{total}}}{P_d} = \frac{\mathbf{m}[k]^T \cdot \mathbf{C}[k,0] \cdot \mathbf{m}[k]}{P_d \left(\mathbf{m}[k]^T \cdot \mathbf{D} \right)^2} = \frac{1}{P_d \left(\mathbf{m}[k]^T \cdot \mathbf{D} \right)}$$

where P_d is the power of the signal desired.

MATLAB is used to simulate the ATA, then ported to the C6x for real-time implementation. Figure 10.67 shows the test setup using a fixed desired signal of 1416 Hz and an undesired signal of 1784 Hz (which can be varied). From MATLAB, and optimal value of τ is found to minimize UDR. This is confirmed in real time, since for that value of τ (varying τ with a GEL file), the undesired signal (initially displayed from an HP3561A analyzer) is greatly attenuated.

10.23.5 FSK Modem

This project implements a digital modulator/demodulator. It generates 8-ary FSK carrier tones. The following steps are performed in the program.

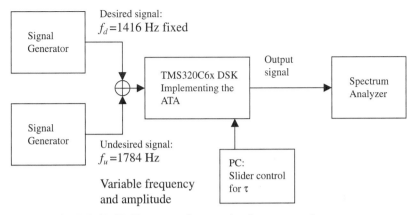

FIGURE 10.67. Test setup for an adaptive temporal attenuator.

1. The sampled data are acquired as input.

2. The 6 MSBs are separated into two 3-bit samples.

3. The most significant portion of the sample data selects an FSK tone.

4. The FSK tone is sent to a demodulator.

5. The FSK tone is windowed using the Hanning window function.

6. DFT (16-point) results are obtained for the windowed FSK tone.

7. DFT results are sent to the function that selects the frequency with the highest amplitude, corresponding to the upper 3 bits of the sampled data.

8. The process is repeated for the lower 3 bits of the sampled data.

9. The bits are combined and sent to the codec.

10. The gel program allows for an option to interpolate or up-sample the reconstructed data for a smoother output waveform.

10.23.6 Image Processing

This project implements various schemes used in image processing:

1. *Edge detection:* for enhancing edges in an image using Sobe's edge detection

2. *Median filtering:* nonlinear filter for removing noise spikes in an image

3. *Histogram equalization:* to make use of the image spectrum

4. *Unsharp masking:* spatial filter to sharpen the image, emphasizing its high-frequency components

5. *Point detection:* for emphasizing single-point features in the image

A major issue was using/loading the images as .*h* files in lieu of using real-time images (due to the course's one-semester time constraint). During the course of this project, the following evolved: a code example for additive noise with a Gaussian distribution, with adjustable variance and mean, and a code example of histogram transformation to map the distribution of one set of numbers to a different distribution (used in image processing).

10.23.7 Filter Design and Implementation Using a Modified Prony's Method

This project designs and implements a filter based on a modified Prony's method [71–74]. The method is based on the correlation property of the filter's representation and does not require computation of any derivatives or an initial guess of the coefficient vector. The filter's coefficients are calculated recursively to obtain the filter's impulse response.

10.23.8 PID Controller

Both nonadaptive and adaptive controllers using the proportional, integral, and derivative (PID) control algorithm have been implemented [17,75,76].

10.23.9 Four-Channel Multiplexer for Fast Data Acquisition

A four-channel multiplexer module was designed and built for this project, implemented in C [8]. It includes an 8-bit flash ADC, a FIFO, a MUX, and a crystal oscillator (2 or 20 MHz). An input is acquired through one of the four channels. The FFT of the input signal is displayed in real time on the PC monitor.

10.23.10 Video Line Rate Analysis

This project is discussed in [8,77] and implemented using C and C30 code. It analyzes a video signal at the horizontal (line) rate. Interactive algorithms commonly used in image processing for filtering, averaging, and edge enhancement using C code are utilized for this analysis. The source of the video signal is a charge-coupled device (CCD) camera as input to a module designed and built for this project. This module includes flip-flops, logic gates, and a clock. Displays on the PC monitor illustrate various effects on one horizontal video line signal from either a 500-kHz or a 3-MHz IIR lowpass filter and from an edge enhancement algorithm.

Acknowledgments

I owe a special debt to all the students who have made this chapter possible. They include students from Roger Williams University, the University of

Massachusetts–Dartmouth, and the Worcester Polytechnic Institute (WPI) who have contributed to my general background in real-time DSP applications over the last 20 years. In particular, the undergraduate and graduate students at WPI who have recently taken my two courses on real-time DSP. Many projects and mini-projects from these students are included in this chapter. A special thanks to the following students: N. Alsindi, E. Boron, A Buchholz, J. Chapman, G. Colangelo, J. Coyne, H. Daempfling, T. Daly, D. Debiasio, A. Dupont, J. Elbin, J. Gaudette, E. Harvey, K. Krishna, M. Lande, M. Lauer, E. Laurendo, R. Lemdiasov, M. Marcantonio, A. Nadkarni, S. Narayanan, A. Navalekar, A. Obi, P. Phadnis, J. Quartararo, V. Rangan, D. Sebastian, M. Seward, D. Tulsiani, and K. Yuksel.

REFERENCES

1. R. Chassaing, *DSP Applications Using C and the TMS320C6x DSK,* Wiley, New York, 2002.

2. J. H. McClellan, R. W. Schafer, and M. A. Yoder, *DSP First: A Multimedia Approach*, Prentice Hall, Upper Saddle River, NJ, 1998.

3. N. Kehtarnavaz and M. Keramat, *DSP System Design Using the TMS320C6000*, Prentice Hall, Upper Saddle River, NJ, 2001.

4. N. Dahnoun, *DSP Implementation Using the TMS320C6x Processors*, Prentice Hall, Upper Saddle River, NJ, 2000.

5. S. Tretter, *Communication System Design Using DSP Algorithms—With Laboratory Experiments for the TMS320C6701 and TMS320C6711*, Kluwer Academic, Boston, 2003.

6. M. Morrow, T. Welch, C. Cameron, and G. York, Teaching real-time beamforming with the C6211 DSK and MATLAB, *Proceedings of the Texas Instruments DSPS Fest Annual Conference*, 2000.

7. R. Chassaing, *Digital Signal Processing Laboratory Experiments Using C and the TMS320C31 DSK*, Wiley, New York, 1999.

8. R. Chassaing, *Digital Signal Processing with C and the TMS320C30*, Wiley, New York, 1992.

9. C. Marven and G. Ewers, *A Simple Approach to Digital Signal Processing*, Wiley, New York, 1996.

10. J. Chen and H. V. Sorensen, *A Digital Signal Processing Laboratory Using the TMS320C30*, Prentice Hall, Upper Saddle River, NJ, 1997.

11. S. A. Tretter, *Communication System Design Using DSP Algorithms*, Plenum Press, New York, 1995.

12. R. Chassaing et al., Student projects on digital signal processing with the TMS320C30, *Proceedings of the 1995 ASEE Annual Conference*, June 1995.

13. J. Tang, Real-time noise reduction using the TMS320C31 digital signal processing starter kit, *Proceedings of the 2000 ASEE Annual Conference*, 2000.

14. C. Wright, T. Welch III, M. Morrow, and W. J. Gomes III, Teaching real-world DSP using MATLAB and the TMS320C31 DSK, *Proceedings of the 1999 ASEE Annual Conference*, 1999.

15. J. W. Goode and S. A. McClellan, Real-time demonstrations of quantization and prediction using the C31 DSK, *Proceedings of the 1998 ASEE Annual Conference*, 1998.

16. R. Chassaing and B. Bitler, Signal processing chips and applications, *The Electrical Engineering Handbook*, CRC Press, Boca Raton, FL, 1997.

17. R. Chassaing et al., Digital signal processing with C and the TMS320C30: Senior projects, *Proceedings of the 3rd Annual TMS320 Educators Conference*, Texas Instruments, Dallas, TX, 1993.

18. R. Chassaing et al., Student projects on applications in digital signal processing with C and the TMS320C30, *Proceedings of the 2nd Annual TMS320 Educators Conference*, Texas Instruments, Dallas, TX, 1992.

19. R. Chassaing, TMS320 in a digital signal processing lab, *Proceedings of the TMS320 Educators Conference*, Texas Instruments, Dallas, TX, 1991.

20. P. Papamichalis, ed., *Digital Signal Processing Applications with the TMS320 Family: Theory, Algorithms, and Implementations*, Vols. 2 and 3, Texas Instruments, Dallas, TX, 1989, 1990.

21. *Digital Signal Processing Applications with the TMS320C30 Evaluation Module: Selected Application Notes*, Texas Instruments, Dallas, TX, 1991.

22. R. Chassaing and D. W. Horning, *Digital Signal Processing with the TMS320C25*, Wiley, New York, 1990.

23. I. Ahmed, ed., *Digital Control Applications with the TMS320 Family*, Texas Instruments, Dallas, TX, 1991.

24. A. Bateman and W. Yates, *Digital Signal Processing Design*, Computer Science Press, New York, 1991.

25. Y. Dote, *Servo Motor and Motion Control Using Digital Signal Processors*, Prentice Hall, Upper Saddle River, NJ, 1990.

26. R. Chassaing, A senior project course in digital signal processing with the TMS320, *IEEE Transactions on Education*, Vol. 32, 1989, pp. 139–145.

27. R. Chassaing, Applications in digital signal processing with the TMS320 digital signal processor in an undergraduate laboratory, *Proceedings of the 1987 ASEE Annual Conference*, June 1987.

28. K. S. Lin, ed., *Digital Signal Processing Applications with the TMS320 Family: Theory, Algorithms, and Implementations*, Vol. 1, Prentice Hall, Upper Saddle River, NJ, 1988.

29. G. Goertzel, An algorithm for the evaluation of finite trigonometric series, *American Mathematics Monthly*, Vol. 65, Jan. 1958.

30. ScenixSemiconductors available at http://www.electronicsweekly.com/toolkits/system/feature3.asp

31. A. Si, Implementing DTMF detection using the Silicon Laboratories Data Access Arrangement (DAA), Scenix Semiconductors, Sept. 1999.

32. www.gamedev.net/reference/programming/features/beatdetection/

33. S. Qian, *Introduction to Time-Frequency and Wavelet Transform,* Prentice-Hall, Upper Saddle River, NJ, 2002.

34. B. Boashah, *Time-Frequency Signal Analysis: Methods and Applications,* Wiley Halsted Press, 1992.

35. U. Zoler, *Digital Audio Signal,* Wiley, Chichester, England, 1995.

36. J. Proakis and M. Salehi, *Communication Systems Engineering,* Prentice-Hall, Upper Saddle River, NJ, 1994.

37. S. Haykin, *Communication Systems,* Wiley, New York, 2001.

38. B. Sklar, *Digital Communications: Fundamentals and Applications,* Prentice-Hall, Upper Saddle River, NJ, 2001.

39. http://www.physics.gmu.edu/~amin/phys251/Topics/NumAnalysis/Approximation/polynomialInterp.html

40. S. Lin, D. J. Costello, *Error Control Coding, Fundamentals and Applications,* Prentice-Hall, Upper Saddle River, NJ, 1983.

41. C. Fleming, A tutorial on convolutional encoding with viterbi decoding. Available at http://home.netcom.com/~chip.f/viterbi/algrthms2.html

42. J. Flanagan and L. Rabiner, *Speech Synthesis,* Dowden, Hutchinson & Ross, Stroudsburg, PA, 1973.

43. R. Rabiner and R. W. Schafer, *Digital Signal Processing of Speech Signals,* Prentice-Hall, Englewood Cliffs, NJ, 1978.

44. R. Deller, J. G. Proakis, and J. H. Hansen, *Discrete-Time Processing of Signals,* Macmillan, New York, 1993.

45. B. Gold and N. Morgan, *Speech and Audio Signal Processing,* Wiley, New York, 2000.

46. R. P. Ramachandran and R. J. Mammoce, eds., *Modern Methods of Speech Processing,* Kluwer Academic, Boston, 1995.

47. M. N. Do, An automatic speaker recognition system, Audio Visual Communications Lab, Swiss Federal Institute of Technology, Lausanne.

48. X. Huang et al., *Spoken Language Processing,* Prentice-Hall, Upper Saddle River, NJ, 2001.

49. R. P. Ramchandran and Peter Kabal, Joint solution for formant and speech predictors in speech processing, *Proceedings of the IEEE International Conference on Acoustics, Speech, Signal Processing,* Apr. 1988, pp. 315–318.

50. L. B. Rabiner and B. H. Juang, *Fundamentals of Speech Recognition,* Prentice-Hall, Upper Saddle River, NJ, 1993.

51. *ITU-T Recommendation G.722 Audio Coding with 64 kbits/s.*

52. P. M. Embree, *C Algorithms for Real-Time DSP,* Prentice Hall, Upper Saddle River, NJ, 1995.

53. ECB Mode (Native DES), Frame Technology, 1994. Available at http://www.cs.nps.navy.mil/curricula/tracks/security/notes/chap04_38.html

54. S. Hallyn, *DES: The Data Encryption Standard,* last modified June 27, 1996. Available at `http://www.cs.wm.edu/~hallyn/des`

55. N. Nicolicim, *Data Encryption Standard (DES) History of DES,* McMaster University, lecture notes, October 9, 2001. Available at `www.ece.mcmaster.ca/faculty/ nicolici/ coe4oi4/2001/lecture10.pdf`

56. B. Sunar, interview and lecture notes. Available at `http://www.ece.wpi.edu/ ~sunar`

57. Roland E. Best, *Phase-Locked Loops Design, Simulation, and Applications,* 4th ed., McGraw-Hill, New York, 1999.

58. W. Li and J. Meiners, *Introduction to Phase Locked Loop System Modeling,* SLTT015, Texas Instruments, Dallas, TX, May 2000.

59. J. P. Hein and J. W. Scott, Z-domain model for discrete-time PLL's, *IEEE Transactions on Circuits and Systems,* Vol. CS-35, Nov. 1988, pp. 1393–1400.

60. R. Chassaing, P. Martin, and R. Thayer, Multirate filtering using the TMS320C30 floating-point digital signal processor, *Proceedings of the 1991 ASEE Annual Conference,* June 1991.

61. R. E. Crochiere and L. R. Rabiner, *Multirate Digital Signal Processing,* Prentice-Hall, Upper Saddle River, NJ, 1983.

62. R. W. Schafer and L. R. Rabiner, A digital signal processing approach to interpolation, *Proceedings of the IEEE,* Vol. 61, 1973, pp. 692–702.

63. R. E. Crochiere and L. R. Rabiner, Optimum FIR digital filter implementations for decimation, interpolation and narrow-band filtering, *IEEE Transactions on Acoustics, Speech, and Signal Processing,* Vol. ASSP-23, 1975, pp. 444–456.

64. R. E. Crochiere and L. R. Rabiner, Further considerations in the design of decimators and interpolators, *IEEE Transactions on Acoustics, Speech, and Signal Processing,* Vol. ASSP-24, 1976, pp. 296–311.

65. M. G. Bellanger, J. L. Daguet, and G. P. Lepagnol, Interpolation, extrapolation, and reduction of computation speed in digital filters, *IEEE Transactions on Acoustics, Speech, and Signal Processing,* Vol. ASSP-22, 1974, pp. 231–235.

66. R. Chassaing, W. A. Peterson, and D. W. Horning, A TMS320C25-based multirate filter, *IEEE Micro,* Oct. 1990, pp. 54–62.

67. R. Chassaing, Digital broadband noise synthesis by multirate filtering using the TMS320C25, *Proceedings of the 1988 ASEE Annual Conference,* Vol. 1, June 1988.

68. B. Widrow and R. Winter, Neural nets for adaptive filtering and adaptive pattern recognition, *Computer,* Mar. 1988, pp. 25–39.

69. D. E. Rumelhart, J. L. McClelland, and the PDP Research Group, *Parallel Distributed Processing: Explorations in the Microstructure of Cognition,* Vol. 1, MIT Press, Cambridge, MA, 1986.

70. I. Progri and W. R. Michalson, Adaptive spatial and temporal selective attenuator in the presence of mutual coupling and channel errors, *ION GPS-2000,* 2000.

71. F. Brophy and A. C. Salazar, Recursive digital filter synthesis in the time domain, *IEEE Transactions on Acoustics, Speech, and Signal Processing,* Vol. ASSP-22, 1974.

72. W. H. Press, S. A. Teukolsky, W. T. Vetterling, and B. P. Flannery, *Numerical Recipes in C: The Art of Scientific Computing*, Cambridge University Press, New York, 1992.

73. J. Borish and J. B. Angell, An efficient algorithm for measuring the impulse response using pseudorandom noise, *Journal of the Audio Engineering Society*, Vol. 31, 1983.

74. T. W. Parks and C. S. Burrus, *Digital Filter Design*, Wiley, New York, 1987.

75. J. Tang, R. Chassaing, and W. J. Gomes III, Real-time adaptive PID controller using the TMS320C31 DSK *Proceedings of the 2000 Texas Instruments DSPS Fest Conference*, 2000.

76. J. Tang and R. Chassaing, PID controller using the TMS320C31 DSK for real-time motor control, *Proceedings of the 1999 Texas Instruments DSPS Fest Conference*, 1999.

77. B. Bitler and R. Chassaing, Video line rate processing with the TMS320C30, *Proceedings of the 1992 International Conference on Signal Processing Applications and Technology (ICSPAT)*, 1992.

78. *MATLAB, The Language of Technical Computing, Version 6.3*, MathWorks, Natick, MA.

A

TMS320C6x Instruction Set

A.1 INSTRUCTIONS FOR FIXED- AND FLOATING-POINT OPERATIONS

Table A.1 shows a listing of the instructions available for the C6x processors. The instructions are grouped under the functional units used by these instructions. These instructions can be used with both fixed- and floating-point C6x processors. Some additional instructions are available for the fixed-point C64x processor [2].

A.2 INSTRUCTIONS FOR FLOATING-POINT OPERATIONS

Table A.2 shows a listing of additional instructions available with the floating-point processor C67x. These instructions handle floating-point type of operations and are grouped under the functional units used by these instructions (see also Table A.1).

REFERENCES

1. *TMS320C6000 CPU and Instruction Set*, SPRU189F, Texas Instruments, Dallas, TX, 2000.
2. *TMS320C6000 Programmer's Guide*, SPRU198G, Texas Instruments, Dallas, TX, 2002.

TABLE A.1   Instructions for Fixed- and Floating-Point Operations

.L Unit	.M Unit	.S Unit	.D Unit
ABS	MPY	ADD	ADD
ADD	MPYH	ADDK	ADDAB
ADDU	MPYHL	ADD2	ADDAH
AND	MPYHLU	AND	ADDAW
CMPEQ	MPYHSLU	B disp	LDB
CMPGT	MPYHSU	B IRP[a]	LDBU
CMPGTU	MPYHU	B NRP[a]	LDH
CMPLT	MPYHULS	B reg	LDHU
CMPLTU	MPYHUS	CLR	LDW
LMBD	MPYLH	EXT	LDB (15-bit offset)[b]
MV	MPYLHU	EXTU	LDBU (15-bit offset)[b]
NEG	MPYLSHU	MV	LDH (15-bit offset)[b]
NORM	MPYLUHS	MVC[a]	LDHU (15-bit offset)[b]
NOT	MPYSU	MVK	LDW (15-bit offset)[b]
OR	MPYU	MVKH	MV
SADD	MPYUS	MVKLH	STB
SAT	SMPY	NEG	STH
SSUB	SMPYH	NOT	STW
SUB	SMPYHL	OR	STB (15-bit offset)[b]
SUBU	SMPYLH	SET	STH (15-bit offset)[b]
SUBC		SHL	STW (15-bit offset)[b]
XOR		SHR	SUB
ZERO		SHRU	SUBAB
		SSHL	SUBAH
		SUB	SUBAW
		SUBU	ZERO
		SUB2	
		XOR	
		ZERO	

[a]S2 only.   [b]D2 only.

Source: Courtesy of Texas Instruments [1,2].

TABLE A.2   Instructions for Floating-Point Operations

.L Unit	.M Unit	.S Unit	.D Unit
ADDDP	MPYDP	ABSDP	ADDAD
ADDSP	MPYI	ABSSP	LDDW
DPINT	MPYID	CMPEQDP	
DPSP	MPYSP	CMPEQSP	
DPTRUNC		CMPGTDP	
INTDP		CMPGTSP	
INTDPU		CMPLTDP	
INTSP		CMPLTSP	
INTSPU		RCPDP	
SPINT		RCPSP	
SPTRUNC		RSQRDP	
SUBDP		RSQRSP	
SUBSP		SPDP	

Source: Courtesy of Texas Instruments [1,2].

B

Registers for Circular Addressing and Interrupts

A number of special-purpose registers available on the C6x processor are shown in Figures B.1 to B.8 [1].

1. Figure B.1 shows the address mode register (AMR) that is used for the circular mode of addressing. It is used to select one of eight register pointers (A4 through A7, B4 through B7) and two blocks of memories (BK0, BK1) that can be used as circular buffers.
2. Figure B.2 shows the control status register (CSR) with bit 0 for the global interrupt enable (GIE) bit.
3. Figure B.3 shows the interrupt enable register (IER).
4. Figure B.4 shows the interrupt flag register (IFR).
5. Figure B.5 shows the interrupt set register (ISR).
6. Figure B.6 shows the interrupt clear register (ICR).
7. Figure B.7 shows the interrupt service table pointer (ISTP).
8. Figure B.8 shows the serial port control register (SPCR).

In Section 3.7.2 we discuss the AMR register and in Section 3.14 the interrupt registers.

REFERENCE

1. *C6000 CPU and Instruction Set*, SPRU189F, Texas Instruments, Dallas, TX, 2000.

Digital Signal Processing and Applications with the C6713 and C6416 DSK   By Rulph Chassaing
ISBN 0-471-69007-4   Copyright © 2005 by John Wiley & Sons, Inc.

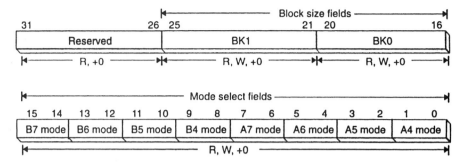

FIGURE B.1. Address mode register (AMR). (Courtesy of Texas Instruments)

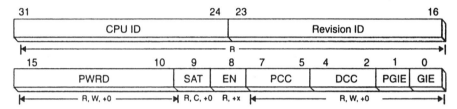

FIGURE B.2. Control status register (CSR). (Courtesy of Texas Instruments)

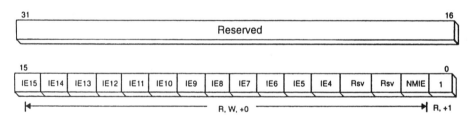

FIGURE B.3. Interrupt enable register (IER). (Courtesy of Texas Instruments)

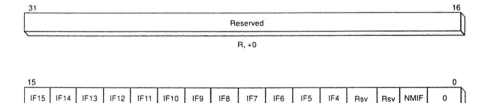

FIGURE B.4. Interrupt flag register (IFR). (Courtesy of Texas Instruments)

31
16
Reserved
15
0
IS15
IS14
IS13
IS12
IS11
IS10
IS9
IS8
IS7
IS6
IS5
IS4
Rsv
Rsv
Rsv
Rsv
W

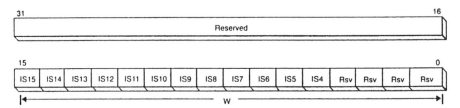

FIGURE B.5. Interrupt set register (ISR). (Courtesy of Texas Instruments)

31
16
Reserved
15
0
IC15
IC14
IC13
IC12
IC11
IC10
IC9
IC8
IC7
IC6
IC5
IC4
Rsv
Rsv
Rsv
Rsv
W

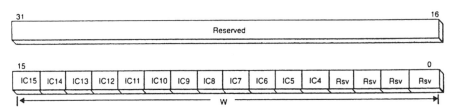

FIGURE B.6. Interrupt clear register (ICR). (Courtesy of Texas Instruments)

31
10 9
5 4
0
ISTB
HPEINT
0
0
0
0
0
R, W, +0
R, +0

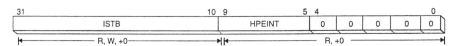

FIGURE B.7. Interrupt service table pointer (ISTP). (Courtesy of Texas Instruments)

31
24
23
22
21
20
19
18
17
16
reserved†
FRST
GRST
XINTM
XSYNCERR‡
XEMPTY
XRDY
XRST
R, +0
RW, +0
RW, +0
RW, +0
RW, +0
R, +0
R, +0
RW, +0
15
14
13
12
10
9
8
7
6
5
4
3
2
1
0
DLB
RJUST
CLKSTP
reserved
reserved
reserved
RINTM
RSYNCERR
RFULL
RRDY
RRST
RW,+0
RW, +0
RW,+0
R, +0
R, +0
R, +0
RW, +0
RW, +0
R, +0
R, +0
RW, +0§

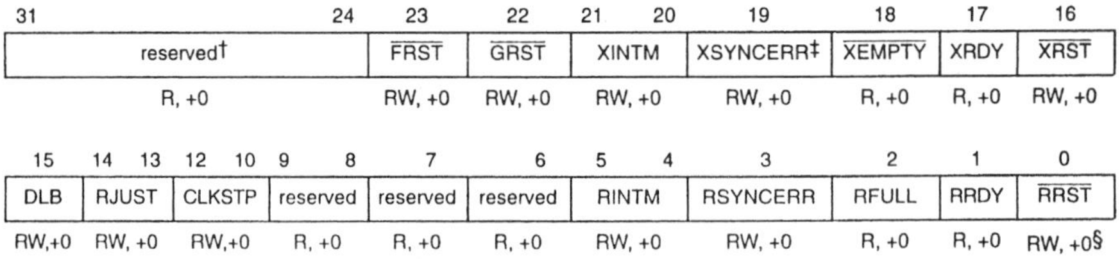

FIGURE B.8. Serial port control register (SPCR). (Courtesy of Texas Instruments)

C

Fixed-Point Considerations

The C6713 is a floating-point processor capable of performing both integer and floating-point operations. Both the C6713 and the A1C23 codec support 2's-complement arithmetic. It is thus appropriate here to review some fixed-point concepts [1].

In a fixed-point processor, numbers are represented in integer format. In a floating-point processor, both fixed- and floating-point arithmetic can be handled. With the floating-point processor C6713, a much greater range of numbers can be represented than with a fixed-point processor.

The dynamic range of an N-bit number based on 2's-complement representation is between $-(2^{N-1})$ and $(2^{N-1} - 1)$, or between $-32,768$ and $32,767$ for a 16-bit system. By normalizing the dynamic range between -1 and 1, the range will have 2^N sections, where $2^{-(N-1)}$ is the size of each section starting at -1 up to $1 - 2^{-(N-1)}$. For a 4-bit system, there would be 16 sections, each of size $\frac{1}{8}$ from -1 to $\frac{7}{8}$.

C.1 BINARY AND TWO'S-COMPLEMENT REPRESENTATION

To make illustrations more manageable, a 4-bit system is used rather than a 32-bit word length. A 4-bit word can represent the unsigned numbers 0 through 15, as shown in Table C.1.

The 4-bit unsigned numbers represent a modulo (mod) 16 system. If 1 is added to the largest number (15), the operation wraps around to give 0 as the answer. Finite bit systems have the same modulo properties as number wheels on combination locks. Therefore, a number wheel graphically demonstrates the addition

Digital Signal Processing and Applications with the C6713 and C6416 DSK   By Rulph Chassaing
ISBN 0-471-69007-4   Copyright © 2005 by John Wiley & Sons, Inc.

TABLE C.1   Unsigned Binary Number

Binary	Decimal
0000	0
0001	1
0010	2
0011	3
.	.
.	.
.	.
1110	14
1111	15

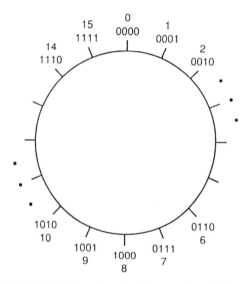

FIGURE C.1. Number wheel for unsigned integers.

properties of a finite bit system. Figure C.1 shows a number wheel with the numbers 0 through 15 wrapped around the outside. For any two numbers x and y in the range, the operation amounts to the following procedure:

1. Find the first number x on the wheel.

2. Step off y units in the clockwise direction, which brings you to the answer.

For example, consider the addition of the two numbers $(5 + 7)$ mod 16, which yields 12. From the number wheel, locate 5, then step 7 units in the clockwise direction to arrive at the answer, 12. As another example, $(12 + 10)$ mod 16 = 6. Starting with 12 on the number wheel, step 10 units clockwise, past zero, to 6.

Negative numbers require a different interpretation of the numbers on the wheel. If we draw a line through 8 cutting the number wheel in half, the right half will rep-

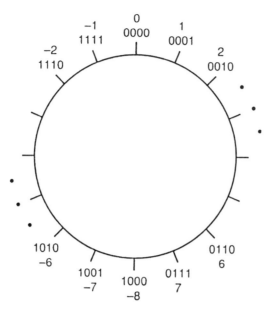

FIGURE C.2. Number wheel for signed integers.

resent the positive numbers and the left half the negative numbers, as shown in Figure C.2. This representation is the 2's-complement system. The negative numbers are the 2's complement of the positive numbers, and vice versa.

A 2's-complement binary integer,

$$B = b_{n-1} \cdots b_1 b_0$$

is equivalent to the decimal integer

$$I(B) = -b_{n-1} \times 2^{n-1} + \cdots + b_1 \times 2^1 + b_0 \times 2^0$$

where the b's are binary digits. The sign bit has a negative weight; all the others have positive weights. For example, consider the number -2,

$$1110 = -1 \times 2^3 + 1 \times 2^2 + 1 \times 2^1 + 0 \times 2^0 = -8 + 4 + 2 + 0 = -2$$

To apply the graphical technique to the operation $6 + (-2)$ mod16 = 4, locate 6 on the wheel, then step off (1110) units clockwise to arrive at the answer 4.

The binary addition of these same numbers,

$$
\begin{array}{r}
0110 \\
1110 \\
\hline
10100 \\
C
\end{array}
$$

shows a carry in the most significant bit, which in the case of finite register arithmetic, will be ignored. This carry corresponds to the wraparound through zero on the number wheel. The addition of these two numbers results in correct answers, by ignoring the carry in the most significant bit position, provided that the answer is in the range of representable numbers -2^{n-1} to $(2^{n-1} - 1)$ in the case of an n-bit number, or between -8 and 7 for the 4-bit number wheel example. When -7 is added to -8 in the 4-bit system, we get an answer of $+1$ instead of the correct value of -15, which is out of range. When two numbers of like sign are added to produce an answer with opposite sign, overflow has occurred. Subtraction with 2's-complement numbers is equivalent to adding the 2's complement of the number being subtracted to the other number.

C.2 FRACTIONAL FIXED-POINT REPRESENTATION

Rather than using the integer values just discussed, a fractional fixed-point number that has values between $+0.99\ldots$ and -1 can be used. To obtain the fractional n-bit number, the radix point must be moved $n - 1$ places to the left. This leaves one sign bit plus $n - 1$ fractional bits. The expression

$$F(B) = -b_0 \times 2^0 + b_1 \times 2^{-1} + b_2 \times 2^{-2} + \cdots + b_{n-1} \times 2^{-(n-1)}$$

converts a binary fraction to a decimal fraction. Again, the sign bit has a weight of negative 1 and the weights of the other bits are positive powers of $\frac{1}{2}$. The number wheel representation for the fractional 2's-complement 4-bit numbers is shown in Figure C.3. The fractional numbers are obtained from the 2's-complement integer numbers of Figure C.2 by scaling them by 2^3. Because the number of bits in a 4-bit system is small, the range is from -1 to 0.875. For a 16-bit word, the signed integers range from $-32,768$ to $+32,767$. To get the fractional range, scale those two signed integers by 2^{-15} or $32,768$, which results in a range from -1 to 0.999969 (usually taken as 1).

C.3 MULTIPLICATION

If one multiplies two n-bit numbers, the common notion is that a $2n$-bit operand will result. Although this is true for unsigned numbers, it is not so for signed numbers. As shown before, sign numbers need one sign bit with a weight of -2^{n-1}, followed by positive weights that are powers of 2. To find the number of bits needed for the result, multiply the two largest numbers together:

$$P = (-2^{n-1})(-2^{n-1}) = 2^{2n-2}$$

This number is a positive number representable in $(2n - 1)$ bits. The MSB of this result occupies the $(2n - 2)$ bit position counting from 0. Since this number is pos-

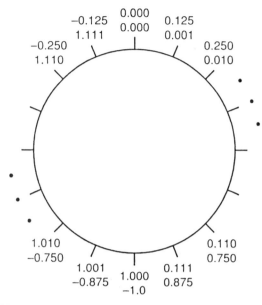

FIGURE C.3. Number wheel for fixed-point representation.

itive, its sign bit, which would show up as a negative number (a power of 2), does not appear. This is an exceptional case, which is treated as an overflow in fractional representation. Since the fractional representation requires that both operand and resultant occupy the same range, $-1 \geqslant$ range $< +1$, the operation $(-1) \times (-1)$ produces an unrepresentable number, $+1$.

Consider the next larger combination:

$$P = \left(-2^{n-1}\right)\left(-2^{n-1} + 1\right) = 2^{2n-2} - 2^{n-1}$$

Since the second number subtracts from the first, the product will occupy up to the $(2n - 3)$ bit position, counting from 0. Thus, it is representable in $(2n - 2)$ bits. With the exceptional case ruled out, this makes the bit position $(2n - 2)$ available for the sign bit of the resultant. Therefore, $(2n - 1)$ bits are needed to support an $(n \times n)$-bit signed multiplication.

To clarify the preceding equation, consider the 4-bit case, or

$$P = \left(-2^3\right)\left(-2^3 + 1\right) = 2^6 - 2^3$$

The number 2^6 occupies bit position 6. Since the second number is negative, the summation of the two is a number that will occupy only bit positions less than bit position 6, or

$$2^6 - 2^3 = 64 - 8 = 56 = 00111000$$

Thus bit position 6 is available for the sign bit. The 8-bit equivalent would have 2 sign bits (bits 6 and 7). The C6x supports signed and unsigned multiplies and therefore provides $2n$ bits for the product.

Consider the multiplication of two fractional 4-bit numbers, with each number consisting of 3 fractional bits and 1 sign bit. Let the product be represented by an 8-bit number. The first number is −0.5 and the second number is 0.75; the multiplication is as follows:

$$-0.50 = 1.100$$
$$\times 0.75 = 0.110$$

$$\underline{11}111000$$
$$\underline{11}1000$$

$$111.101000$$
$$C$$
$$= -2^1 + 2^0 + 2^{-1} + 2^{-3} = -0.375$$

The underlined bits of the multiplicand indicate sign extension. When a negative multiplicand is added to the partial product, it must be sign-extended to the left up to the limit of the product in order to give the proper larger bit version of the same number. To demonstrate that sign extension gives the correct expanded bit number, scan around the number wheel in Figure C.2 in the counterclockwise direction from 0. Write the codes for 5-bit, 6-bit, 7-bit, . . . negative numbers. Notice that they would be derived correctly by sign-extending the existing 4-bit codes; therefore, sign extension gives the correct expanded bit number. The carry-out will be ignored; however, the numbers 111.101000 (9-bit word), 11.101000 (8-bit word), and 1.101000 (7-bit word) all represent the same number: −0.375. Thus, the product of the preceding example could be represented by $(2n − 1)$ bits, or 7 bits for a 4-bit system.

When two 16-bit numbers are multiplied to produce a 32-bit result, only 31 bits are needed for the multiply operation. As a result, bit 30 is sign-extended to bit 31. The extended bits are frequently called *sign bits*.

Consider the following example: to multiply $(0101)_2$ by $(1110)_2$, which is equivalent to multiplying 5 by −2 in decimal, which would result in −10. This result is outside the dynamic range $\{-8,7\}$ of a 4-bit system. Using a Q-3 format, this corresponds to multiplying 0.625 by −0.25, yielding a result of −0.15625, which is within the fractional range.

When two Q-15 format numbers (each with a sign bit) are multiplied, the result is a Q-30 format number with one extra sign bit. The MSB is the extra sign bit. One can shift right by 15 to retain the MSBs and only one of the 2 sign bits. By shifting right by 15 (dividing by 2^{15}) to be able to store the result into a 16-bit system, this discards the 15 LSBs, thereby losing some precision. One is able to retain high precision by keeping the most significant 15 bits. With a 32-bit system, a left shift by 1 bit would suffice to get rid of the extra sign bit.

Note that when two Q-15 numbers, represented with a range of −1 to 1, are multiplied, the resulting number remains within the same range. However, the addition of two Q-15 numbers can produce a number outside this range, causing overflow. Scaling would then be required to correct this overflow.

REFERENCE

1. R. Chassaing and D. W. Horning, *Digital Signal Processing with the TMS320C25*, Wiley, New York, 1990.

D

MATLAB Support Tools

Several support tools using MATLAB [1,2] are described in this appendix:

1. The filter designer **SPTool** and the filter design and analysis tool **FDATool** for FIR and IIR filter design using a graphical user interface (GUI).
2. FIR and IIR filter design using functions available with the Student Version of MATLAB
3. Bilinear transformation
4. FFT and IFFT

D.1 SPTool AND FDATool FOR FIR FILTER DESIGN

MATLAB provides GUIs with both the filter designer **SPTool** and the filter design and analysis **FDATool** to design FIR filters (and IIR in the next section).

Example D.1: SPTool and FDATool for FIR Filter Design

SPTool
1. From MATLAB, type the following:

```
>>SPTool
```

Digital Signal Processing and Applications with the C6713 and C6416 DSK   By Rulph Chassaing
ISBN 0-471-69007-4   Copyright © 2005 by John Wiley & Sons, Inc.

to access MATLAB's GUI filter designer SPTool for the design of both FIR and IIR filters.

2. From the startup window `startup.spt`, select a new design and use the characteristics shown in Figure D.1*a* to design an FIR bandstop filter centered at 2700 Hz. The filter contains $N = 89$ coefficients (MATLAB shows order as $N - 1$) and uses the Kaiser window function. The real-time implementation of this filter is tested in Example 4.1.

3. Access the `startup` window again. Select → Edit → Name. Change the name (enter new variable name) to *bs2700*.

4. Select File → Export → Export to Workspace the *bs2700* design.

5. Access MATLAB's workspace and type the following two commands

```
>>bs2700.tf.num;
>>round(bs2700.tf.num*2^15)
```

to find the numerator coefficients of the transfer function. These coefficients are in a float format. The second command scales these coefficients by 2^{15} so that they can be used for a fixed-point implementation. The scaled coefficients of the FIR bandstop filter should be listed within the workspace as

```
-14  23  -9...23  -14
```

FDATool

MATLAB's filter design and analysis tool can be invoked in a similar fashion, typing

```
>>FDATool
```

Figure D.1*b* shows the corresponding graphical window for the design of the same FIR bandstop filter. Select File → Export → to Workspace. Export as coefficients with a corresponding variable name bs2700. Within the workspace, type bs2700 to list the coefficients in a float format. Scale these coefficients by typing

```
>>bs2700 = int16(bs2700*2^15)
```

to obtain these coefficients in a 16-bit integer format. These coefficients can then be saved in the `.MAT` file `FIRcoeffs.mat` using

```
>>save FIRcoeffs.mat bs2700
```

These coefficients are contained in the file *bs2700.cof*, shown in Figure D.2 and used in Example 4.1.

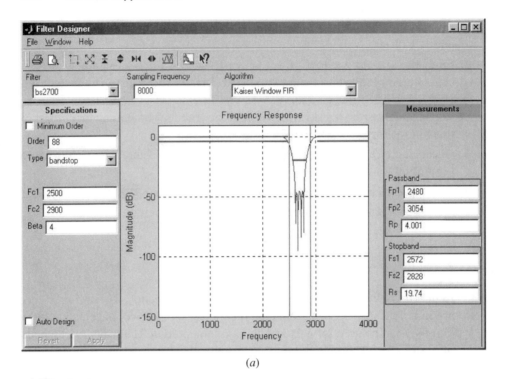

(*a*)

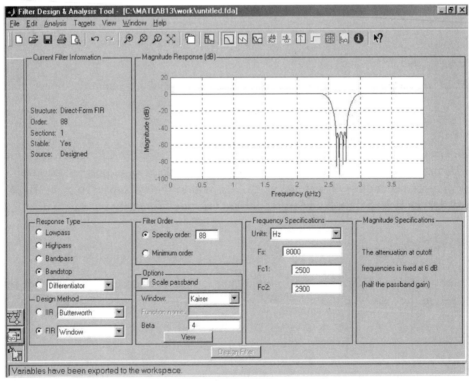

(*b*)

FIGURE D.1. Characteristics of a FIR bandstop filter centered at 2700 Hz using (*a*) SPTool; (*b*) FDATool.

```
//BS2700.cof FIR bandstop coefficients designed with MATLAB

#define N 89                              //number of coefficients

short h[N]={-14,23,-9,-6,0,8,16,-58,50,44,-147,119,67,-245,200,72,
-312,257,53,-299,239,20,-165,88,0,105,-236,33,490,-740,158,932,-1380,
392,1348,-2070,724,1650,-2690,1104,1776,-3122,1458,1704,29491,1704,
1458,-3122,1776,1104,-2690,1650,724,-2070,1348,392,-1380,932,158,-740,
490,33,-236,105,0,88,-165,20,239,-299,53,257,-312,72,200,-245,67,119,
-147,44,50,-58,16,8,0,-6,-9,23,-14};
```

FIGURE D.2. Coefficient file for an FIR bandstop filter centered at 2700 Hz designed using MATLAB's filter designer SPTool (bs2700.cof).

Real-Time SPTool (RTSPTool)

Real-time SPTool (RTSPTool) provides a direct interface for the DSK [3–5] for filter design and implementation (within the MATLAB environment) on the DSK in real time. RTSPTool's window is similar to SPTool's filter designer window, with additional toolbars to run the filter in real time on the DSK. Upon pressing an appropriate toolbar, the filter is designed and the coefficients are scaled and saved in an appropriate file that is included in a generic FIR program. MATLAB's file filtdes.m was modified to provide that interface to the DSK. A MATLAB(.m) function accesses CCS code generation tools to compile/assemble, link, and load/run the resulting executable file on the DSK (load/run using dsk6xldr filename.out).

D.2 SPTool AND FDATool FOR IIR FILTER DESIGN

Section D.1 illustrates the design of FIR filters using MATLAB's **SPTool** and **FDATool**. Some of the same procedures are used for the design of IIR filters as well.

Example D.2: SPTool and FDATool for IIR Filter Design

SPTool

Figure D.3a shows MATLAB's filter designer SPTool displaying the characteristics of a 10th-order IIR bandstop filter centered at 1750 Hz. MATLAB shows the order as 5, which represents the number of second-order sections. Save it as bs1750 (see Example D.1). Export the coefficients to the workspace, as with the previous FIR design. From MATLAB's workspace, type the following commands:

```
>>[z,p,k] = tf2zp(bs1750.tf.num, bs1750.tf.den);
>>sec_ord_sec = zp2sos(z,p,k);
>>sec_ord_sec = round(sec_ord_sec*2^15)
```

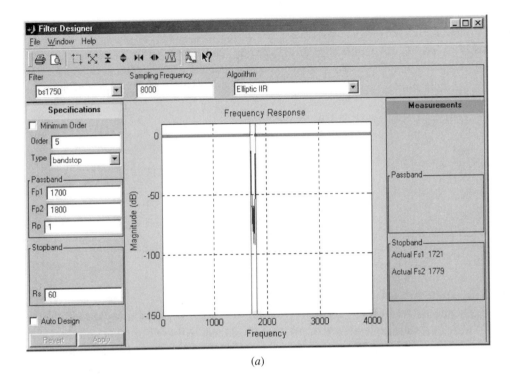

(a)

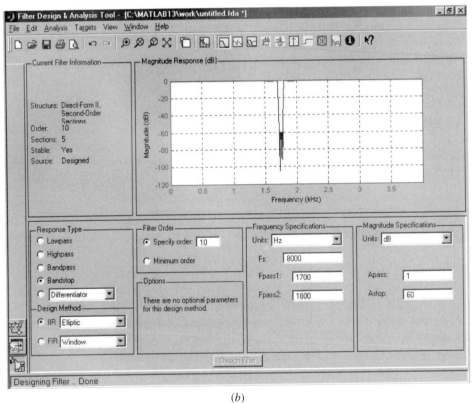

(b)

FIGURE D.3. Characteristics of a 10th-order IIR bandstop filter centered at 1750 Hz using (a) SPTool; (b) FDATool.

The first command finds the roots of the numerator and the denominator (zeros and poles). The second command converts the resulting floating-point coefficients into a format for implementation as second-order sections. The third command scales these coefficients for a fixed-point implementation. The resulting numerator and denominator coefficients should be listed as

```
27940  -10910  27940  32768  -11417  25710
.

.

.
32768  -14239  32768  32768  -15258  32584
```

These 30 coefficients represent the numerator coefficients a_0, a_1, and a_2 and the denominator coefficients b_0, b_1, and b_2. They represent six coefficients per stage, with b_0 normalized to 1 and scaled by $2^{15} = 32,768$.

FDATool

Figure D.3*b* shows the corresponding **FDATool** window for the design of the same IIR bandstop filter. Note that a 10th order is specified with FDATool. Export the coefficients to the workspace as bs1750. Within the workspace, type bs1750 to obtain the coefficients in a float format. These coefficients can be scaled into a 16- or 32-bit integer format.

The coefficients using SPTool are contained in the file *bs1750.cof*, listed in Figure D.4 and used in Example 5.1. Figure D.4 shows 25 coefficients (in lieu of 30). Since the coefficient b_0 is always normalized to 1, it is not used in the program. Note

```
//bs1750.cof  IIR bandstop coefficient file, centered at 1,750 Hz

#define stages 5                    //number of 2nd-order stages

int a[stages][3]=       {       //numerator coefficients
{27940,  -10910,  27940},        //a10, a11, a12 for 1st stage
{32768,  -11841,  32768},        //a20, a21, a22 for 2nd stage
{32768,  -13744,  32768},        //a30, a31, a32 for 3rd stage
{32768,  -11338,  32768},        //a40, a41, a42 for 4th stage
{32768,  -14239,  32768}  };

int b[stages][2]=       {       //*denominator coefficients
{-11417,  25710},                //b11, b12 for 1st stage
{-9204,  31581},                 //b21, b22 for 2nd stage
{-15860,  31605},                //b31, b32 for 3rd stage
{-10221,  32581},                //b41, b42 for 4th stage
{-15258,  32584}          };     //b51, b52 for 5th stage
```

FIGURE D.4. Coefficient file for an IIR bandstop filter centered at 1750 Hz designed using MATLAB's filter designer SPTool (BS1750.cof).

that the a's and b's used by MATLAB correspond to the b's and a's, respectively, used in this book, with the b's (in this book) representing the denominator coefficients of the transfer function.

As with the FIR design, this IIR bandstop filter can be implemented in real time with a push of a button within RTSPTool [3,4].

D.3 MATLAB FOR FIR FILTER DESIGN USING THE STUDENT VERSION

FIR filters can be designed using the Student Version [2] of the MATLAB software package [1]. See also Section D.1 for the design of FIR filters using MATLAB's GUI filter designer SPTOOL.

Example D.3: FIR Filter Design Using MATLAB's Student Version

Figure D.5 shows a listing of a MATLAB program mat33.m to design a 33-coefficient FIR bandpass filter. The function remez uses the Parks–McClellan algorithm based on the Remez exchange algorithm and Chebyshev's approximation theory. The desired filter has a center frequency of 1 kHz with a sampling frequency of 10 kHz. The frequency v represents the normalized frequency variable, defined as $v = f/F_N$, where F_N is the Nyquist frequency. The bandpass filter is represented by three bands:

1. The first band (stopband) has normalized frequencies between 0 and 0.1 (0 to 500 Hz), with a corresponding magnitude of 0.
2. The second band (passband) has normalized frequencies between 0.15 and 0.25 (750 to 1250 Hz), with a corresponding magnitude of 1.
3. The third band (stopband) has normalized frequencies between 0.3 and the Nyquist frequency of 1 (1500 to 5000 Hz), with a corresponding magnitude of 0.

Run this program from MATLAB and verify the magnitude response of the ideal desired filter plotted within MATLAB in Figure D.6. Note that the frequencies 750

```
%Mat33.m  MATLAB program for FIR Bandpass with 33 coefficients Fs=10 kHz

nu= [0 0.1 0.15 0.25 0.3 1];     %normalized frequencies
mag= [0 0 1 1 0 0];              %magnitude at normalized frequencies
c=remez (32,nu,mag);            %invoke remez algorithm for 33 coeff
bp33=c';                         % coeff values transposed
save matpb33.cof bp33 -ascii;    %save in ASCII file with coefficients
[h,w] =freqz (c,1,256);          %frequency response with 256 points
plot(5000*nu,mag,w/pi,abs(h))    %plot ideal magnitude response
```

FIGURE D.5. MATLAB program for FIR filter design (mat33.m).

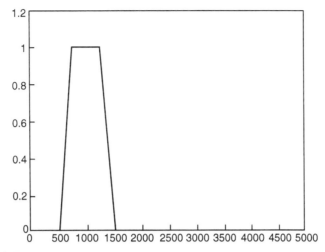

FIGURE D.6. Frequency response of the FIR bandpass filter design obtained with MATLAB.

and 1250 Hz represent passband frequencies with normalized frequencies of 0.15 and 0.25, respectively, and associated magnitudes of 1. The frequencies 500 and 1500 Hz represent stopband frequencies with normalized frequencies of 0.1 and 0.3, respectively, and associated magnitudes of 0. The last normalized frequency value of 1 corresponds to the Nyquist frequency of 5000 Hz and has a magnitude of zero. The program generates a set of 33 coefficients saved in the file `matbp33.cof` in ASCII format.

Example D.4: Multiband FIR Filter Design Using MATLAB

This example extends the preceding three-band example to a five-band design in order to obtain two passbands. The program `mat63.m` (Figure D.7) is similar to the preceding MATLAB program, `mat33.m`. This filter with two passbands is represented by a total of five bands: the first band (stopband) has normalized frequencies between 0 and 0.1 (0 to 500 Hz), with corresponding magnitude of 0; the second band (passband) has normalized frequencies between 0.12 and 0.18 (600 to 900 Hz), with a corresponding magnitude of 1, and so on. This is summarized as follows:

Band	Frequency (Hz)	Normalized f/F_N	Magnitude
1	0–500	0–0.1	0
2	600–900	0.12–0.18	1
3	1000–1500	0.2–0.3	0
4	1600–1900	0.32–0.38	1
5	2000–5000	0.4–1	0

```
%Mat63.m  MATLAB program for two passbands, 63 coefficients Fs=10 kHz

nu= [0 0.1 0.12 0.18 0.2 0.3 0.32 0.38 0.4 1]; %normalized frequencies
mag= [0 0 1 1 0 0 1 1 0 0];     %magnitude at normalized frequencies
c=remez (62,nu,mg);             %invoke remez algorithm for 63 coeff
bp63=c';                        % coeff values transposed
save mat2bp.cof bp63 -ascii;    %save in ASCII file with coefficients
[h,w] =freqz (c,1,256);         %frequency response with 256 points
plot (500*nu,mag,w/pi,abs(h))   %plot ideal magnitude response
```

FIGURE D.7. MATLAB program for a two-passband FIR filter design (mat63.m).

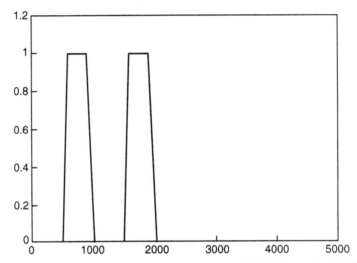

FIGURE D.8. Frequency response of a two-passband FIR filter using MATLAB.

Run this program from MATLAB and verify the magnitude response of the ideal two-passband filter in Figure D.8. This program generates a set of 63 coefficients saved in the coefficient file mat2bp.cof in ASCII format.

D.4 MATLAB FOR IIR FILTER DESIGN USING THE STUDENT VERSION

MATLAB can also be used for the design of IIR filters using the Student Edition of MATLAB. See also Section D.2 for the design of IIR filters using MATLAB's GUI filter designer SPTOOL.

Example D.5: IIR Filter Design Using MATLAB's Student Version

The function *yulewalk*, available in MATLAB, allows for the design of recursive filters based on a best least squares fit [1,2]. Consider again the MATLAB program `mat33.m` in Figure D.5 to obtain a 33-coefficient FIR bandpass filter centered at 1000 Hz. In lieu of the *remez* function for an FIR design, the MATLAB command

```
>>[a,b] = yulewalk (n,nu,mag)
```

returns the *a* and *b* coefficients in the general input–output equation in Chapter 5, associated with an IIR filter. The filter's order n represents the number of second-order sections. The C program in Example 5.1 implements an IIR filter with cascaded second-order sections, as is most commonly done. For example, if n = 6 in the *yulewalk* function, the general transfer function in Chapter 5 in terms of the resulting *a* and *b* coefficients from MATLAB needs to be reduced to one in terms of three cascaded sections.

D.5 BLT USING MATLAB AND SUPPORT PROGRAMS ON CD

This section expands on the discussion of BLT in Section 5.3.

Exercise D.1: First-Order IIR Lowpass Filter

Given a first-order lowpass analog transfer function $H(s)$, a corresponding discrete-time filter with transfer function $H(z)$ can be obtained. Let the bandwidth or cutoff frequency $B = 1$ *r/s* and the sampling frequency $F_s = 10$ Hz.

1. Choose an appropriate transfer function

$$H(s) = \frac{1}{s+1}$$

which represents a lowpass filter with a bandwidth of 1 *r/s*.

2. Prewarp ω_D using

$$\omega_A = \tan\frac{\omega_D T}{2} = \tan\left(\frac{1}{20}\right) \cong \frac{1}{20}$$

where $\omega_D = B = 1$ *r/s* and $T = \frac{1}{10}$.

3. Scale $H(s)$ to obtain

$$H(s/\omega_A) = \frac{1}{20s+1}$$

4. Obtain the desired transfer function $H(z)$, or

$$H(z) = H(s/\omega_A)|_{s=(z-1)/(z+1)} = \frac{z+1}{21z - 19}$$

Exercise D.2: First-Order IIR Highpass Filter

Given a highpass transfer function $H(s) = s/(s+1)$, obtain a corresponding transfer function $H(z)$. Let the bandwidth or cutoff frequency be 1 r/s and the sampling frequency be 5 Hz. From the preceding procedure, $H(z)$ is found to be

$$H(z) = \frac{10(z-1)}{11z - 9}$$

Exercise D.3: Second-Order IIR Bandstop Filter

Given a second-order analog transfer function $H(s)$ for a bandstop filter, a corresponding discrete-time transfer function $H(z)$ can be obtained. Let the lower and upper cutoff frequencies be 950 and 1050 Hz, respectively, with a sampling frequency F_s of 5 kHz.

The transfer function selected for a bandstop filter is

$$H(s) = \frac{s^2 + \omega_r^2}{s^2 + sB + \omega_r^2}$$

where B and ω are the bandwidth and center frequencies, respectively. The analog frequencies are

$$\omega_{A1} = \tan\frac{\omega_{D1}T}{2} = \tan\frac{2\pi \times 950}{2 \times 5000} = 0.6796$$

$$\omega_{A2} = \tan\frac{\omega_{D2}T}{2} = \tan\frac{2\pi \times 1050}{2 \times 5000} = 0.7756$$

The bandwidth $B = \omega_{A2} - \omega_{A1} = 0.096$ and $\omega_r^2 = (\omega_{A1})(\omega_{A2}) = 0.5271$. The transfer function $H(s)$ becomes

$$H(s) = \frac{s^2 + 0.5271}{s^2 + 0.096s + 0.5271} \tag{D.1}$$

and the corresponding transfer function $H(z)$ can be obtained with $s = (z-1)/(z+1)$, or

$$H(z) = \frac{\{(z-1)/(z+1)\}^2 + 0.5271}{[(z-1)/(z+1)]^2 + 0.096(z-1)/(z+1) + 0.5271}$$

which can be reduced to

$$H(z) = \frac{0.9408 - 0.5827 z^{-1} + 0.9408 z^{-2}}{1 - 0.5827 z^{-1} + 0.8817 z^{-2}} \tag{D.2}$$

As shown later, $H(z)$ can be verified using the program BLT.BAS (on the accompanying CD), or MATLAB, which calculates $H(z)$ from $H(s)$ using the BLT technique, as we will illustrate. This can be quite useful in applying this procedure for higher-order filters.

Exercise D.4: Fourth-Order IIR Bandpass Filter

A fourth-order IIR bandpass filter can be obtained using the BLT procedure. Let the upper and lower cutoff frequencies be 1 and 1.5 kHz, respectively, and the sampling frequency be 10 kHz.

1. The transfer function $H(s)$ of a fourth-order Butterworth bandpass filter can be obtained from the transfer function of a second-order Butterworth lowpass filter, or

$$H(s) = H_{LP}(s)\Big|_{s=(s^2+\omega_r^2)/sB}$$

where $H_{LP}(s)$ is the transfer function of a second-order Butterworth lowpass filter. $H(s)$ then becomes

$$H(s) = \frac{1}{s^2 + \sqrt{2}s + 1}\Bigg|_{s=(s^2+\omega_r^2)/SB}$$

$$= \frac{s^2 B^2}{s^4 + \sqrt{2}Bs^3 + (2\omega_r^2 + B^2)s^2 + \sqrt{2}B\omega_r^2 s + \omega_r^4} \tag{D.3}$$

2. The analog frequencies ω_{A1} and ω_{A2} are

$$\omega_{A1} = \tan\frac{\omega_{D1}T}{2} = \tan\frac{2\pi \times 1050}{2 \times 10,000} = 0.3249$$

$$\omega_{A2} = \tan\frac{\omega_{D2}T}{2} = \tan\frac{2\pi \times 1500}{2 \times 10,000} = 0.5095$$

3. The center frequency ω_r and the bandwidth B can now be found:

$$\omega_r^2 = (\omega_{A1})(\omega_{A2}) = 0.1655$$
$$B = \omega_{A2} - \omega_{A1} = 0.1846$$

4. The analog transfer function $H(s)$ is (D.3) reduces to

$$H(s) = \frac{0.03407 s^2}{s^4 + 0.26106 s^3 + 0.36517 s^2 + 0.04322 s + 0.0274} \tag{D.4}$$

5. The corresponding $H(z)$ becomes

$$H(z) = \frac{0.02008 - 0.04016 z^{-2} + 0.02008 z^{-4}}{1 - 2.5495 z^{-1} + 3.2021 z^{-2} - 2.0359 z^{-3} + 0.64137 z^{-4}} \tag{D.5}$$

which is in the form of (5.4). This can be verified using the program BLT.BAS (on the CD).

Exercise D.5: H(z) from H(s) Using Bilinear Function in MATLAB

Using Exercise D.3 with the second-order IIR bandstop filter, the transfer function in the analog s-plane [from (D.1)],

$$H(s) = \frac{s^2 + 0.5271}{s^2 + 0.096 s + 0.5271}$$

can be converted to an equivalent transfer function in the digital z-plane using the bilinear function from MATLAB with the following commands:

```
>>num = [1, 0, 0.5271];          %numerator coefficients
>>den = [1, 0.096, 0.5271];      %denominator coefficients
>>T = 2; Fs = 1/T;               %K=1 from bilinear equation
>>[a,b]=bilinear (num, den, Fs)  %invoke bilinear function
```

to obtain the coefficients a and b associated with the transfer function in (5.4), or

$$H(z) = \frac{0.9409 - 0.5827 z^{-1} + 0.9409 z^{-2}}{1 - 0.5827 z^{-1} + 0.8817 z^{-2}}$$

which is the same transfer function (D.2) as that found in Exercise D.3. Note that $T = 2$ was chosen with MATLAB since the constant $K = 2/T$ in the bilinear equation in Chapter 5 was set to 1 for convenience. Note that MATLAB uses the following notation in the general input–output equation:

$$y(n) = b_0 x(n) + b_1 x(n-1) + b_2 x(n-2) + \cdots - a_1 y(n-1) - a_2 y(n-2) - \cdots$$

which yields a transfer function of the form

```
Enter the # of numerator coefficients (30 = Max, 0 = Exit) --> 3
     Enter a(0)s^2  --> 1
     Enter a(1)s^1  --> 0
     Enter a(2)s^0  --> 0.5271

Enter the # of denominator coefficients --> 3
     Enter b(0)s^2  --> 1
     Enter b(1)s^1  --> 0.096
     Enter b(2)s^0  --> 0.5271

Are the above coefficients correct ? (y/n) y
```
(a)

```
     a(0)z^-0 =   0.94085        b(0)z^-0 =   1.00000
     a(1)z^-1 =  -0.58271        b(1)z^-1 =  -0.58271
     a(2)z^-2 =   0.94085        b(2)z^-2 =   0.88171
```
(b)

FIGURE D.9. Use of BLT.BAS program for bilinear transformations: (a) coefficients in the s-plane; (b) coefficients in the z-plane.

$$H(z) = \frac{b_0 + b_1 z^{-1} + b_2 z^{-2} + \cdots}{1 + a_1 z^{-1} + a_2 z^{-2} + \cdots}$$

which shows that MATLAB's a and b coefficients are the reverse of the notation used in (5.1).

Exercise D.6: Utility Program BLT.BAS to Find H(z) from H(s)

The utility program BLT.BAS (on the CD), written in BASIC, converts an analog transfer function $H(s)$ into an equivalent transfer function $H(z)$ using the bilinear equation $s = (z - 1)/(z + 1)$. To verify the results in (D.1) found in Exercise D.3 for the second-order bandstop filter, run GWBASIC, then load and run BLT.BAS. The prompts and the associated data for the a and b coefficients associated with $H(s)$ are shown in Figure D.9a, and the a and b coefficients associated with the transfer function $H(z)$ are shown in Figure D.9b, which verifies (D.1). Run BLT.BAS again to verify (D.5) using the data in (D.4).

Exercise D.7: Utility Program AMPLIT.CPP to Find Magnitude and Phase

The utility program AMPLIT.CPP (on the CD), written in C++, can be used to plot the magnitude and phase responses of a filter for a given transfer function $H(z)$ with a maximum order of 10. Compile (using Borland's C++ compiler) and run this program. Enter the coefficients of the transfer function associated with the second-order IIR bandstop filter (D.2) in Exercise D.3, as shown in Figure D.10a. Figures D.10b and D.10c show the magnitude and phase of the second-order bandstop filter.

FILTER COEFFICIENTS	
NUMERATOR	DENOMINATOR
z-0 .9408 z-1 −.5827 z-3 .9408 z-4 z-5 z-6 z-7 z-8 z-9 z-10	z-0 1 z-3 −.5827 z-4 .8817 z-5 z-6 z-7 z-8 z-9 z-10
F1 HELP F5 QUIT F10 PLOT	

(*a*)

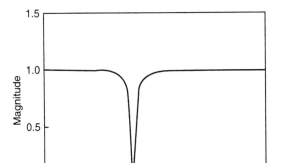

F1 for PRINTOUT   ENTER to continue

(*b*)

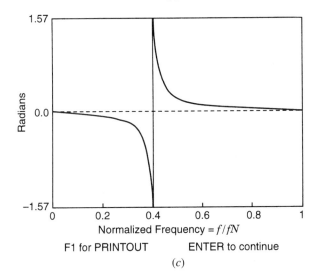

F1 for PRINTOUT   ENTER to continue

(*c*)

FIGURE D.10. Use of the `AMPLIT.CPP` program for plotting magnitude and phase: (*a*) coefficients in the *z*-plane; (*b*) normalized magnitude; (*c*) normalized phase.

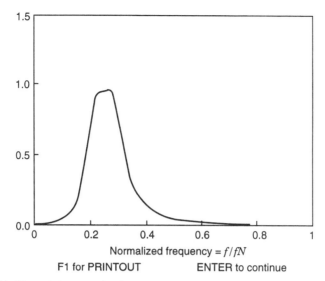

FIGURE D.11. Plot of the magnitude response of a fourth-order IIR bandpass filter using AMPLIT.CCP.

From the plot of the magnitude response of $H(z)$, the normalized center frequency is shown at $v = f/F_N = 1000/2500 = 0.4$.

Run this program again to plot the magnitude response associated with the fourth-order IIR bandpass filter in Exercise D.4. Verify the plot shown in Figure D.11. The normalized center frequency is shown at $v = 1250/5000 = 0.25$.

A utility program MAGPHSE.BAS (on the CD), written in BASIC, can be used to tabulate the magnitude and phase responses.

D.6 FFT AND IFFT

MATLAB can be used to find both the fast Fourier transform FFT of a sequence of numbers and the inverse Fourier transform IFFT.

Exercise D.8: Eight-Point FFT and IFFT Using MATLAB

The eight-point FFT in Exercise 6.1 can readily be verified with MATLAB, with the following commands:

```
>>x = [1 1 1 1 0 0 0 0];
>>y = fft(x)
>>magy = abs(y)
>>plot (magy)
```

The resulting output magnitude transform is also plotted.

Similarly, the inverse FFT can also be verified. Given the output sequence X's in Exercise 6.1, the inverse FFT or IFFT can be found:

```
>>X = [4 1-2.414*i 0 1-0.414+i 0 1+0.414*i 0 1+2.414*i];
>>y = ifft(X)
```

where y is the resulting rectangular sequence.

REFERENCES

1. *MATLAB, The Language of Technical Computing,* Math Works, Natick, MA, 2003.

2. *MATLAB Student Version,* MathWorks, Natick, MA, 2000.

3. W J. Gomes III and R. Chassaing, Filter design and implementation using the TMS320C6x interfaced with MATLAB, *Proceedings of the 1999 ASEE Annual Conference,* 1999.

4. W J. Gomes III and R. Chassaing, Real-time FIR and IIR filter design using MATLAB interfaced with the TMS320C31 DSK, *Proceedings of the 1999 ASEE Annual Conference,* 1999.

5. R. Chassaing, *Digital Signal Processing Laboratory Experiments Using C and the TMS320C31 DSK,* Wiley, New York, 1999.

E

Additional Support Tools

The following additional support tools are available (see also Appendix D for MATLAB support):

1. Goldwave utility for signal generation, virtual instrument, and so on
2. FIR and IIR filter design using digifilter from MultiDSP
3. Homemade filter development package
4. Visual Application Builder (VAB) and LabVIEW
5. Codec support from integrated DSP

E.1 GOLDWAVE SHAREWARE UTILITY AS A VIRTUAL INSTRUMENT

Goldwave is a shareware utility software program that can turn a PC with a sound card into a virtual instrument. It can be downloaded from the Web [1]. One can create a function generator to generate different signals such as a sine wave and random noise. It can also be used as an oscilloscope, as a spectrum analyzer, and to record/edit a speech signal. Effects such as echo and filtering can be obtained. Lowpass, highpass, bandpass, and bandstop filters can be implemented on a sound card with Goldwave and their effects on a signal illustrated readily.

Goldwave was used to obtain an input voice (`TheForce.wav`, on the CD) added with two sinusoidal signals of frequencies 900 and 2700 Hz, respectively. This corrupted voice signal, shown in Figure 4.24, is used in Example 4.7 to illustrate removal of the two sinusoidal signals.

Digital Signal Processing and Applications with the C6713 and C6416 DSK   By Rulph Chassaing
ISBN 0-471-69007-4   Copyright © 2005 by John Wiley & Sons, Inc.

One can use two copies of Goldwave running under Windows: one to generate a signal as input to the DSK, another to use the DSK's output into the sound card as a spectrum analyzer.

Other shareware utility programs, such as Cool Edit [2] or Spectrogram [3], also can be used as virtual spectrum analyzers.

E.2 FILTER DESIGN USING DIGIFILTER

DigiFilter is a filter design package for the design of both FIR and IIR filters [4]. Currently, it interfaces to the C31 DSK for real-time implementation. It can still be used for the design of FIR and IIR filters.

E.2.1 FIR Filter Design

Figure E.1 shows a plot of the log magnitude response of a 61-coefficient FIR bandpass filter centered at 2 kHz using the Kaiser window function. For a specific design,

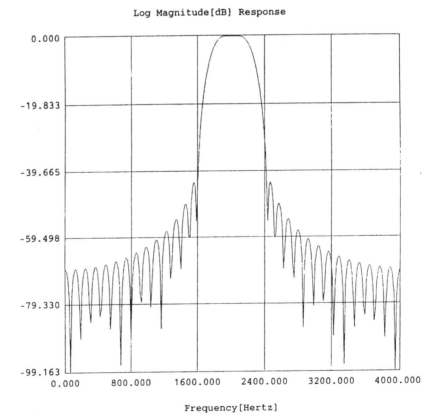

FIGURE E.1. Magnitude response of an FIR bandpass filter using DigiFilter.

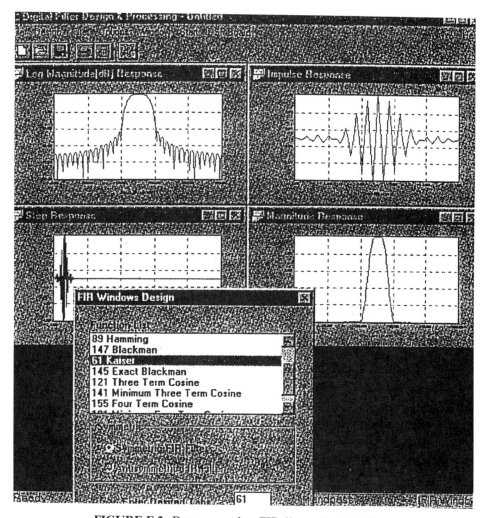

FIGURE E.2. Responses of an FIR filter using DigiFilter.

the user can select among several window functions, with the specification of the number of taps (coefficients) associated with each window (rectangular, Hamming, etc.). Impulse as well as step responses can also be obtained, as shown in Figure E.2. Note that an implementation with a Hamming window function would require 89 coefficients, whereas a Kaiser window would require 61 coefficients (Figure E.2).

E.2.2 IIR Filter Design

An IIR filter can readily be designed with the filter package DigiFilter. One can choose among several designs using the following functions: Butterworth, Chebyshev, elliptic, and Bessel, each associated with a specific filter order. A plot of the magnitude response similar to an FIR design, as well as a plot of the poles and zeros of $H(z)$, can be obtained.

E.3 FIR FILTER DESIGN USING A FILTER DEVELOPMENT PACKAGE

A noncommercial filter development package appears on the accompanying CD. The program *FIRprog.bas*, written in BASIC, calculates the coefficients of an FIR filter. This program is discussed in Refs. 5 to 7. It allows for the design of lowpass, highpass, bandpass, and bandstop FIR filters using the rectangular, Hanning, Hamming, Blackman, and Kaiser window functions. The resulting coefficients can be generated in integer or float format. This file with the coefficients needs to be modified and incorporated into one of the generic FIR programs.

E.3.1 Kaiser Window

1. Run BASIC (GWBASIC) and load/run the program `FIRprog.bas`. Figures E.3*a* and E.3*b* show a display of available window functions and the frequency-selective filters that can be designed. Select the Kaiser window option and a bandpass filter. A separate module for the Kaiser window (`FIRproga.bas`) is called from *FIRprog.bas*.

```
                        Main Menu
                        _____


                        1. . . .RECTANGULAR
                        2. . . .HANNING
                        3. . . .HAMMING
                        4. . . .BLACKMAN
                        5. . . .KAISER
                        6. . . .Exit to DOS

        Enter window desired (number only) -> 5
                           (a)

             Selections:
                        1. . . .LOWPASS
                        2. . . .HIGHPASS
                        3. . . .BANDPASS
                        4. . . .BANDSTOP
                        5. . . .Exit back to Main Menu

        Enter desired filter type (number only) -> 3
                           (b)
```

FIGURE E.3. FIR filter design with a filter development package (on CD): (*a*) choice of windows; (*b*) type of filter; (*c*) filter specifications; (*d*) menu for coefficients format.

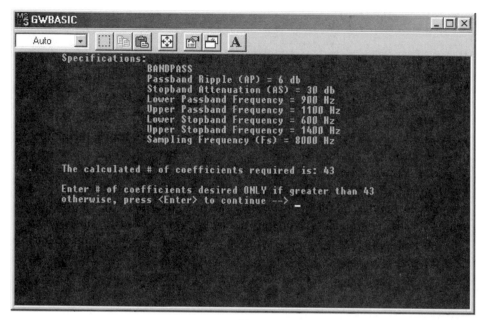

(c)

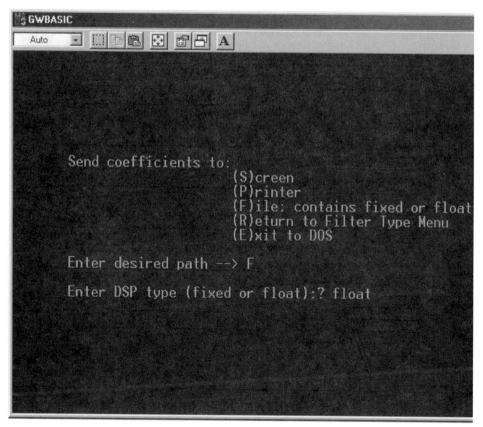

(d)

FIGURE E.3. (*Continued*)

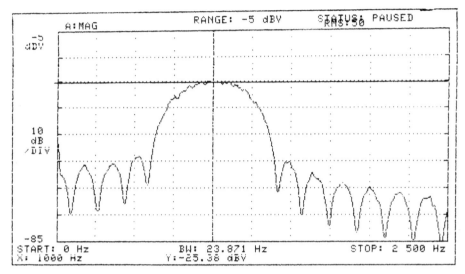

FIGURE E.4. Frequency response of FIR bandpass filter using coefficient file BP43K.cof generated with filter package on CD.

2. Enter the specifications shown in Figure E.3c. Choose the float option (Figure E.3d) to save the 43 resulting coefficients into a file in a float format (the fixed option saves the coefficients in hexadecimal). Save it as BP43K.cof.

3. Edit it (an edited version is on the CD). Include it in the program FIRPRN.c in Example 4.4. Build/run and verify the frequency response of the FIR bandpass filter centered at 1000 Hz shown in Figure E.4, obtained with an HP analyzer. An internally generated noise sequence becomes the input to the FIR filter in the program FIRPRN.c. This filter was designed so that the center frequency is at 1000 Hz with a sampling frequency of 8 kHz.

E.3.2 Hamming Window

Repeat this procedure for a Hamming window function. Enter 900 and 1100 for the lower and upper cutoff frequencies. Enter 5.2 (ms) for the duration D of the impulse response, since the number of coefficients N is

$$N = (D \times F_s) + 1$$

This will yield a design with 53 coefficients. Save the resulting coefficient file as BP53H.cof. Edit it as with the Kaiser window, test it using the program FIRPRN.C, and verify an FIR bandpass filter with a narrower mainlobe.

E.4 VISUAL APPLICATION BUILDER AND LABVIEW

The Visual Application Builder (VAB), available from National Instruments or Hyperception [8,9], is a component-based virtual design tool that can be used to implement DSP algorithms. VAB uses a methodology of developing DSP algorithms and systems graphically simply by connecting functional components together with a mouse. The user only needs to choose the desired functions, place them onto a worksheet, select their parameters interactively, and describe the data flow using line connections. The method of design is quite similar to drawing a block diagram of the system being designed. DSP-based design implementations can be created and executed on DSP hardware without having to write any source code at all.

VAB contains a wide range of functional block components for FFT, filtering, and so on, and supports the C6713 DSK. One can design and test a DSP system that includes functional blocks such as signal generators, A/D and D/A, filters, FFT, image processing components, and so on. Results can be displayed on the PC monitor as the algorithm is executing or to an external device such as an oscilloscope.

See also LabVIEW introduced in Section 9.5 [9].

E.5 ALTERNATIVE INPUT/OUTPUT

A Daughter card, based on the AD77 stereo codec that interfaces to the C6x DSK, is available from Integrated-DSP [10]. It plugs on the DSK and can provide an alternative input and output.

REFERENCES

1. Goldwave. Available at www.goldwave.com.

2. Cool Edit. Available at www.syntrillium.com.

3. Gram412.zip from Spectrogram, address from shareware utility with the database address www.simtel.net.

4. DigiFilter, from MultiDSP. Available at multidsp@aol.com.

5. R. Chassaing. *Digital Signal Processing Laboratory Experiments Using C and the TMS32OC31 DSK*, Wiley, New York, 1999.

6. R. Chassaing, *Digital Signal Processing with C and the TMS320C30*, Wiley, New York, 1992.

7. R. Chassaing and D. W. Horning, *Digital Signal Processing with the TMS320C25*, Wiley, New York, 1990.

8. Hyperception. Available at info@hyperception.com.

9. National Instruments, www.ni.com.

10. Integrated DSP. Available at www.integrated-dsp.com.

F

Fast Hartley Transform

Whereas complex additions and multiplications are required for an FFT, the Hartley transform [1–8] requires only real multiplications and additions. The FFT maps a real function of time into a complex function of frequency, whereas the fast Hartley transform (FHT) maps the same real-time function into a real function of frequency. The FHT can be particularly useful in cases where the phase is not a concern.

The discrete Hartley transform (DHT) of a time sequence $x(n)$ is defined as

$$H(k) = \sum_{n=0}^{N-1} x(n)\text{cas}\left(\frac{2\pi nk}{N}\right), \quad k = 0, 1, \ldots, N-1 \tag{F.1}$$

where

$$\text{cas } u = \cos u + \sin u \tag{F.2}$$

In a similar development to the FFT, (F.1) can be decomposed as

$$H(k) = \sum_{n=0}^{(N/2)-1} x(n)\text{cas}\left(\frac{2\pi nk}{N}\right) + \sum_{n=N/2}^{N-1} x(n)\text{cas}\left(\frac{2\pi nk}{N}\right) \tag{F.3}$$

Let $n = n + N/2$ in the second summation of (F.3),

Digital Signal Processing and Applications with the C6713 and C6416 DSK   By Rulph Chassaing
ISBN 0-471-69007-4   Copyright © 2005 by John Wiley & Sons, Inc.

$$H(k) = \sum_{n=0}^{(N/2)-1} \left\{ x(n)\text{cas}\left(\frac{2\pi nk}{N}\right) + x\left(n+\frac{N}{2}\right)\text{cas}\left(\frac{2\pi k[n+N/2]}{N}\right) \right\} \qquad \text{(F.4)}$$

Using (F.2) and the identities

$$\sin(A+B) = \sin A \cos B + \cos A \sin B$$
$$\cos(A+B) = \cos A \cos B - \sin A \sin B, \qquad \text{(F.5)}$$

For odd k,

$$\begin{aligned}
\text{cas}\left(\frac{2\pi k[n+N/2]}{N}\right) &= \cos\left(\frac{2\pi nk}{N}\right)\cos(\pi k) - \sin\left(\frac{2\pi nk}{N}\right)\sin(\pi k) \\
&\quad + \sin\left(\frac{2\pi nk}{N}\right)\cos(\pi k) + \cos\left(\frac{2\pi nk}{N}\right)\sin(\pi k) \\
&= -\cos\left(\frac{2\pi nk}{N}\right) - \sin\left(\frac{2\pi nk}{N}\right) \\
&= -\text{cas}\left(\frac{2\pi nk}{N}\right) \qquad \text{(F.6)}
\end{aligned}$$

and, for even k,

$$\text{cas}\left(\frac{2\pi k[n+N/2]}{N}\right) = \cos\left(\frac{2\pi nk}{N}\right) + \sin\left(\frac{2\pi nk}{N}\right) = \text{cas}\left(\frac{2\pi nk}{N}\right) \qquad \text{(F.7)}$$

Using (F.6) and (F.7), (F.4) becomes

$$H(k) = \sum_{n=0}^{(N/2)-1} \left[x(n) + x\left(n+\frac{N}{2}\right) \right] \text{cas}\left(\frac{2\pi nk}{N}\right), \quad \text{for even } k \qquad \text{(F.8)}$$

and

$$H(k) = \sum_{n=0}^{(N/2)-1} \left[x(n) - x\left(n+\frac{N}{2}\right) \right] \text{cas}\left(\frac{2\pi nk}{N}\right), \quad \text{for odd } k \qquad \text{(F.9)}$$

Let $k = 2k$ for even k, and let $k = 2k+1$ for odd k. Equations (F.8) and (F.9) become

$$H(2k) = \sum_{n=0}^{(N/2)-1} \left[x(n) + x\left(n+\frac{N}{2}\right) \right] \text{cas}\left(\frac{2\pi n 2k}{N}\right) \qquad \text{(F.10)}$$

$$H(2k+1) = \sum_{n=0}^{(N/2-1)} \left[x(n) - x\left(n+\frac{N}{2}\right) \right] \text{cas}\left(\frac{2\pi n[2k+1]}{N}\right) \qquad \text{(F.11)}$$

Furthermore, using (F.5)

$$\mathrm{cas}\left(\frac{2\pi n[2k+1]}{N}\right) = \cos\left(\frac{2\pi n}{N}\right)\left\{\cos\left(\frac{2\pi n2k}{N}\right) + \sin\left(\frac{2\pi n2k}{N}\right)\right\}$$
$$+ \sin\left(\frac{2\pi n}{N}\right)\left\{\cos\left(\frac{2\pi n2k}{N}\right) - \sin\left(\frac{2\pi n2k}{N}\right)\right\}$$

and

$$\sin\left(\frac{2\pi kn}{N}\right) = -\sin\left(\frac{2\pi k[N-n]}{N}\right)$$
$$\cos\left(\frac{2\pi kn}{N}\right) = \cos\left(\frac{2\pi k[N-n]}{N}\right)$$

Equation (F.11) becomes

$$H(2k+1) = \sum_{n=0}^{(N/2)-1}\left\{\left[x(n) - x\left(n + \frac{N}{2}\right)\right]\cos\left(\frac{2\pi n}{N}\right)\mathrm{cas}\left(\frac{2\pi n2k}{N}\right)\right.$$
$$\left. + \sin\left(\frac{2\pi n}{N}\right)\mathrm{cas}\left(\frac{2\pi 2k[N-n]}{N}\right)\right\}$$

(F.12)

Substituting $N/2 - n$ for n in the second summation, (F.12) becomes

$$H(2k+1) = \sum_{n=0}^{(N/2)-1}\left\{\left[x(n) - x\left(n + \frac{N}{2}\right)\right]\cos\left(\frac{2\pi n}{N}\right)\right.$$
$$\left. + \left[x\left(\frac{N}{2} - n\right) - x(N - n)\right]\sin\left(\frac{2\pi n}{N}\right)\right\}\mathrm{cas}\left(\frac{2\pi n2k}{N}\right)$$

(F.13)

Let

$$a(n) = x(n) + x\left(n + \frac{N}{2}\right)$$
$$b(n) = \left[x(n) - x\left(n + \frac{N}{2}\right)\right]\cos\left(\frac{2\pi n}{N}\right)$$
$$+ \left[x\left(\frac{N}{2} - n\right) - x(N - n)\right]\sin\left(\frac{2\pi n}{N}\right)$$

Equations (F.10) and (F.13) become

$$H(2k) = \sum_{n=0}^{(N/2)-1} a(n)\mathrm{cas}\left(\frac{2\pi n2k}{N}\right)$$

(F.14)

$$H(2k+1) = \sum_{n=0}^{(N/2)-1} b(n)\text{cas}\left(\frac{2\pi n 2K}{N}\right) \qquad (\text{F.15})$$

A more complete development of the FHT can be found in [3]. We now illustrate the FHT with two exercises: an 8-point FHT and a 16-point FHT. We will then readily verify these results from the FFT exercises in Chapter 6.

Exercise F.1: Eight-Point Fast Hartley Transform

Let the rectangular sequence $x(n)$ be represented by $x(0) = x(1) = x(2) = x(3) = 1$, and $x(4) = x(5) = x(6) = x(7) = 0$. The flow graph in Figure F.1 is used to find $X(k)$. We will now use $X(k)$ instead of $H(k)$. The sequence is first permuted and the intermediate results after the first two stages are as shown in Figure F.1. The coefficients Cn and Sn are (with $N = 8$)

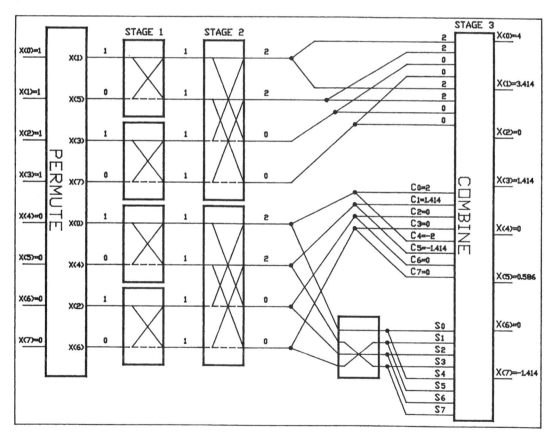

FIGURE F.1. Eight-point FHT flow graph.

$$Cn = \cos(2\pi n/N)$$
$$Sn = \sin(2\pi n/N)$$

The output sequence $X(k)$ after the final stage 3 is also shown in Figure F.1. For example,

$$X(0) = 2 + 2C0 + 2S0 = 2 + 2(1) + 2(0) = 4$$
$$X(1) = 2 + 2C1 + 2S1 = 2 + 1.414 + 0 = 3.41$$
$$\vdots$$
$$X(7) = 0 + 0(C7) + 2S7 = -1.414 \qquad \text{(F.16)}$$

This resulting output sequence can be verified from the $X(k)$ obtained with the FFT, using

$$\text{DHT}\{x(n)\} = \text{Re}\{\text{DFT}[x(n)]\} - \text{Im}\{\text{DFT}[x(n)]\} \qquad \text{(F.17)}$$

For example, from the eight-point FFT in Exercise 6.1, $X(1) = 1 - j2.41$, and

$$\text{Re}\{X(1)\} = 1$$
$$\text{Im}\{X(1)\} = -2.41$$

Using (F.17),

$$\text{DHT}\{x(1)\} = X(1) = 1 - (-2.41) = 3.41$$

as in (F.16). Conversely, the FFT can be obtained from the FHT using

$$\text{Re}\{\text{DFT}[x(n)]\} = \tfrac{1}{2}\{\text{DHT}[x(N-n)] + \text{DHT}[x(n)]\}$$
$$\text{Im}\{\text{DFT}[x(n)]\} = \tfrac{1}{2}\{\text{DHT}[x(N-n)] - \text{DHT}[x(n)]\} \qquad \text{(F.18)}$$

For example, using (F.18) to obtain $X(1) = 1 - j2.41$ from the FHT,

$$\text{Re}\{X(1)\} = \tfrac{1}{2}\{X(7) + X(1)\} = \tfrac{1}{2}\{-1.41 + 3.41\} = 1$$
$$\text{Im}\{X(1)\} = \tfrac{1}{2}[X(7) - X(1)] = \tfrac{1}{2}\{-1.41 - 3.41\} = -2.41 \qquad \text{(F.19)}$$

where the left-hand side of (F.18) is associated with the FFT and the right-hand side with the FHT.

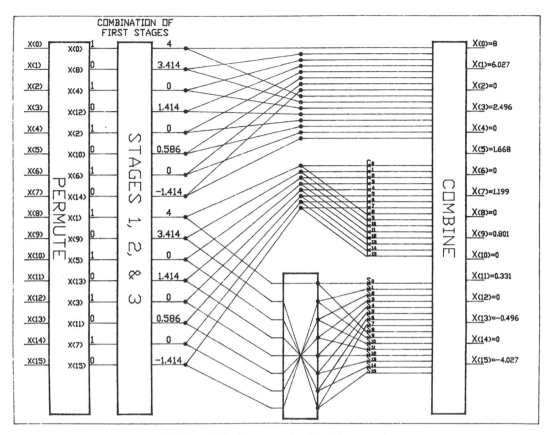

FIGURE F.2. Sixteen-point FHT flow graph.

Exercise F.2: 16-Point Fast Hartley Transform

Let the rectangular sequence $x(n)$ be represented by $x(0) = x(1) = \ldots = x(7) = 1$, and $x(8) = x(9) = \ldots = x(15) = 0$. A 16-point FHT flow graph can be arrived at, building on the 8-point FHT. The permutation of the input sequence before the first stage is as follows for the first (upper) eight-point FHT: $x(0)$, $x(8)$, $x(4)$, $x(12)$, $x(2)$, $x(10)$, $x(6)$, $x(14)$ and for the second (lower) eight-point FHT: $x(1)$, $x(9)$, $x(5)$, $x(13)$, $x(3)$, $x(11)$, $x(7)$, $x(15)$. After the third stage, the intermediate output results for the upper and the lower eight-point FHTs are as obtained in the previous eight-point FHT example. Figure F.2 shows the flow graph of the fourth stage for the 16-point FHT. The intermediate output results from the third stage become the input to the fourth stage in Figure F.2. The output sequence $X(0)$, $X(1)$, $\ldots$, $X(15)$ from Figure F.2 can be verified using the results obtained with the 16-point FFT in Exercise 6.2. For example, using

$$Cn = \cos\frac{2\pi n}{N} = \cos\frac{\pi n}{8}$$

$$Sn = \sin\frac{2\pi n}{N} = \sin\frac{\pi n}{8}$$

with $N = 16$, $X(1)$ can be obtained from Figure F.2:

$$X(1) = 3.414 + 3.414C1 - 1.414S1 = 3.414 + 3.154 - 0.541 = 6.027$$

as in Figure F.2. Equation (F.18) can be used to verify $X(1) = 1 - j5.028$, as obtained using the FFT in Example 6.2. Note that, for example,

$$
\begin{aligned}
X(15) &= -1.414 + (-1.414C15) + (3.414S15) \\
&= -1.414 - 1.306 - 1.306 \\
&= -4.0269
\end{aligned}
$$

as shown in Figure 6.15.

REFERENCES

1. R. N. Bracewell, The fast Hartley transform, *Proceedings of the IEEE*, Vol. 72, Aug. 1984, pp. 1010–1018.

2. R. N. Bracewell, Assessing the Hartley transform, *IEEE Transactions on Acoustics, Speech, and Signal Processing*, Vol. ASSP-38, 1990, pp. 2174–2176.

3. R. N. Bracewell, *The Hartley Transform*, Oxford University Press, New York, 1986.

4. R. N. Bracewell, *The Fourier Transform and its Applications*, McGraw Hill, New York, 2000.

5. H. V. Sorensen, D. L. Jones, M. T. Heidman, and C. S. Burrus, Real-valued fast Fourier transform algorithms, *IEEE Transactions on Acoustics, Speech, and Signal Processing*, Vol. ASSP-35, 1987, pp. 849–863.

6. H. S. Hou, The fast Hartley transform algorithm, *IEEE Transactions on Computers*, Vol. C-36, Feb. 1987, pp. 147–156.

7. H. S. Hou, Correction to "The fast Hartley transform algorithm," *IEEE Transactions on Computers*, Vol. C-36, Sept. 1987, pp. 1135–1136.

8. A. Zakhor and A. V. Oppenheim, Quantization errors in the computation of the discrete Hartley transform, *IEEE Transactions on Acoustics, Speech, and Signal Processing*, Vol. ASSP-35, Oct. 1987, pp. 1592–1601.

G

Goertzel Algorithm

Goertzel's algorithm performs a DFT using an IIR filter calculation. Compared to a direct N-point DFT calculation, this algorithm uses half the number of real multiplications, the same number of real additions, and requires approximately $1/N$ the number of trigonometric evaluations. The biggest advantage of the Goertzel algorithm over the direct DFT is the reduction of the trigonometric evaluations. Both the direct method and the Goertzel method are more efficient than the FFT when a "small" number of spectrum points is required rather than the entire spectrum. However, for the entire spectrum, the Goertzel algorithm is an N^2 effort, just as is the direct DFT.

G.1 DESIGN CONSIDERATIONS

Both the first-order and the second-order Goertzel algorithms are explained in several books [1–3] and in Ref. [4]. A discussion of them follows. Since

$$W_N^{-kN} = e^{j2\pi k} = 1$$

both sides of the DFT in (6.1) can be multiplied by it, giving

$$X(k) = W_N^{-kN} \sum_{r=0}^{N-1} x(r) W_N^{+kr} \tag{G.1}$$

Digital Signal Processing and Applications with the C6713 and C6416 DSK   By Rulph Chassaing
ISBN 0-471-69007-4   Copyright © 2005 by John Wiley & Sons, Inc.

which can be written as

$$X(k) = \sum_{r=0}^{N-1} x(r) W_N^{-k(N-r)} \tag{G.2}$$

Define a discrete-time function as

$$y_k(n) = \sum_{r=0}^{N-1} x(r) W_N^{-k(n-r)} \tag{G.3}$$

The discrete transform is then

$$X(k) = y_k(n)\big|_{n=N} \tag{G.4}$$

Equation (G.3) is a discrete convolution of a finite-duration input sequence $x(n)$, $0 < n < N - 1$, with the infinite sequence W_N^{-kn}. The infinite impulse response is therefore

$$h(n) = W_N^{-kn} \tag{G.5}$$

The Z-transform of $h(n)$ in (G.5) is

$$H(z) = \sum_{n=0}^{\infty} h(n) z^{-n} \tag{G.6}$$

Substituting (G.5) into (G.6) gives

$$H(z) = \sum_{n=0}^{\infty} W_N^{-kn} z^{-n} = 1 + W_N^{-k} z^{-1} + W_N^{-2k} z^{-2} + \cdots = \frac{1}{1 - W_N^{-2k} z^{-1}} \tag{G.7}$$

Thus equation (G.7) represents the transfer function of the convolution sum in equation (G.3). Its flow graph represents the first-order Goertzel algorithm and is shown in Figure G.1. The DFT of the kth frequency component is calculated by

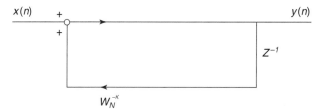

FIGURE G.1. First-order Goertzel algorithm.

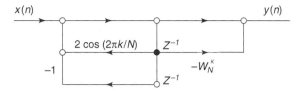

FIGURE G.2. Second-order Goertzel algorithm.

starting with the initial condition $y_k(-1) = 0$ and running through N iterations to obtain the solution $X(k) = y_k(N)$. The $x(n)$'s are processed in time order, and processing can start as soon as the first one comes in. This structure needs the same number of real multiplications and additions as the direct DFT but $1/N$ the number of trigonometric evaluations.

The second-order Goertzel algorithm can be obtained by multiplying the numerator and denominator of (G.7) by $1 - W_N^{-k}z^{-1}$ to give

$$H(z) = \frac{1 - W_N^{+k}z^{-1}}{1 - 2\cos(2\pi k/N)z^{-1} + z^{-2}} \tag{G.8}$$

The flow graph for this equation is shown in Figure G.2. Notice that the left half of the graph contains feedback flows and the right half contains only feedforward terms. Therefore, only the left half of the flow graph must be evaluated each iteration. The feedforward terms need only be calculated once for $y_k(N)$. For real data, there is only one real multiplication in this graph and only one trigonometric evaluation for each frequency. Scaling is a problem for fixed-point arithmetic realizations of this filter structure; therefore, simulation is extremely useful.

The second-order Goertzel algorithm is more efficient than the first-order Goertzel algorithm. The first-order Goertzel algorithm (assuming a real input function) requires approximately $4N$ real multiplications, $3N$ real additions, and two trigonometric evaluations per frequency component as opposed to N real multiplications, $2N$ real additions, and two trigonometric evaluations per frequency component for the second-order Goertzel algorithm. The direct DFT requires approximately $2N$ real multiplications, $2N$ real additions, and $2N$ trigonometric evaluations per frequency component.

This Goertzel algorithm is useful in situations where only a few points in the spectrum are necessary, as opposed to the entire spectrum. Detection of several discrete frequency components is a good example. Since the algorithm processes samples in time order, it allows the calculation to begin when the first sample arrives. In contrast, the FFT must have the entire frame in order to start the calculation.

Section 10.1 describes a DTMF project. It is implemented using the Goertzel's algorithm on the C6416 DSK (see Appendix H) and can be readily transported to the C6713 DSK.

REFERENCES

1. G. Goertzel, An Algorithm for the Evaluation of Finite Trigonometric Series, *American Mathematics Monthly*, 65, January 1958.

2. A. V. Oppenheim and R. Schafer, *Discrete-Time Signal Processing*, Prentice Hall, Upper Saddle River, NJ, 1989.

3. C. S. Burus and T. W. Parks, *DFT/FFT and Convolution Algorithms: Theory and Implementation*, Wiley, New York, 1988.

4. `http://ptolemy.eecs.berkeley.edu/papers/96/dtmf_ict/www/node3.html`

H

TMS320C6416 DSK

H.1 TMS320C64x PROCESSOR

Another member of the C6000 family of processors is the C64x, which can operate at a much higher clock rate. The C6416 DSK operates at 720 MHz for a 1.39 ns instruction cycle time. With eight instructions per cycle, this translates into 5760 million instructions per second (MIPS). Features of the C6416 architecture include: four 16×16-bit multiplier (each .M unit can perform two multiplies per cycle), sixty-four 32-bit general-purpose registers, more than 1 MB of internal memory consisting of 1 MB of L2 RAM/cache, 16 kB of each L1P program cache and L1D data cache [1–7].

The C64x is based on the architecture VELOCITI.2, which is an extension of VELOCITI [2]. The extra registers allow for packed data types to support four 8-bit or two 16-bit operations associated with one 32-bit register, increasing parallelism [3]. For example, the instruction MPYU4 performs four 8-bit multiplications within a single instruction cycle time. Several special-purpose instructions have also been added to handle many operations encountered in wireless and digital imaging applications, where 8-bit data processing is common. In addition, the .M unit (for multiply operations) can also handle shift and rotate operations. Similarly, the .D unit (for data manipulation) can also handle logical operations.

The C64x is a fixed-point processor. Existing instructions are available to more units. Double-word load (LDDW) and store (STDW) instructions can access 64 bits of data, with up to a two double-word load or store instructions per cycle (read or write 128 bits per cycle).

Digital Signal Processing and Applications with the C6713 and C6416 DSK By Rulph Chassaing
ISBN 0-471-69007-4 Copyright © 2005 by John Wiley & Sons, Inc.

A few instructions have been added for the C64x processor. For example, the instruction

```
BDEC LOOP,B0
```

decrements a counter B0 and performs a conditional branch to LOOP based on B0. The branch decision is *before* the decrement; with the branch decision based on a negative number (not on whether the number is zero). This multitask instruction resembles the syntax used in the C3x and C4x family of processors.

Furthermore, with the intrinsic C function _dotp2, it can perform two 16×16 multiplies and adds the products together to further reduce the number of cycles. This intrinsic function in C has the corresponding assembly function DOTP2. With two multiplier units, four 16×16 multiplies per cycle can be performed, double the rate of the C62x or C67x. At 720 MHz, this corresponds to 2.88 billion multiply operations per second, or 5.76 billion 8×8 multiplies per second.

H.2 PROGRAMMING EXAMPLES USING THE C6416 DSK

A 720-MHz 6416-based DSK is currently available. Most of the programs associated with the C6713 DSK can be transported readily to the C6416DSK. Note that the examples in Chapters 1–5 can be transported almost "as is." The BSL and CSL utilities between the two types of DSKs are very similar. An equivalent init/comm file C6416dskinit.c and the corresponding header file are included in the folder **DSK6416**. This is obtained primarily by replacing the occurrences 6713 with 6416 in c6713dskinit.c (and .h) files. Otherwise, these two "black box" init/comm files are very similar.

The C6416 DSK package includes CCS with a tutorial on the C6416 DSK. Subfolders associated with several examples illustrate the similarities between the C6713 and C6416 DSKs. The appropriate support files for these examples are in the folder **DSK6416**, including the init file C6416dskinit.c (with the corresponding header file c6416dskinit.h), the vector file, and the linker command file.

Example H.1: Sine Generation with DIP Switch Control Using the C6416 DSK (*sine8_LED*)

Figure H.1 shows the C source program sine8_LED.c that generates a sine wave when the user accessible dip switch sw0 is pressed. Verify similar results shown in Example 1.1. From the Build Options (linker tab), the corresponding library support files are included: rts6400.lib, dsk6416bsl.lib, csl6416.lib. The support files for this project are in the folder DSK6416/sine8_LED.

Example 1.2 illustrates the generation of a sine wave and plotting in both time-and-frequency domains using CCS. Verify similar results with the C6416 DSK. The appropriate files are in the folder **DSK6416/sine8_buf**.

```
//Sine8_LED.c   Sine generation with DIP switch control

#include "dsk6416_aic23.h"              //support file for codec,DSK
Uint32 fs = DSK6416_AIC23_FREQ_8KHZ;   //set sampling rate
short loop = 0;                        //table index
short gain = 10;                       //gain factor
short sine_table[8]={0,707,1000,707,0,-707,-1000,-707};//sine values

void main()
{
 comm_poll();                          //init DSK,codec,McBSP
 DSK6416_LED_init();                   //init LED from BSL
 DSK6416_DIP_init();                   //init DIP from BSL
 while(1)                              //infinite loop
 {
  if(DSK6416_DIP_get(0)==0)            //=0 if DIP switch #0 pressed
  {
   DSK6416_LED_on(0);                  //turn LED #0 ON
   output_sample(sine_table[loop]*gain);//output sine values
   if (++loop > 7)  loop = 0;          //check for end of table
  }
  else DSK6416_LED_off(0);             //turn LED off if not pressed
 }                                     //end of while(1) infinite loop
}                                      //end of main
```

FIGURE H.1. Sine generation using the C6416 DSK (sine8_LED.c).

Example H.2: Loop Program Using the C6416 DSK (loop_intr)

Figure H.2 shows the C source program loop_intr.c that implements a loop program. Compare this program with that shown in Figure 2.4 in Example 2.1. As with the C6713, input and output default to the left channel of the AIC23 codec.

Build this project as **loop_intr** and verify that the results are similar to those in Example 2.1. The appropriate files are in the folder **DSK6416/loop_intr**.

Example H.3: FIR/IIR Implementation Using the C6416 DSK (FIR/IIR)

Examples 4.1 and 5.1 implement an FIR and IIR filter, respectively, using the C6713 DSK. Using the C6416 DSK, verify similar results as in those examples.

Using the program FIR.c, verify that the coefficient file lp1500_256.cof represents a 256-tap FIR lowpass filter with a bandwidth of 1500 Hz, sampling at 48 kHz. Verify also that lp1500_768.cof represents a 768-tap FIR lowpass filter with the same bandwidth, but sampling at 8 kHz.

Figure H.3 shows the output of the C6416 DSK as an IIR bandpass filter, centered at 2000 Hz (as in Example 5.1). This plot was obtained with an HP analyzer using noise as input. Support files for both FIR and IIR are within the folder **DSK6416**.

```
//Loop_intr.c Loop program with the DSK6416. Output = delayed input

#include "dsk6416_aic23.h"              //codec-DSK support file
Uint32 fs=DSK6416_AIC23_FREQ_8KHZ;     //set sampling rate

interrupt void c_int11()               //interrupt service routine
{
    short sample_data;

    sample_data = input_sample();      //input data
    output_sample(sample_data);        //output data
    return;
}

void main()
{
    comm_intr();                       //init DSK, codec, McBSP
    while(1);                          //infinite loop
}
```

FIGURE H.2. Loop program using the C6416 DSK (loop_intr).

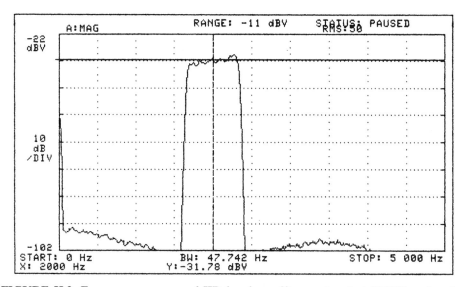

FIGURE H.3. Frequency response of IIR bandpass filter centered at 2000 Hz using the C6416 DSK.

Example H.4: FFT with C-Coded FFT Function Using the C6416 DSK (FFT256c)

Example 6.2 illustrates the implementation of a 256-point FFT. The interrupt-driven C source file for this project is FFT256c.c, that calls a C-coded FFT function. Figure H.4 shows the output of the C6416 DSK which represents the FFT of a 2 kHz input sinusoidal signal (as in Example 6.2). The support files are within the folder **DSK6416**.

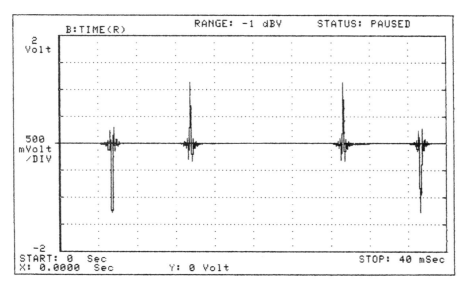

FIGURE H.4. Output of 256-point FFT using the C6416 DSK.

The fixed-point C6416-based DSK, operating at 720 MHz, executes floating-point operations much *slower* than the floating-point C6713-based DSK, operating at 225 MHz. This can be verified using the polling-version of the 256-point FFT (FFT256c_poll.c) with the C6416 DSK, since the distance between the two negative spikes (used as reference) is approximately 35 ms (not 32 ms as with the C6713-based DSK).

Example H.5: Adaptive FIR filter Implementation Using the C6416 DSK (adaptnoise)

Example 7.2 illustrates an adaptive FIR filter for the cancellation of a sinusoidal noise. Verify similar results using the C6416 DSK. The appropriate files are in the folder **DSK6416/adaptnoise**.

Example H.6: DTMF Implementation on the C6416 DSK Using the Goertzel Algorithm and the FFT, with RTDX Using Visual C++ (DTMF_goertzel, DTMF_FFT)

Section 10.1 describes a DTMF project using different methods. The necessary support files, including the Visual C++ files for RTDX (see also Section 9.3), are within the folder **DSK6416**.

Using Goertzel Algorithm
The Goertzel's algorithm, described in Appendix G, is used to implement the DTMF project on the C6416 DSK. It can be tested as in Section 10.1. The files are in the folder **DSK6416/DTMF_goertzel**.

Using Radix-4 FFT

The DTMF was also implemented on the C6416 DSK using the radix-4 FFT. The appropriate files are in the folder **DSK6416/DTMF_FFT**.

REFERENCES

1. *TMS320C6416, TMS320C6415, TMS320C6416 Fixed-Pont Digital Signal Processors*, SPRS146, Texas Instruments, Dallas, TX, 2003.

2. *TMS320C6000 Programmer's Guide*, SPRU198G, Texas Instruments, Dallas, TX, 2002.

3. *TMS320C6000 CPU and Instruction Set*, SPRU189F, Texas Instruments, Dallas, TX, 2000.

4. *TMS320C64x Technical Overview*, SPRU395, Texas Instruments, Dallas, TX, 2003.

5. *How to Begin Development Today with the TMS320C6416, TMS320C6415, and TMS320C6416 DSPs Application Report*, SPRA718, Texas Instruments, Dallas, TX, 2003.

6. *TMS320C6000 Chip Support Library API User's Guide*, SPRU401, Texas Instruments, Dallas, TX, 2003.

7. *TMS320C6000 DSK Board Support Library API User's Guide*, SPRU432, Texas Instruments, Dallas, TX, 2001.

I

TMS320C6711 DSK

Dozens of examples are included in Ref 1. Most of the examples in Chapters 1–8 are also included in Ref. 1. The following interrupt-driven example ilustrates the differences in the implementations between the C6713 and C6711 DSKs. Note that illustrating input and output, one can insert a specific algorithm between the lines of code for input and output. A corresponding pooling-based program can readily be obtained.

Example I.1: Loop Program Using the C6711 DSK

This example is included to show the similarities between the DSKs. See also Examples 2.1 and 2.2. Figure I.1 shows the C source program `loop_intr.c`, included in the folder **DSK6711** with the necessary support files (`C6xdskinit.c`, `C6xdskinit.h`, etc.), and is discussed in Ref. 1.

REFERENCE

1. R. Chassaing, *DSP Applications Using C and the TMS320C6x DSK*, Wiley, New York, 2002.

```
//Loop_intr.c Loop program using interrupt, output= delayed input
//Comm routines and support files included in C6xdskinit.c

interrupt void c_int11()           //interrupt service routine
{
   short sample_data;

   sample_data = input_sample(); //input data
   output_sample(sample_data);    //output data
   return;
}

void main()
{
   comm_intr();                    //init DSK, codec, McBSP
   while(1);                       //infinite loop
}
```

FIGURE I.1 Loop program using the C6711 DSK (`loop_intr`).

Index

Note: Page numbers followed by f refer to figures, page numbers followed by t refer to tables.

Digital Signal Processing and Applications with the C6713 and C6416 DSK By Rulph Chassaing
ISBN 0-471-69007-4 Copyright © 2005 by John Wiley & Sons, Inc.